Rye Cereal

The book represents a comprehensive and authoritative compilation aimed at addressing the growing demand for in-depth knowledge on rye cereal. Rye, often overshadowed by other cereal grains, has witnessed a resurgence in interest due to its unique nutritional profile, resilience to adverse environmental conditions and diverse industrial applications. This handbook endeavors to fill a critical gap in the existing literature by offering a multidisciplinary perspective that covers every facet of rye, from cultivation and production to its industrial applications. The project's key components include an exhaustive exploration of rye cultivation practices, encompassing optimal growing conditions, pest management and sustainable farming techniques.

Key Features

- Contains comprehensive information on rye cereal.
- Discusses the extraction and characterization of rye cereal.
- Discusses the economic importance of rye cereal.
- Contains brief information on safety and toxicity of rye cereal.
- Sheds light on relationship between rye cereal intake and health benefits.

Rye holds significant agricultural and economic value in regions like Scandinavia, Germany, Poland and the United States, where it is a staple crop. The proposed handbook addresses various aspects of rye cultivation, offering insights into optimal production methods, crop characterization and its diverse industrial applications. The handbook targets students, academics, farmers, food researchers and nutritionists offering comprehensive scientific and practical insight into rye production, characterization, sustainability and industrial application.

Rye Cereal

Production, Cultivation, Characterization, and Industrial Applications

Edited by
Ali Ikram, Tabussam Tufail
and Gulzar Ahmad Nayik

CRC Press
Taylor & Francis Group
Boca Raton London New York

CRC Press is an imprint of the
Taylor & Francis Group, an **informa** business

Designed cover image: Getty

First edition published 2027
by CRC Press
2385 NW Executive Center Drive, Suite 320, Boca Raton FL 33431

and by CRC Press
4 Park Square, Milton Park, Abingdon, Oxon, OX14 4RN

CRC Press is an imprint of Taylor & Francis Group, LLC

ISBN: 9781032956244 (hbk)
ISBN: 9781032956275 (pbk)
ISBN: 9781003585817 (ebk)

DOI: 10.1201/9781003585817

Typeset in Times
by KnowledgeWorks Global Ltd.

Dedication

All the editors dedicate this book
to their beloved family.

Contents

Preface

Rye (*Secale cereale* L.) is one of the most resilient and historically significant cereal crops cultivated by humankind. Despite its long-standing role in ensuring food security in temperate and cold regions, rye has often remained underrepresented in mainstream cereal science literature when compared with wheat, rice and maize. In recent years, however, renewed global interest in rye has emerged due to its remarkable adaptability to marginal environments, superior dietary fiber content, rich phytochemical profile and expanding applications in food, feed, pharmaceutical and bio-industrial sectors. This resurgence underscores the urgent need for a comprehensive and authoritative reference that consolidates contemporary knowledge on rye from field to fork and beyond.

The handbook of *Rye Cereal: Production, Cultivation, Characterization, and Industrial Applications* has been conceived to address this need. This volume brings together contributions from leading researchers, academicians and industry experts to provide a holistic and multidisciplinary overview of rye cereal. The book systematically covers fundamental aspects such as botanical classification, agronomic practices, varietal diversity and post-harvest handling, while also delving into advanced topics, including physical and chemical characterization, processing technologies, quality and safety aspects, health-endorsing properties, allergens and toxicity concerns. By integrating both traditional knowledge and recent scientific advancements, this handbook aims to bridge the gap between basic research, applied science and industrial utilization.

Special emphasis has been placed on processing techniques, industrial applications and by-product utilization to highlight rye's growing importance in value-added food products, brewing and distilling industries, animal nutrition, bioenergy production and functional foods. Furthermore, the book explores the evolving role of food technologists and quality managers in maintaining grain quality, ensuring safety and meeting regulatory standards in an increasingly globalized food system.

The structure of this handbook has been carefully designed to serve a diverse readership. Graduate and postgraduate students will find it an accessible yet rigorous academic resource, while researchers and academicians will benefit from its depth, breadth and extensive references. Agricultural practitioners, food scientists, nutritionists and industry professionals will appreciate the practical insights and applied perspectives offered throughout the volume. Additionally, policymakers and stakeholders interested in sustainable cereal production and food security will find valuable guidance within these pages.

We sincerely acknowledge the invaluable contributions of all chapter authors whose expertise and dedication have shaped this volume. Their collective efforts reflect a shared commitment to advancing rye cereal research and promoting its wider adoption across agricultural and industrial domains. We are also grateful to the reviewers and editorial team for their constructive feedback and support during the development of this handbook.

It is our hope that this book will serve not only as a definitive reference on rye cereal but also as a catalyst for future research, innovation and collaboration. By highlighting the untapped potential of rye, this handbook aspires to contribute meaningfully to sustainable agriculture, improved human health and the development of novel rye-based products worldwide.

Dr. Ali Ikram
Dr. Tabussam Tufail
Dr. Gulzar Ahmad Nayik

About the Editors

Ali Akram earned his PhD from the Government College University Faisalabad, Pakistan in September 2022. His research expertise lies in cereal sciences, functional foods and food product development. Dr. Ali is working as Assistant Professor at University Institute of Food Science and Technology (UIFST). He has published 53 research articles (15 as first author) in Scopus/WOS-indexed journals, having a 100+ impact factor, 325 citations and an h-index of 11. Dr. Ali also participated in and presented his research work at seven different international conferences in New York, Australia, Turkey and Azerbaijan.

Tabussam Tufail is currently working as Assistant Professor in University Institute of Diet and Nutritional Sciences, Faculty of Allied Health Sciences, The University of Lahore, Lahore, Punjab, Pakistan. He completed his PhD in Food Science and Technology and MS in Food Science and Technology from Government College University Faisalabad. He has completed 25 international and national trainings as well as courses from different organizations. He has published over 80 peer-reviewed research/review papers, 20 book chapters, 1 book and presented papers in 25 international and national conferences as well as attended a number of conferences, seminars, workshops and webinars. Tabussam Tufail is serving as an editorial board member in Acta Scientific Publications. He is a frequent reviewer of several reputed journals in the area of Food & Nutrition as well as Food Science and Technology. He is a lifetime member of the Scientific and Technical Research Association (STRA), Healthcare & Biological Sciences Research Association (HBSRA), Teaching and Education Research Association (TERA), Social Science and Humanities Research Association (SSHRA), and the Pakistan Society of Food Scientists & Technologist (PSFST).

Gulzar Ahmad Nayik completed his Master's degree in Food Technology from the Islamic University of Science & Technology, Awantipora, Jammu and Kashmir, India, and earned his PhD from Sant Longowal Institute of Engineering & Technology, Longowal, Sangrur, Punjab, India. He has published more than 160 peer-reviewed research and review papers, contributed over 80 book chapters and edited 22 books with leading publishers. He has delivered presentations at numerous national and international conferences, seminars, workshops and webinars. Dr. Nayik was shortlisted twice (2017 and 2018) for the Inspire-Faculty Award by the Indian National Science Academy, New Delhi, and was nominated for the INSA Medal for Young Scientists (2019–20). In 2025, Dr. Nayik was featured among the top 2% cited scientists published on Elsevier Data Repository. He serves as editor, associate editor, assistant editor and reviewer for several reputable journals in Food Science and Technology. He has received multiple awards, appreciations and recognitions, and is an active member of various international scientific societies and organizations.

Contributors

Gholamreza Abdi
Department of Biotechnology, Persian Gulf Research Institute, Persian Gulf University
Bushehr, Iran

Zunair Abdullah
University Institute of Food Science and Technology, The University of Lahore
Lahore, Pakistan

B.S. Adithya
Department of Food Technology and Nutrition, Lovely Professional University
Phagwara, Punjab, India

Muhammad Afzaal
Department of Food Science, Government College University Faisalabad
Faisalabad, Pakistan

Aneeq Ahmad
University Institute of Food Science and Technology, The University of Lahore
Lahore, Pakistan

Muhammad Ahmad
University Institute of Food Science and Technology, The University of Lahore
Lahore, Pakistan

Samreen Ahsan
Institute of Food Science and Technology, Khwaja Fareed University of Engineering and Information Technology
Rahim Yar Khan, Pakistan

Ubaida Akbar
Department of Food Technology and Nutrition, Lovely Professional University
Phagwara, Punjab, India

Shazia Akhter
Division of Food Science and Technology, Sher-e-Kashmir University of Agricultural Sciences and Technology
Srinagar, India

Muhammad Tayyab Arshad
University Institute of Food Science and Technology, The University of Lahore
Lahore, Pakistan

Mehvish Mushtaq Banday
Division of Food Science and Technology, Sher-e-Kashmir University of Agricultural Sciences and Technology
Srinagar, India

Muhammad Farhan Jahangir Chughtai
Institute of Food Science and Technology, Khwaja Fareed University of Engineering and Information Technology
Rahim Yar Khan, Pakistan

Aalia Farooq
Division of Food Science and Technology, Sher-e-Kashmir University of Agricultural Sciences and Technology
Srinagar, India

Adil Farooq
Institute of Food Science and Technology, Khwaja Fareed University of Engineering and Information Technology
Rahim Yar Khan, Pakistan

Hareem Fatima
Department of Food Science, Government College University Faisalabad
Faisalabad, Pakistan

Muhammad Hammad ul Hassan
Department of Food Science and Technology, Muhammad Nawaz Shareef University of Agriculture
Multan, Pakistan

Mahreen Faqeer Hussain
University Institute of Diet and Nutritional Sciences, The University of Lahore
Lahore, Pakistan

Muhammad Nouman Ijaz
Institute of Chemistry, Khwaja Fareed University of Engineering & Information Technology
Rahim Yar Khan, Pakistan

Ali Ikram
University Institute of Food Science and Technology, The University of Lahore
Lahore, Pakistan

Mariam Islam
Department of Food Science, Government College University Faisalabad
Faisalabad, Pakistan

Ghulam Mustafa Kamal
Institute of Chemistry, Khwaja Fareed University of Engineering & Information Technology
Rahim Yar Khan, Pakistan

Muhammad Khalid
Institute of Chemistry, Khwaja Fareed University of Engineering & Information Technology
Rahim Yar Khan, Pakistan

Adnan Khaliq
Institute of Food Science and Technology, Khwaja Fareed University of Engineering and Information Technology
Rahim Yar Khan, Pakistan

Mujahid Hassan Khan
Food Chemistry and Technology Department, Teagasc Food Research Centre, Moorepark, Fermoy, Co. Cork
Ireland

Atif Liaqat
Institute of Food Science and Technology, Khwaja Fareed University of Engineering and Information Technology
Rahim Yar Khan, Pakistan

Aaruba Maqbool
Division of Food Science and Technology, Sher-e-Kashmir University of Agricultural Sciences and Technology
Srinagar, India

Sammra Maqsood
National Institute of Food Science and Technology, University of Agriculture Faisalabad
Faisalabad, Pakistan

Tariq Mehmood
Institute of Food Science and Technology, Khwaja Fareed University of Engineering and Information Technology
Rahim Yar Khan, Pakistan

Ushna Momal
Department of Food Science and Technology, Muhammad Nawaz Shareef University of Agriculture
Multan, Pakistan

Md Sabir Ahmed Mondol
Department of Agricultural Biochemistry, Bidhan Chandra Krishi Viswavidyalaya
Nadia, West Bengal, India

Tuba Mustafa
Department of Food Science, Government College University Faisalabad
Faisalabad, Pakistan

Aliza Naeem
Department of Food Science, Government College University Faisalabad
Faisalabad, Pakistan

Hammad Naeem
Department of Food Science and Technology, Muhammad Nawaz Shareef University of Agriculture
Multan, Pakistan

Muhammad Adnan Nasir
Department of Allied Health Sciences, Faculty of Allied Health Sciences, The University of Chenab
Gujrat, Pakistan

Feroza Naveed
University Institute of Food Science and Technology, The University of Lahore
Lahore, Pakistan

Gulzar Ahmad Nayik
Marwadi University Research Centre, Department of Microbiology, Marwadi University
Rajkot, Gujarat, India

Bushra Niaz
Department of Food Science, Government College University Faisalabad
Faisalabad, Pakistan

Asfa Perween
Department of Human Nutrition and Dietetics, Muhammad Nawaz Shareef University of Agriculture
Multan, Pakistan

Dayanand Peter
Division of Food Processing Technology, Karunya Institute of Technology and Sciences
Coimbatore, Tamil Nadu, India

Sidharth Pradeep
Division of Food Processing Technology, Karunya Institute of Technology and Sciences
Coimbatore, Tamil Nadu, India

Charalampos Proestos
Food Chemistry Laboratory, Department of Chemistry, National and Kapodistrian University of Athens, Panepistimiopolis Zographou
Athens, Greece

T V Ranganathan
Division of Food Processing Technology, Karunya Institute of Technology and Sciences
Coimbatore, Tamil Nadu, India

Amara Rasheed
Department of Food Science, Government College University Faisalabad
Faisalabad, Pakistan

Jahangir Ahmad Rather
Department of Food Science and Technology, University of Kashmir
Jammu & Kashmir, India

Awais Raza
University Institute of Food Science and Technology, The University of Lahore
Lahore, Pakistan

Nighat Raza
Department of Food Science and Technology, Muhammad Nawaz Shareef University of Agriculture
Multan, Pakistan

Tehseena Sabir
Institute of Chemistry, Khwaja Fareed University of Engineering & Information Technology
Rahim Yar Khan, Pakistan

Farhan Saeed
Department of Food Science, Government College University Faisalabad
Faisalabad, Pakistan

Nimra Sameed
Institute of Food Science and Technology, Khwaja Fareed University of Engineering and Information Technology
Rahim Yar Khan, Pakistan

Muhammad Shahbaz
Department of Food Science and Technology, Faculty of Food and Home Sciences, Muhamamd Nawaz Shareef University of Agriculture
Multan, Pakistan

Aanchal Sharma
University Centre for Research and Development, Chandigarh University
Mohali, Punjab, India

Ayesha Siddiqa
School of Agriculture, Food and Ecosystem Sciences, Faculty of Science, The University of Melbourne
Parkville, VIC, Australia
Institute of Food Science and Technology, Khwaja Fareed University of Engineering and Information Technology
Rahim Yar Khan, Pakistan

Assam Bin Tahir
State Key Laboratory, School of Food Science and Resources, Jiangnan University
Wuxi, Jiangsu, China
The University Institute of Food Science & Technology, University of Lahore
Lahore, Pakistan

Maryam Abbasi Tarighat
Department of Chemistry, Faculty of Nano and Bio Science and Technology, Persian Gulf University
Bushehr, Iran

Ruqaya Tariq
Division of Food Science and Technology, Sher-e-Kashmir University of Agricultural Sciences and Technology
Srinagar, Jammu & Kashmir, India

Johann Koshy Thampan
Division of Food Processing Technology, Karunya Institute of Technology and Sciences
Coimbatore, Tamil Nadu, India

Samra Yasmin
Institute of Chemistry, Khwaja Fareed University of Engineering & Information Technology
Rahim Yar Khan, Pakistan

Hamna Zafar
Department of Food Science, Government College University Faisalabad
Faisalabad, Pakistan

1 Rye Cereals
An Overview

Ali Ikram, Muhammad Tayyab Arshad, Sammra Maqsood, Mahreen Faqeer Hussain, Aneeq Ahmad, Awais Raza, Zunair Abdullah and Muhammad Ahmad

1.1 INTRODUCTION

The oldest and hardiest grain grown in northern climes where other grains cannot thrive well, rye (*Secale cereale*), was an essential crop used in the world's food and agricultural systems (Jonsson et al., 2018). Rye was discovered first in the Near East and quickly spread over northern and Eastern Europe, Scandinavia and parts of Asia due to its high tolerance for cold temperatures and poor soils (El-Mahis et al., 2023). For thousands of years, rye has been farmed to be a grain with significant cultural value; in the past, traditional bread was made using the grain, especially in Germany, Russia and Nordic nations (Engeset et al., 2015). Today, rye's agricultural and economic importance extends beyond traditional regions. Rye is a farming crop that sustains global agrarian resilience and food security by providing an alternative in drought-prone or excessively hot regions (Ghafoor et al., 2024). In the cold parts of North America and Northern Europe, where poor soil makes wheat and barley crops not worthwhile, its tolerance for lower-than-ideal growing conditions has made rye an essential crop in these areas (Wrigley, 2019). Rye supports crop diversification in challenging environments, such as helping farmers differentiate between crop rotations that could mean healthier soils and reduced input dependency on fertilizers while boosting the regional economy.

In addition to its durability, rye is also widely consumed in both traditional and modern diets because of its nutritional profile. Rye flour is a raw material for whole-grain bread, is rich in vitamins and bioactive substances and has a high dietary fiber content. Recent nutritional studies have shown possible health benefits (Kaur et al., 2021). Studies further show that daily whole-grain rye products can improve weight control, cardiovascular health and glycemic control (Jonsson et al., 2018). Rye has also been proven to positively impact gut flora, contributing to better digestive health and addressing the increasing requirement for functional food products (Table 1.1) (Koecher et al., 2019).

Many traditional food products reflect the cultural role and the culinary application of rye. For example, rich, nutritious rye bread with a low glycemic index (GI) is still a staple food for people of all ages in Nordic countries to have more prolonged

DOI: 10.1201/9781003585817-1

TABLE 1.1
Nutritional Profile of Rye

Component	Description	Value (%)	Reference
Moisture	The water content of the rye grain.	10–15	Wrigley (2019)
Ash	Mineral content, including calcium, iron and magnesium.	1.5–2.5	El-Mahis et al. (2023)
Protein	Total protein content, including gluten proteins.	10–15	Kaur et al. (2021)
Fat	Total fat content, primarily unsaturated fats.	1.5–2.5	Jonsson et al. (2018)
Carbohydrates	Total carbohydrate content, mainly starch and sugars.	70–75	Ghafoor et al. (2024)
Dietary fiber	Total fiber content, including soluble and insoluble fibers.	10–15	Koecher et al. (2019)

energy release (Sandvik et al., 2017). This keeps demanding consumers as rye-based products are growing the interest in whole grains, mostly about serving the rich earthy flavors and the culture they first originated from with traditional meals. Rye is such a vital cereal, primarily due to its versatility, cultural value and nutritional importance, and this trait can be extended into both ancient and modern agriculture. Because of the resilience of rye to poor growth conditions and many health benefits, it holds strategic importance in food security and sustainability in agriculture for the future (Litwinek et al., 2022).

This chapter offers an in-depth review of rye grains with information from their botanical, agricultural and nutritional perspectives. The chapter tries to give detailed knowledge of why rye is an essential cereal crop, especially for regions facing challenging growing conditions, through the historical, economic and cultural facets of rye.

1.2 BOTANICAL AND NUTRITIONAL ASPECTS OF RYE

1.2.1 Botanical Characteristics

The resilient cereal crop rye (*S. cereale* L.) of the *Poaceae* grasses family grows and has botanical characteristics that differ from all other grains. Taxonomy classifies rye in the same tribe as wheat and barley, the *Triticeae*. This is shown in the rye plant structure that shares various morphological features with other grains but still exhibits certain features that have been exceptionally well suited for harsher soil and climatic conditions. Rye is essential for food security in particular climates, where it is as resilient and applicable in areas where other cereal crops cannot grow because of cold and poor soils (Ikram et al., 2023). The anatomy of the rye plant facilitates resistance to drought and frost due to its strong roots, long, thin stalks and narrow leaves. Being a crop that thrives well in difficult weather conditions, rye does exceptionally well because its long, flexible stalks enable the plant to persevere under strong winds. The dense spike produced by the spikelets of this plant is usually two or three florets in length, thereby leading to good pollination and seed development. Rye is more heterogeneous and plastic than wheat in terms of genetics since it undergoes most of its breeding through cross-pollination (Schreiber et al., 2021).

Like most other cereals, rye goes through a cycle of growth phases—from germination to maturity. Still, its tillering stage is relatively more prolonged, enabling it to withstand harsh growing conditions other cereals cannot resist. Rye plants experience some substages in maturity, namely germination, tillering, stem elongation, heading, flowering and grain-filling. Since the right time for planting and harvesting maximizes yield and quality, this growth sequence must be crucial for optimal productivity. Again, the utility of rye as a reliable food grain crop in northern regions is further justified by its ability to thrive under low light conditions and temperatures that would provide a longer growing season (Rodehutscord et al., 2016). Rye has a specific nutritional profile with high dietary fiber, mainly arabinoxylans, to support its health benefits, including better glycemic control and digestion. This makes rye a good grain for controlling metabolic health due to its high fiber content, which helps regulate blood sugar levels (Jonsson et al., 2018). Besides, rye is high in B vitamins, indispensable for energy metabolism and other minerals like potassium, phosphorus and magnesium (Kaur et al., 2021). Though it is not ideal for those with celiac disease, rye still has a lower gluten level than wheat, making it easier for most people to digest (Rani et al., 2021).

Additionally, rye has bioactive compounds, like lignans and phenolic acids with antioxidant activity, related to potential health benefits like reducing the risk of cardiovascular diseases and certain cancers. Bioactive compounds, high fiber content and thus the status of being a functional food with health benefits characterize rye. However, rye can also produce mycotoxins under some conditions, particularly under moisture stress. It is, therefore, worthy of careful monitoring of the food chain (Khodaei et al., 2021). Overall, rye's nutritional and botanical properties give it value as a strong and nutrient-dense cereal crop best suited for both traditional diets and diets with health needs. More so in climates and regions that favor its growth, the variety and unique profile of its nutrients make it an essential grain with massive potential for food security and health uses (Dziki, 2022; Papageorgiou & Skendi, 2018).

1.2.2 Nutritional Profile

Rye's nutritional complex is characterized by a complex set of macro- and microelements, significant dietary fiber and antioxidants and bioactive materials, which enhance its nutraceutical properties. The most considerable macronutrient component consists of carbohydrates, primarily consisting of complex polysaccharides. These carbohydrates trigger a more sustained energy flow and have a lower impact on GI effects compared to those of a few other grains (Ikram et al., 2022). Although rye has a relatively lower protein value than wheat, it is still valuable, especially in regions where it is a staple meal. Rye proteins are unique because they have less gluten, which limits their use in leavened breads but may be advantageous for specific dietary requirements (Rani et al., 2021). High in fiber, rye contains high amounts of arabinoxylans, which make it essential for digestive well-being. Rye fibers are known to regulate glucose levels, control cholesterol levels and maintain gut well-being, besides regulating digestion. In addition, the prebiotic properties of fiber from rye enhance gut-beneficial microbiota growth for better metabolic and immunological health (Jonsson et al., 2018).

Rye micronutrients are iron, magnesium and phosphorus. These minerals play a vital role in the body through several physiological functions, such as the metabolism of energy and the maintenance of bones. On the other hand, it is a good source of B vitamins, such as folate and niacin. These B vitamins support production in red blood cells and cellular processes (El-Mahis et al., 2023). Rye is a grain of interest in nutrient-dense diets since it contains higher levels of these micronutrients than other cereals (Kataria et al., 2024). In addition to those factors, rye is composed of bioactive compounds such as lignans and phenolic acids that demonstrate anti-inflammatory activities and possess high antioxidant capacities. Such compounds in rye are associated with lowered chronic disease risks, such as heart disease and multiple cancers (Deleu et al., 2020). Apart from its health properties, rye antioxidants ensure the sustainability of products produced from rye through the hindrance of the oxidation processes, ensuring a product's quality remains long (El-Mahis et al., 2023). However, mycotoxins can be present in rye, especially in wet regions. If not appropriately regulated, the mycotoxins ergot alkaloids may accumulate in rye grains and harm human health (Khodaei et al., 2021). Careful farming practices and effective surveillance are needed to minimize these risks and ensure that rye products are safe (Figure 1.1). In conclusion, its complete nutritional profile with great fiber and vital nutrients alongside bioactive substances places it as a functional food with many health advantages. In these regions, where it is mainly a staple grain, this nutritional richness and antioxidant value make it an essential complement to well-balanced diets (Papageorgiou & Skendi, 2018; Rodehutscord et al., 2016).

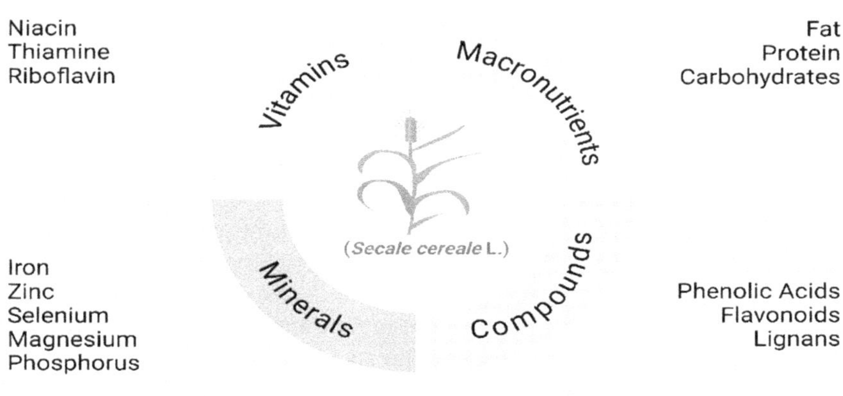

FIGURE 1.1 Composition of rye.

1.3 HEALTH BENEFITS

Rye has several health benefits because of its unique mixture of dietary fiber, antioxidants and bioactive chemicals (Figure 1.2). The health benefits that rye gives people include good heart health, digestive health and regulating blood sugar levels. Some of the health benefits that rye is known for are its heart health effects. According to Jonsson et al. (2018), soluble fibers in rye have high amounts of arabinoxylans, which result in the lowering of low-density lipoprotein (LDL) cholesterol, which is essential for good cardiovascular health. Because lignans can reduce inflammation and prevent lipid oxidation, their presence in rye has been associated with lower risks of heart disease and several types of cancer (Deleu et al., 2020). Lignans are a form of phytoestrogen. Rye also provides significant advantages concerning digestive health. High dietary fiber in rye promotes the maintenance of gut health through regular bowel movements and increased stool volume. The prebiotic fiber in rye increases healthy gut flora growth, enhancing immunity and minimizing gastrointestinal discomfort (Rani et al., 2021). This consumption of regular rye may also improve the gut microbiota, which is essential for upholding the best digestive health possible (Ikram et al., 2022). Rye is an excellent cereal for diabetic or insulin-resistant clients since it is incredibly efficient in regulating blood glucose. Rye's low GI, caused by its complex carbohydrate and high fiber, slows digestion and causes an incremental release of glucose into the bloodstream (Kaur et al., 2021). This controlled glucose release prevents sudden blood sugar spikes, which means better glycemic control and reduced chances of

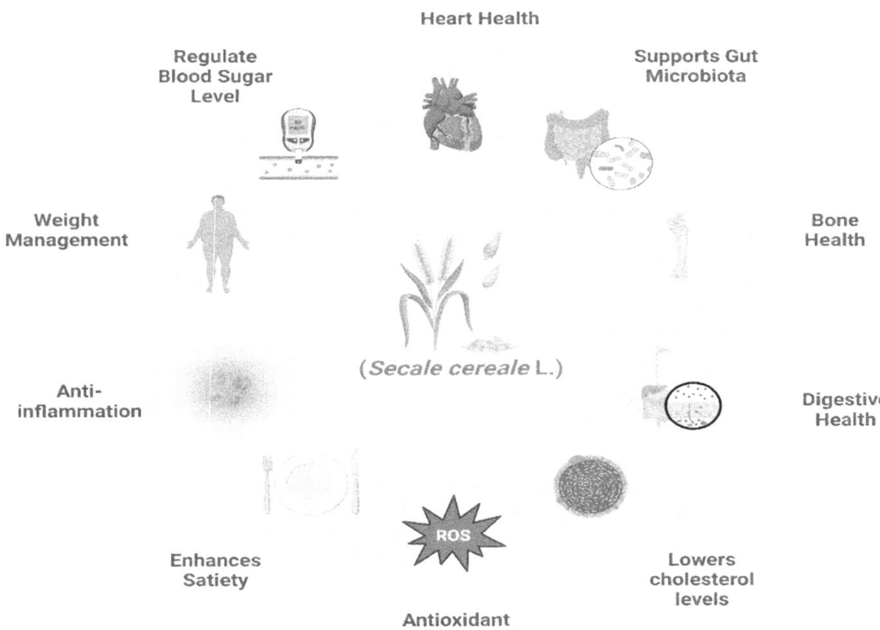

FIGURE 1.2 Health perspectives of rye.

TABLE 1.2
Health Benefits of Rye

Benefit	Description	Citations
Heart health	Rye's high fiber content and presence of antioxidants contribute to lowering cholesterol levels, reducing inflammation and improving overall cardiovascular health.	Jonsson et al. (2018) and Rani et al. (2021)
Digestive health	Rich in dietary fiber, rye promotes gut health by supporting beneficial gut microbiota, improving bowel regularity and reducing the risk of colorectal cancer.	Koecher et al. (2019), Jonsson et al. (2018) and Ikram et al. (2022)
Blood sugar regulation	Rye's low glycemic index and high fiber content slow glucose absorption, helping regulate blood sugar levels and reducing postprandial spikes in blood glucose.	El-Mahis et al. (2023) and Deleu et al. (2020)
Weight management	The high fiber content in rye increases satiety, which can help with weight management by reducing overall calorie intake and promoting more prolonged periods of fullness.	Sandvik et al. (2017) and Ghafoor et al. (2024)
Antioxidant activity	Rye contains bioactive compounds like phenolic acids and lignans that have antioxidant properties, protecting cells from oxidative stress and lowering chronic disease risks.	Jonsson et al. (2018) and Daly et al. (2022)

type 2 diabetes. Furthermore, rye products have been related to higher satiety, thus helping with weight management by reducing calorie intake (El-Mahis et al., 2023). Besides, rye's antioxidants, including phenolic acids, support its immunoprotective and anti-inflammatory features in the fight against chronic diseases like cancer and neurodegenerative diseases (Johansson et al., 2018). As these bioactive compounds decrease cell damage and potentially delay aging, they detoxify free radicals (Table 1.2). Put together, the nutritional content of rye can benefit several positive effects on human health, such as enhanced digestive performance and lesser risk to cardiac health, with its effectiveness in bettering sugar content in the body's blood cells. Functional food recognized and well-discussed for its benefits to an individual's health is often sought. Therefore, it appears beneficial in their diet plan (Daly et al., 2022; Kataria et al., 2024).

1.4 PROCESSING AND PRODUCTION

1.4.1 Agronomic Practices

Rye (*S. cereale* L.) is an adaptable cereal crop with peculiar agronomic requirements that could thrive in various soil types and temperatures. It is an essential crop for areas where other cereals would not have grown due to flexibility, such as in Northern and Central European cold climates, and for their tolerance to heavy winters (Jonsson et al., 2018). Winter rye varieties are widely used for their cold hardness; they are planted in the fall and harvested in the summer (Kaur et al., 2021). Furthermore, rye is an ideal crop for places with little rainfall or regions lacking adequate irrigation due to its superior drought tolerance (El-Mahis et al., 2023). It is relatively free of soil requirements but grows readily on low-fertility, sandy and acidic soils where other grains do not. It prefers soils of high organic matter and good drainage, however. Rye tolerates somewhat acidic soils because of an optimum pH range of 5.5–7.0. Rye is an excellent crop for soil conservation and erosion prevention due to its extensive and efficient root system, which enables it to reach deep into the soil in pursuit of water and nutrients (Papageorgiou & Skendi, 2018). Deep rye roots support growth on marginal soils by slowly improving the health of the soil by bringing in organic matter. Concerning land preparation techniques, rye requires minimal tillage, which, in turn, supports soil conservation practices. It is planted from seed and is usually done either by conducting the operation straight into the land or by operating with traditional techniques depending on what is needed or the particular growing conditions. The ideal planting depth of winter rye is between 2 and 5 cm, and seeding rates go about 150–200 kg/ha (Kaur et al., 2021). As a challenging crop, rye requires fewer chemical inputs of fertilizers or pesticides than other grains such as wheat. Such a lower requirement is partly contributed by the extreme resilience it displays for common diseases associated with most grains, particularly rust and powdery mildew, according to Daly et al. (2022).

Although fertilization is necessary for maximum production, rye needs less nitrogen input than crops such as maize and wheat—high levels of nitrogen result in lodging. Lodging is when plants fall in great numbers and make harvestability difficult. Instead, it provides moderate amounts of nitrogen to promote growth but does not favor hefty top growth (Wrigley & Bushuk, 2017). The crops also depend on the soil type, whereby different fertilizers, such as potassium or phosphorus, are used under the assumption that the mineral is inadequate within the soils since most farmlands mostly have such soils. As a highly competitive crop, rye aids practical organic farming because it slows the growth of weed spread; hence, weeds are rarely controlled (Dziki, 2022). Climate change has affected rye agronomy during the past years in a manner that requires adaptation techniques to be adopted to ensure production yields remain unchanged due to changes in climatic factors. Rial-Lovera et al. (2017) highlight that the methods comprise breeding techniques for developing hybrid rye varieties, thereby providing enhanced resistance capabilities for tolerating increased temperatures and variations in rainfall patterns. For example, hybrid rye breeding became of great interest to improve grain quality, yield and resistance to disease while leaving the crop capacity to thrive under adverse growing conditions, according to Miedaner and Laidig (2019). Rye is an incredibly resilient crop that can do exceptionally well in stressful situations with little assistance. Its positive impacts on the health of the soil make it one of the most important cereal crops that should be grown in sustainable agriculture. Farmers can have high-quality grain with lower environmental impact through agronomic techniques that utilize the resilience inherent in rye.

1.4.2 Harvesting and Post-Harvest Handling

The harvest and post-harvest handling of rye (*S. cereale* L.) are thus fundamental for retaining nutrients and quality. Early maturity harvests may lead to lowered yields and lower seed germination, and full maturity is the best maturity at which rye is harvested to ensure high-quality grains (Jonsson et al., 2018). Timing is essential because if rye is harvested later than the scheduled time, it might be subjected to less favorable weather conditions that favor mycotoxin growth and fungal contamination (Khodaei et al., 2021). Rye requires adequate drying after harvesting to decrease its moisture content; to discourage microbial growth, it should usually be kept below 14% (Deleu et al., 2020). Low-temperature drying is very much used for preservation and retaining nutritional characteristics, especially the fiber and bioactive compounds of rye; the method of drying should be selected taking into account the local climate as well as the infrastructure accessible and at a cost-effective manner (El-Mahis et al., 2023). Drying must be done on the rye grains; after that, they must be kept in a controlled environment for no spoilage. Optimal storage jeopardizes productivity and safety because it controls the temperature and humidity, preventing mold and insect infestation (Ghafoor et al., 2024). There must be proper ventilation so that the moisture in those storage facilities is minimized at all times; such is found to lower the probability of spoilage of the grain (Papageorgiou & Skendi, 2018). In addition, for the product to be safe and high-quality, rye grains kept in long-term storage can be sampled and fumigated at intervals (El-Mahis et al., 2023).

1.5 INDUSTRIAL PROCESSING

Many vital processes in the industrial production of rye, such as milling, refining and manufacturing its derivatives, can affect the quality and use value of the products based on rye. Generally, if rye is processed into flour, most of the bran and germ remain intact, contributing much of the nutritional content, including dietary fiber, B vitamins and bioactive compounds. Rye contains different baking properties due to its higher levels of pentosan and less gluten than wheat. These properties influence the taste and texture of the bread and come with some unique health benefits attributed to the high fiber content in rye (Sandvik et al., 2017). Rye flour is usually milled so as not to refine it too much so that some nutritional value in the grain is retained. The use of whole-grain rye flour, for example, is ideal for old-time rye bread and health-conscious products (Litwinek et al., 2022). Apart from flour, industrial processing also produces rye bran, flakes and other by-products that are greatly valued for utilization in dietary supplements and functional meals because of their high antioxidant and diet fiber (Kaur et al., 2021). Therefore, rye processing improvements aim to increase its nutritional content and shelf life and research new applications of rye derivatives in various food systems (Johansson et al., 2018).

1.6 QUALITY CONTROL AND STANDARDS

The quality control of rye during its production is essential so that it is safe and up to legal standards. Rye is very prone to infections by fungi, like the *Fusarium* species, that produce mycotoxins, toxic products. Thus, it would be necessary to monitor such pollutants as mycotoxins (Khodaei et al., 2021). Routine evaluation for moisture content, foreign material and mycotoxins are usually incorporated in the quality evaluations; permitted limits are based on regional and international standards (Korzun et al., 2021). Good Agricultural Practices (GAPs) are also promoted for rye production to reduce the risk of contamination and increase crop quality (Daly et al., 2022). Other physical properties include grain size, color and kernel hardness, which indicate the grain's milling potential and the final product's quality (Laidig et al., 2017). In the food trade, regulatory frameworks guarantee that rye meets specific pre-determined safety and quality requirements to ensure consumer confidence and value for the products in the global market (Engeset et al., 2015). The second is to safeguard rye from harmful substances and guarantee the quality of products for customers. All rye imported into the European Union (EU) must be tested according to EU-specific requirements and pass the checks. In addition to public health protection, these standards make exporting rye into the international market financially viable.

1.7 VARIETIES OF RYE

1.7.1 Overview of Rye Varieties

Rye is a highly hardy cereal crop adaptable to places where other cereals cannot grow well. Such versatility of growth traits in rye varieties, high yield potential and high adaptation to environmental conditions exist among the large variability seen in rye varieties, where some types of rye are far more suited to colder or harsh conditions than others (Daly et al., 2022). For instance, some winter rye varieties are ideal in northern parts since they can tolerate the cold and are often seeded during the fall. Spring rye varieties are typically planted in areas with a relatively mild winter and when the climate is relatively warm (Deleu et al., 2020). Drought resistance is another environmental adaptation characteristic of rye varieties that makes them suitable for poor, dry soils where crops, such as wheat, may fail to grow (Ghafoor et al., 2024). The toughness of rye makes it more resistant and valuable in sustainable agriculture due to unstable weather conditions.

1.7.2 Popular Cultivars

Due to such specific features, productivity and resistance to environmental conditions, several rye cultivars are cultivated globally. For example, in Europe, some of the popular cultivars that people like in cold locations are 'Amilo' and 'Dankowskie Zlote' due to excellent production and hardiness against winter (Laidig et al., 2017). They also have high value among these hybrid cultivars, providing higher yields and disease resistance. This is so, especially in Central and Eastern European countries where rye is largely cultivated (Litwinek et al., 2022). The 'Hancock' and 'Wintergrazer' are widely grown in North America. Winter rye varieties can grow biomass, which is vital in cover cropping. Such varieties show the complexity of rye; as per the needs of the local agriculture and the environment, such types are chosen for qualities like biomass, grain or fodder.

1.7.3 Selection and Breeding

The improvements in rye breeding that have been observed to date involve improvements in disease resistance, yield and nutritional quality to match the present demand for agriculture and consumers. Rye breeders have developed disease-resistant varieties of rye against the most prevalent diseases, such as rusts and powdery mildews, and they have reduced the use of fungicides and increased crop sustainability. Hybrid breeding programs, particularly in Europe, have resulted in hybrid rye cultivars that offer higher yields, nutrient profile superiority and improved disease resistance, according to Hackauf et al. (2022). In addition, these hybrids are less susceptible to lodging or breaking down in the rain because they have enhanced lodging resistance (El-Mahis et al., 2023). Breeding efforts have recently targeted the nutritional quality of rye to meet the ever-growing demands of more health-conscious consumer markets, especially in terms of its dietary fiber and bioactive chemicals, which are believed to have health benefits (Jonsson et al., 2018). With continuous breeding advancement, rye is proving to be an even more complicated, nutritionally beneficial crop for farmers to be more commercially successful globally.

1.8 DIETARY USES AND CULINARY APPLICATIONS

1.8.1 Traditional Uses

For many years, rye has been an essential grain among European nations and is also culturally significant in those nations. Traditional recipes with rye bread are crucial characteristics of Scandinavian and Eastern European food, like bread, which is very strong and dense. Over time, rye porridge has been valued as a dish with high fiber content and nutritional benefits. Because rye thrives well in cold-weather conditions, such a dish is very much preferred by people (Engeset et al., 2015). The applications from the ancient era denote all characteristics of rye, including versatility, nutritional richness and cultural significance across multiple population categories.

1.8.2 Modern Culinary Applications

Lately, rye has been used innovatively in modern cooking mainly because of the increased consumer demand for sustainable and plant-based diets. Chefs and food producers have come up with rye pasta, crackers and even non-dairy drinks to try and meet the demand for whole grains in modern diets (Kaur et al., 2021). Moreover, rye is a commonly used ingredient in gourmet and artisanal foods and plant-based products that gain consumers' attention because it is healthy and has acquired taste and nutritional properties (El-Mahis et al., 2023). This increase in rye's culinary applications demonstrates its adaptability and popularity in modern diets.

1.8.3 Functional Foods

Considering its highly established health properties, rye is more predominantly applied in fortified products and functional foods. Rich in fiber and some bioactive chemical content, it can manage blood sugar, digestive health and possibly cancer

TABLE 1.3
Different Aspects of Rye

Aspect	Description	References
Whole grain rye products	Rye is rich in dietary fiber, vitamins and minerals, making whole-grain rye products like bread and porridge excellent functional foods that promote health benefits.	El-Mahis et al. (2023) and Wrigley and Bushuk (2017)
Fortification potential	Rye can be fortified with additional vitamins, minerals and probiotics, enhancing its nutritional profile and health benefits in functional foods.	Jonsson et al. (2018) and Poutanen et al. (2014)
Bioactive compounds	Rye contains bioactive compounds such as phenolic acids, antioxidants and lignans that may help reduce the risk of chronic diseases and improve overall health.	Kaur et al. (2021) and Ikram et al. (2023)
Gut health benefits	Fermented rye products can serve as functional foods that support gut health by promoting the growth of beneficial gut microbiota and improving digestion.	Koecher et al. (2019) and Litwinek et al. (2022)
Metabolic health	The low glycemic index of rye helps regulate blood sugar levels and may aid in weight management, making it beneficial in functional food formulations for metabolic health.	Ghafoor et al. (2024) and Rani et al. (2021)
Culinary versatility	Rye flour can be used in various food products, from baked goods to snacks, allowing for innovative functional food development while maintaining taste and texture.	Papageorgiou and Skendi (2018) and Dziki (2022)

prevention. Applying such health features, food scientists formulate functional meals made of rye, such as snack bars and fortified bread (Ikram et al., 2023). It markets these products as offering better nutritional quality to conscious consumers and how rye may become a valued addition for improved dietary quality that benefits well-being (Table 1.3).

1.9 CHALLENGES AND CONSIDERATIONS

1.9.1 Agricultural Challenges

Some of the many agricultural issues rye faces include susceptibility to diseases such as ergot and powdery mildew, which are expected to significantly impact crops' quality and yields (Crespo-Herrera et al., 2017). In addition, though rye is cold and dry tolerant, extreme weather from climate change can easily affect yield consistency and general adaptability (Ghafoor et al., 2024). These traits demand that rye be bred further to make it more resilient to diseases and pests and adapt to varied environmental conditions (Korzun et al., 2021).

1.9.2 Market and Economic Challenges

This means that profitability among farmers is limited by the relatively low demand for rye compared to other grains, like wheat and corn (Laidig et al., 2017). The hectare yield of rye is usually lower than that of other grains, and production costs are relatively costly because rye demands specialized harvesting and processing machinery (Rodehutscord et al., 2016). These financial problems probably deter rye from cultivating since other grains are more productive.

1.9.3 Nutritional and Dietary Concerns

From a nutritional perspective, celiac disease or gluten sensitivity means abstaining from rye due to its gluten content (Ikram et al., 2023). Rye contains antinutrient compounds such as phytic acid, which may interfere with mineral absorption, although these are often minimized in fermentation and other processing mechanisms (Wrigley, 2019). Consumer preferences further weaken the popularity of rye in modern diets, considering that some customers prefer a less-flavored cereal than wheat (Sandvik et al., 2017).

1.10 FUTURE TRENDS AND RESEARCH DIRECTIONS

1.10.1 Innovations in Rye Cultivation

Rye breeding research aims to develop the highest-yielding climate-change-tolerant rye varieties, which may also show tolerance to stress caused by environmental changes. Hybrid breeding and genetic selection must enhance pest and disease resistance in addition to adverse weather. This stabilizes yield throughout growing places (Miedaner & Laidig, 2019). Given that the impacts of climatic change are fast becoming more intense, developing this new production methodology becomes essential to enhance the food supply, which should be resilient and sustainable.

1.10.2 Health and Nutrition Research

Recent research has shifted its focus toward the benefits that rye brings about in health, which includes beneficial effects on metabolic health and gut microbiota. Rye contains dietary fiber and bioactive compounds, and its intake has been positively related to better glucose management and digestion, leading to fewer incidents of cardiovascular diseases and type 2 diabetes (Koecher et al., 2019). Current research continues to shape rye's reputation as a desirable crop for health-conscious customers regarding its functional properties, from promoting beneficial gut flora to expressing antioxidant and other bioactive phytochemicals (Jonsson et al., 2018).

1.10.3 Environmental Sustainability

Rye is famous for its role in sustainable agriculture because it directly improves soil structure, thus holding valuable nutrients in the soil and preventing erosion. Besides, rye grows well in low-input environments, requiring minimal amounts of fertilizer and pesticide inputs that diminish the environmental impact of the crop. As a cover crop, its usage promotes even more biodiversity and richness through sustainable agriculture practices that entail soil preservation and ecosystem stability.

1.10.4 Technological Advancements

Technological advancements in food processing increase the functionality and quality of rye-based products. Fermentation and hydrothermal processing lower the levels of antinutrients, thereby making it more nutritious, tastier and easier to digest (El-Mahis et al., 2023). Rye is becoming increasingly valuable for the kitchen because of such technical advances, and it is possible to produce a whole range of products that appeal to modern consumers while retaining nutritional value (Papageorgiou & Skendi, 2018). In this chapter, rye has been one of the most discussed cereal grains due to its primary historical, nutritional and economic functions in world food systems. Rye offers utility when considering its nutrient makeup and temperature flexibility, thus making it an ideal crop where other cereals cannot grow well. The botanical features of rye make it suitable for farming under adverse climatic conditions. Therefore, its significance as a staple crop in the diets of several cultures is justifiable. Rye contains dietary fiber, carbohydrates and bioactive compounds that contribute to heart health, improve digestion and maintain blood sugar levels. The health benefits resonate well with rye as a preventive health food, more so since consumers now look for grains with excellent health properties.

The successful production of rye requires efficient growing, harvesting and processing technology. Advanced processing techniques ensure that the food is safe and that the range of culinary possibilities for rye is also widened. The chapter highlighted the growth of rye in modern cuisine and its potential for fortified, health-conscious goods and its traditional and contemporary uses, demonstrating cultural relevance in conventional recipes like rye bread and porridge. Rye's versatility in dietary applications allows it to fulfill the growing need for whole grains and functional foods and the shift in customer preferences. Despite its advantages, rye production has disadvantages due to its susceptibility to pests and diseases, lack of tolerance to changed climate conditions and economic reasons such as production costs and market demand. These hardships impact both the accessibility of customers and the viability of producers. Consumer acceptance is also determined by diet-related factors such as gluten intolerance and the presence of particular anti-nutritional compounds. However, it has mentioned encouraging trends in rye farming, like improvements in breeding that are supposed to make it more productive, resilient and nutritious. The value of rye as a nutrient-rich grain is further enhanced by continuing health and nutrition studies, which reveal more benefits from the grain. In a nutshell, rye is such a crop with vast potential in traditional and contemporary food systems. Rye is the crop that improves soil health while reducing inputs. Rye, therefore, meets most of the agricultural environmental goals. Its cultural value, adaptability and nutritional benefit make rye a core crop that addresses the challenge of food security and health development. In the future, rye will increasingly be a part of diets worldwide because of improved processing and product development, becoming part of an even more sustainable and health-oriented food system.

1.11 CONCLUSION

This chapter has provided an extensive introduction to rye as an essential cereal, as well as its botanical features, nutrition and historical and cultural characteristics, all of which were discussed previously. Since rye can be farmed within various climates, this crop is economically and agronomically important for several regions. Nutritional and botanical qualities, such as fiber, antioxidants and bioactive components, put special attention on the health benefits of rye, including blood sugar regulation, digestive support and heart health. Rye production should be conducted using appropriate agronomic practices to enhance the yield and quality of this crop. The rye processing steps, from milling up to post-harvest handling, demonstrate how significant industry standards are to achieving quality, safety and regulatory compliance. There are many rye cultivars, each with specific growing requirements and tolerance to the environment. Rye is becoming a crop in demand as the progress in breeding increases its potential production, disease resistance and nutritional content. Rye also has a vital application in nutrition, that is, in traditional or modern kitchens. The culinary range spans traditional rye bread to functional novel foods, bridging the old-fashioned diet and the most current plant-based products. In particular, there are prominent issues in the rye sector that it has to tackle and face, such as the sensitivity to insects, climate risks, instability of the market and consumers' diet conditions. Research in rye farming is promising in the future since attempts are made to develop cultivars with high yields and resilience to climate conditions. With the public interest in nutrition and well-being on the rise, constant research here continues to underscore the health benefits of rye, such as its positive impact on gastrointestinal and metabolic health. Moreover, rye's contribution to sustainable agriculture makes it an attractive crop for promoting biodiversity and soil quality, especially under adverse climatic conditions. Therefore, it remains well-positioned as one of the most essential grains underpinning traditional and current consumption patterns. Future studies, breeding and developments in processing technology strengthen the position of rye as an agricultural crop, a nutritional support mechanism and an environmental sustainability vehicle further into the global supply.

REFERENCES

Crespo-Herrera, L. A., Garkava-Gustavsson, L., & Åhman, I. (2017). A systematic review of rye (*Secale cereale* L.) as a source of resistance to pathogens and pests in wheat (*Triticum aestivum* L.). *Hereditas*, *154*, 1–9.

Daly, E. J., Hernandez-Ramirez, G., Puurveen, D., Ducholke, C., Kim, K., & Oatway, L. (2022). Perennial rye as a grain crop in Alberta, Canada: Prospects and challenges. *Agronomy Journal*, *114*(1), 471–489.

Deleu, L. J., Lemmens, E., Redant, L., & Delcour, J. A. (2020). The major constituents of rye (*Secale cereale* L.) flour and their role in the production of rye bread, a food product to which a multitude of health aspects are ascribed. *Cereal Chemistry*, *97*(4), 739–754.

Dziki, D. (2022). Rye flour and rye bran: New perspectives for use. *Processes*, *10*(2), 293.

El-Mahis, A., Baky, M. H., & Farag, M. A. (2023). How does rye compare to other cereals? A comprehensive review of its potential nutritional value and better opportunities for its processing as a food-based cereal. *Food Reviews International*, *39*(7), 4288–4311.

Engeset, D., Hofoss, D., Nilsson, L. M., Olsen, A., Tjønneland, A., & Skeie, G. (2015). Dietary patterns and whole grain cereals in the Scandinavian countries–differences and similarities. The HELGA project. *Public Health Nutrition, 18*(5), 905–915.

Ghafoor, A. Z., Karim, H., Studnicki, M., Raza, A., Javed, H. H., & Asghar, M. A. (2024). Climate change and rye (*Secale cereale* L.) production: Challenges, opportunities and adaptations. *Journal of Agronomy and Crop Science, 210*(4), e12725.

Hackauf, B., Siekmann, D., & Fromme, F. J. (2022). Improving yield and yield stability in winter rye by hybrid breeding. *Plants, 11*(19), 2666.

Ikram, A., Saeed, F., Afzaal, M., Abdullah, M., Niaz, B., Asif Khan, M., …, & Siddeeg, A. (2022). Comparative study of biochemical properties, anti-nutritional profile, and antioxidant activity of newly developed rye variants. *International Journal of Food Properties, 25*(1), 608–616.

Ikram, A., Saeed, F., Noor, R. A., Imran, A., Afzaal, M., Rasheed, A., …, & Kinki, A. B. (2023). A comprehensive review on biochemical and technological properties of rye (*Secale cereale* L.). *International Journal of Food Properties, 26*(1), 2212–2228.

Johansson, D. P., Gutiérrez, J. L. V., Landberg, R., Alminger, M., & Langton, M. (2018). Impact of food processing on rye product properties and their in vitro digestion. *European Journal of Nutrition, 57*, 1651–1666.

Jonsson, K., Andersson, R., Knudsen, K. E. B., Hallmans, G., Hanhineva, K., Katina, K., …, & Landberg, R. (2018). Rye and health-Where do we stand and where do we go do we go. *Trends in Food Science & Technology, 79*, 78–87.

Kataria, D., Kaur, J., Singh, J., Rasane, P., Bhadariya, V., & Kaur, S. (2024). Rye. In *Cereals and nutraceuticals* (pp. 107–119). Springer Nature Singapore.

Kaur, P., Sandhu, K. S., Purewal, S. S., Kaur, M., & Singh, S. K. (2021). Rye: A wonder crop with industrially important macromolecules and health benefits. *Food Research International, 150*, 110769.

Khodaei, D., Javanmardi, F., & Khaneghah, A. M. (2021). The global overview of the occurrence of mycotoxins in cereals: A three-year survey. *Current Opinion in Food Science, 39*, 36–42.

Koecher, K. J., McKeown, N. M., Sawicki, C. M., Menon, R. S., & Slavin, J. L. (2019). Effect of whole-grain consumption on changes in fecal microbiota: A review of human intervention trials. *Nutrition Reviews, 77*(7), 487–497.

Korzun, V., Ponomareva, M. L., & Sorrells, M. E. (2021). Economic and academic importance of rye. *The Rye Genome*, 1–12.

Laidig, F., Piepho, H. P., Rentel, D., Drobek, T., Meyer, U., & Huesken, A. (2017). Breeding progress, variation, and correlation of grain and quality traits in winter rye hybrid and population varieties and national on-farm progress in Germany over 26 years. *Theoretical and Applied Genetics, 130*, 981–998.

Litwinek, D., Boreczek, J., Gambuś, H., Buksa, K., Berski, W., & Kowalczyk, M. (2022). Developing lactic acid bacteria starter cultures for wholemeal rye flour bread with improved functionality, nutritional value, taste, appearance and safety. *PLoS One, 17*(1), e0261677.

Miedaner, T., & Laidig, F. (2019). Hybrid breeding in rye (*Secale cereale* L.). *Advances in Plant Breeding Strategies: Cereals, 5*, 343–372.

Papageorgiou, M., & Skendi, A. (2018). Introduction to cereal processing and by-products. In *Sustainable recovery and reutilization of cereal processing by-products* (pp. 1–25). Woodhead Publishing.

Poutanen, K., Katina, K., & Heiniö, R. L. (2014). Rye. In *Bakery products science and technology* (pp. 75–87). Wiley Online Library.

Rani, M., Singh, G., Siddiqi, R. A., Gill, B. S., Sogi, D. S., & Bhat, M. A. (2021). Comparative quality evaluation of physicochemical, technological, and protein profiling of wheat, rye, and barley cereals. *Frontiers in Nutrition, 8*, 694679.

Rial-Lovera, K., Davies, W. P., & Cannon, N. D. (2017). Implications of climate change predictions for UK cropping and prospects for possible mitigation: A review of challenges and potential responses. *Journal of the Science of Food and Agriculture*, *97*(1), 17–32.
Rodehutscord, M., Rückert, C., Maurer, H. P., Schenkel, H., Schipprack, W., Bach Knudsen, K. E., …, & Mosenthin, R. (2016). Variation in chemical composition and physical characteristics of cereal grains from different genotypes. *Archives of Animal Nutrition*, *70*(2), 87–107.
Sandvik, P., Nydahl, M., Marklinder, I., Næs, T., & Kihlberg, I. (2017). Different liking but similar healthiness perceptions of rye bread among younger and older consumers in Sweden. *Food Quality and Preference*, *61*, 26–37.
Schreiber, M., Özkan, H., Komatsuda, T., & Mascher, M. (2021). Evolution and domestication of rye. *The Rye Genome*, 85–100.
Wrigley, C., & Bushuk, W. (2017). Rye: Grain-quality characteristics and management of quality requirements. In *Cereal grains* (pp. 153–178). Woodhead Publishing.
Wrigley, C. W. (2019). Rye. In *Whole grains* (pp. 173–196). CRC Press.

2 Nutritional Comparison of Rye with Other Cereals

Feroza Naveed and Ali Ikram

2.1 INTRODUCTION

Due to a variety of health-promoting benefits, nowadays, a significant percentage of the population is prioritizing diets that are abundant in nutrients and bioactive components (Kaur et al., 2021). Plant-based foods, particularly cereal grains, serve as a primary staple for a large portion of the population, providing over 50% of caloric energy worldwide (Avelar et al., 2024). Globally, researchers are exploring various cereal grains to reveal their hidden potential, emphasizing the nutritional content and health benefits (Purewal et al., 2020). Whole grains are abundant in complex carbohydrates, dietary fiber (DF), fats, proteins, minerals, vitamins and phytochemicals, offering potential health benefits beyond basic nutrition and energy supply (Allai et al., 2022). Whole grains are edible seeds of grass species within the Poaceae family, including crops such as rye, wheat, barley, oats, millet, sorghum, rice, maize and triticale (Ficco et al., 2024). All cereal grains are excellent sources of energy and each cereal plays a unique role in supporting nutritional needs. This review aims to analyze the nutritional profile of rye in comparison to other cereals.

Rye (*Secale cereale* L.) ranks as the second most widely used cereal after wheat in the production of bread and bakery products (Biel et al., 2020). European countries (i.e., Russia, Finland, Poland, Ukraine, Germany, Denmark) account for approximately 90% of global rye production (Deleu et al., 2020). The global rye harvest currently totals around 14.8 million tons (Poutanen et al., 2014). Rye provides substantial amounts of starch, proteins, antioxidants, minerals and vitamins and stands out among other cereals due to its high content of DF, which is the main nutritional benefit of rye. Rye-based foods provide nearly 40% of total DF intake in the western world (Bondia-Pons et al., 2009). Rye also contains a wide variety of bioactive compounds (Németh & Tömösközi, 2021).

2.2 MACRONUTRIENT COMPARISON

2.2.1 Starch Comparison

Starch is a complex carbohydrate and is considered a key source of energy in the human diet. It also contributes to enhancing the functional characteristics of food items (Kaur et al., 2021). Rye, like other cereals, contains starch as its main carbohydrate component (Németh & Tömösközi, 2021). The starch levels of rye vary between 55 and 70%, which is comparable to that of wheat (60–75%) and triticale

DOI: 10.1201/9781003585817-2

TABLE 2.1
Comparison of Total Starch and Composition of Amylose and Amylopectin Between Cereal Grains

Cereals	Total Starch Content (%)	Amylose (%)	Amylopectin (%)	References
Rye	55–70	13.1–25.8	74–86.9	Kaur et al. (2021)
Rice	75–81	22.9–29.0	57.1–65.1	Huang et al. (2024)
Wheat	60–75	25–28	72–75	Li et al. (2023)
Triticale	60–75	12.8–35.1	64.9–87.2	Camerlengo & Kiszonas (2023)
Oats	51–65	25.2–29.4	70.6–74.8	Zhang et al. (2021)
Barley	50.3–72.4	15.9–30.7	30.4–51.6	Geng et al. (2022)
Maize	67–72	2.7–50	59.9–97.2	Reddappa et al. (2022)
Sorghum	67.6	20.1	79.8	Chen et al. (2019)
Millet	51–79	20–30	70–80	Mahajan et al. (2021)

(60–75%) but slightly higher than that of oats (51–65) (Camerlengo & Kiszonas, 2023; Kaur et al., 2021; Li et al., 2023; Zhang et al., 2021), whereas rice has the highest starch content (75–81%) among cereals (Huang et al., 2024). The composition of rye starch includes large (A-type, 60–90%) granules of a diameter ranging from 23 to 40 µm and small (B-type, 10–40%) granules with a diameter less than 10 µm. Rye starch granules have a shape and size similar to those of wheat (Sluková et al., 2021). Starch is composed of a combination of amylose and amylopectin. In cereal starches, amylose typically makes up 22–28% while amylopectin accounts for 72–78% (Gomand et al., 2011). The composition of amylose and amylopectin in different cereal grains is presented in Table 2.1.

2.2.2 Protein Comparison

Proteins constitute about 10–15% (d.w) in cereal grains (Shewry, 2023). Most of the protein content of cereals is found in the endosperm; however, the aleurone and sub-aleurone layers contain the highest concentration of protein (Poutanen et al., 2022). Rye finds worldwide utilization in the food sector for the production of bread, crackers, beer and rusk. This is due to the fact that rye is the only cereal, apart from wheat, that has notable levels of gluten proteins (Kaur et al., 2021). Rye proteins contain high lysine levels, but unlike wheat proteins, they are unable to establish a continuous gluten network (Drakos et al., 2017). The percentage of rye proteins generally varies between 8 and 15%, primarily influenced by growth conditions (Németh & Tömösközi, 2021). Albumins and prolamins are the main proteins in rye grain, comprising 34 and 19%, respectively, followed by globulins at 11% and glutenins at 9%. Wheat, triticale, barley and sorghum have protein levels similar to those of rye. Oats lead all cereals in protein content, which ranges

TABLE 2.2
Total Percentages of Protein and Its Types Across Various Cereals

Cereals	Total Protein Content	Albumins	Globulins	Prolamins	Glutelins	References
Rye	8–15%	29–40%	8–11%	Secalins (17–19%)	9–15%	Németh and Tömösközi (2021)
Wheat	9.8%	21.92%		33.1%	33.9%	Rajnincová et al. (2019)
Oats	15–20%	1–12%	50–80%	Avelins (4–15%)	10%	Kumar et al. (2021)
Barley	7–25%	5.5–5.8%		Hordein (30–50%)	35–45%	Geng et al. (2022); Jaeger et al. (2021)
Rice (brown)	6–7%	5–10%	7–17%	3–6%	75–81%	Jayaprakash et al. (2022)
Sorghum	7–15%	6.7%	7.3%	Kafirin (56%)	22.4%	Selle et al. (2020); Siminiuc and Coşciug (2021)
Triticale	8.6–16.3%	42.7%	24.2%	Secalins (7.1%)	9.3%	Camerlengo and Kiszonas (2023)
Maize	9%	3%	3%	Zein (60%)	34%	Maqbool et al. (2021)

from 15 to 20% (Kumar et al., 2021), whereas rice and maize contain protein in relatively lesser amounts. The main storage proteins in wheat, maize, barley, sorghum and millet are prolamins, whereas the predominant storage proteins in rice and oats are glutelins and globulins, respectively (Poutanen et al., 2022). Table 2.2 highlights the total protein content along with levels of albumins, prolamins, globulins and glutelins in various cereals.

The amino acid profile of rye proteins is marginally more advantageous than wheat due to their greater lysine levels. Nevertheless, they are deficient in tryptophan and isoleucine (Németh & Tömösközi, 2021). Table 2.3 outlines the amino acid concentrations across major cereals.

2.2.3 Lipids Comparison

Although lipids are found in cereals in relatively small quantities, they significantly influence the quality and texture of foods. This is because of their amphipathic nature, which allows them to interact with starch and proteins, forming inclusion complexes (Šramková et al., 2009). The highest percentage of lipids was shown in naked oats (8.3–11.4%), followed by barley (3.12–3.56%) (Geng et al., 2022; Leszczyńska et al., 2023), whereas millet (1.5–5%), sorghum (1–5%), maize (2–4%), rye (2–3%), tritcale

TABLE 2.3
Amino Acid Profiles of Cereals (g/100-g Protein) *(Histidine, Isoleucine, Leucine, Lysine, Threonine, Tryptophan, Valine, Alanine, Arginine, Aspartic Acid, Glutamic Acid, Glycine, Proline, Serine)*

Amino Acids (g/100-g Protein)	Rye	Wheat	Barley	Oats	Rice	Sorghum	Maize	Millet	Triticale
His	2.3	2.3	2.1	2.2	2.3	2.2	2.9	2.3	1.9
Ile	3.2	3.3	3.1	3.6	3.8	3.9	3.4	4.1	3.5
Leu	6.1	6.6	6.3	7.2	7.9	13.0	12.0	10.0	6.3
Lys	3.7	2.8	3.6	4.1	3.5	2.0	3.1	2.8	2.5
Thr	3.3	2.9	3.1	3.4	3.4	3.3	3.6	3.7	2.9
Trp	1.1	1.2	1.3	1.4	1.4	1.1	0.8	-	0.9
Val	4.5	4.2	4.8	4.9	5.5	4.9	4.7	5.3	4.7
Ala	4.2	3.5	3.8	4.7	5.5	9.5	7.4	7.7	3.8
Arg	5.1	4.8	5.0	6.5	7.9	3.9	4.8	4.4	5.4
Asp	7.1	5.1	6.0	7.9	8.8	6.8	6.6	7.8	6.2
Glu	22.1	28.1	24.8	19.2	17.1	21.7	18.1	18.8	29.4
Gly	4.4	4.0	3.9	4.9	4.3	2.8	3.9	3.1	3.9
Pro	9.0	9.8	9.2	5.1	4.5	8.3	8.9	6.1	8.5
Ser	4.3	4.5	4.2	4.7	4.8	3.9	4.8	4.5	3.8
References	Sá & House (2024)	JIANG et al. (2008)	Biel et al. (2020)	Tomičić et al. (2022)	Farooq et al. (2023)	Khan et al. (2023)	Sá & House (2024)	Sá & House (2024)	Biel et al. (2020)

(1.2–2.4%), wheat (1–2%) and rice (0.4–1.9%) contain lipids in very minute quantities (Bhatt et al., 2022; Camerlengo & Kiszonas, 2023; Carcea, 2021; Khalid et al., 2023; Németh & Tömösközi, 2021; Swati et al., 2024; Tanwar et al., 2023). The key constituents of cereal lipids are oleic acid, palmitic acid, linoleic acid, stearic acid and linolenic acid. Compositions of these fatty acids in some cereals are presented in Table 2.4.

TABLE 2.4
Comparison of Fatty Acid Profiles among Cereals

Fatty Acids	Rye (bran)	Wheat (germ)	Oats (bran)	Maize (germ)	Rice (bran)
Linoleic (%)	61.0	57	32.5–32.8	54.3	38.4
Oleic (%)	13.65	15	44.0–46.6	29.4	34.3
Palmitic (%)	13.7	17.5	15.6–16	11.5	19.8
Stearic (%)	–	–	1.7–1.8%	2.8%	–
Linolenic (%)	6.3	6	–	–	–
References (%)	Povilaitis and Venskutonis (2015)	Meriles et al. (2022)	Liu et al. (2020)	Klempová et al. (2020)	Xu et al. (2021)

2.3 MICRONUTRIENT COMPARISON

Micronutrients are considered equally essential as macronutrients as they contribute to sustaining diverse biochemical processes within the body (Kaur et al., 2021). Cereals serve as a significant source of vitamins (A, B complex and E) and minerals (P, Ca, Cu, K, Zn, Fe and Mg) (Garutti et al., 2022). As compared to other cereals, rye grain comprises a natural blend of minerals such as sodium, potassium, zinc, magnesium, iron, calcium and selenium, as well as vitamins such as B1, B2, B3, B5, B6, B9, α-tocopherol and choline, which are essential for regulating various body functions. Moreover, rye also contains ample amounts of α-tocopherol (Rodehutscord et al., 2016). Table 2.5 displays the composition of minerals and vitamins across different cereals.

TABLE 2.5
Minerals and Vitamins Composition (mg/100 g) of Whole Grains Cereals

Minerals (mg)	**Rye**	**Wheat**	**Oats**	**Barley**	**Triticale**	**Maize**	**Rice**	**Sorghum**	**Millet**
Phosphorus	252.6	306	523	264	321	256	94	289	285
Potassium	79.4	284	429	452	466	287	92	363	195
Calcium	7.5	41	54	33	35	15	24	13	8
Magnesium	36.9	138	177	133	153		20	165	114
Iron	2.11	5.3	4.72	3.6	2.59	2.4	0.8	3.36	3.01
Zinc	2.65	2.7	4.0	2.77	2.66	2.21	1.3	1.67	1.68
Manganese	2.58	2.28	4.92	1.94	4.18			1.6	1.63
Copper	0.33	0.68	0.17	0.49	0.55	0.31	0.18	0.28	0.75
Sodium	2	17.1	2	12	2	35	5	2	5
Selenium	0.13					0.155	0.01	0.12	0.02
Vitamins (mg)									
Thiamin	0.31	0.38	0.76	0.64	0.37	0.39	0.34	0.332	0.42
Riboflavin	0.25	0.11	0.13	0.28	0.13	0.2	0.09	0.096	0.29
Niacin	4.27	4.38	0.96	4.6	2.86	3.6	4.62	3.69	4.72
Pantothenic acid	1.46	0.95	1.35	0.28	2.17	0.42	0.92	3.67	0.84
Pyridoxine	0.29	0.36	0.11	0.31	0.4	0.62	1.3	0.44	0.38
Folate		0.56				0.26	0.03		
α-Tocopherol	0.85	1.01–4.05			0.9	2.4–7.9	0.14	0.5	
Vitamin A	0.003			0.007		0.06			
Vitamin K	0.05	0.002				0.0003	0.0001		
References	Manzoor et al. (2024); Garutti et al. (2022); Garg et al. (2021)								

2.4 DIETARY FIBERS COMPARISON

DFs are described as complex carbohydrates, naturally occurring in plant-based foods, with a polymerization degree of 3 or more units. They are non-digestible and resist absorption in the small intestine. DFs are further categorized as soluble DFs and insoluble DFs depending on their water solubility. DFs can be obtained from various plant-based foods, including cereals, fruits and vegetables (Joye, 2020). Cereal grains are major sources of DFs. Approximately half of the DF consumed in Western nations comes from cereals (Cooke, 2017). The DF content of rye, comprising non-starch polysaccharides and lignin, is the highest among cereals (Németh & Tömösközi, 2021). The fiber content of rye varies between 14 and 21%, depending on its position inside the grain. The inner endosperm has a lower DF content of 12%, whereas the outer endosperm has a higher content of approximately 22% (Joye, 2020). Soluble and insoluble DF levels of rye are 2.5 and 13.2%, respectively. DF composition of other cereals such as wheat, oats, barley and triticale is relatively lower than that of rye, as shown in Table 2.6. However, barley contains the highest percentage of soluble DF among these cereals. The composition of DF of rye shows arabinoxylans as the predominant component, with significant proportions of fructans, β-glucan and cellulose (Kaur et al., 2021).

2.4.1 Arabinoxylans (AX)

Arabinoxylans are pentosans that are polysaccharides located in the cell walls of plants and are made up of xylose and arabinose units (Zannini et al., 2022). These units form the backbone of AX, with a-L-arabino-furanosyl units substituting the (1-4)-b-D-xylo-pyranosyl residues at varying degrees, generally at O_2 (2mXyl), O_3 (3mXyl) or both (dXyl) positions. Moreover, small amounts of ferrulate residues

TABLE 2.6
Dietary Fibers and Their Specific Components Composition in Various Cereals

Dietary Fibers	Rye	Wheat	Oats	Barley	Triticale
Total DF content (%)	14–21	11.5–15.5	11.8–19.1	8.0–16.2	13.0–16.0
Soluble DF (%)	2.5	3.6	3.5–12.3	4.1	2.6–3.2
Insoluble DF (%)	13.2	9.4–12	6.8–8.3	12.1	11.4–12
Arabinoxylans (%)	8–12	5.53–7.42	1.1–2.0	5.2	6.8
β-glucan (%)	1.3–2.2	0.51–0.96	3.9–7.1	3.2–4.6	0.7
Fructans (%)	4.6–6.6	0.84–1.85	1.4	1.6	2.3
Cellulose (%)	1–1.7	1.67–3.05	1.3	1.9	2.1
References	Ikram et al. (2023)	Shewry and Hey (2015)	Németh and Tömösközi (2021)	Németh and Tömösközi (2021)	Rakha et al. (2011)

are also found that are linked to arabinose via ester bonds at their O_2 position (Ikram et al., 2023). AXs, major DF components, are hemicelluloses found in peripheral layers and endosperm cell walls of cereal grains (Long et al., 2024). AXs play a vital role in enhancing viscosity, forming hydrogels, absorbing moisture, stabilizing emulsions and foams, forming films and serving as substitutes for fats. They are considerably important in cereal processing such as flour milling, brewing, bread making and production of animal feed (Izydorczyk, 2021). Rye grains supply the highest levels of AXs, followed by triticale, wheat and barley. AX is the most abundant component of DF in rye grain, significantly enhancing flour's water retention capabilities (Poutanen et al., 2014). The AX content of rye cereal fluctuates between 7.5 and 11.5%, while the extractable amount of AX varies between 2.1 and 4.0% (Németh & Tömösközi, 202)). In rye flour, the average amount of AX and water-extractable AX (WE-AX) ranges from 3.2 to 3.6% and 1.2 to 1.3%, respectively. Whereas in bran, these values are higher, ranging from 12.9 to 13.4% for AX and 1.1 to 1.3% for WE-AX (Brzozowski et al., 2023). Wheat AXs, ranging from 5.53 to 7.42%, are concentrated at the outer region of the grain, accounting for 60–70% in the bran cell walls, 10.9–26% in bran and 3–5% in endosperm. On the other hand, barley AXs (5.2%) constitute approximately 10–14% of the bran function, 1.2–1.3% of the endosperm and about 25–40% of the barley cell walls (Zannini et al., 2022). Triticale has AXs levels similar to those of wheat. Other types of cereals such as oats, maize, sorghum, rice and millet contain relatively low levels of AXs. As major components of DF, AX contribute to the nutritional profile of rye-based foods, providing significant nutritional and health benefits (Izydorczyk, 2021).

2.4.2 Fructans

Fructans are types of oligosaccharides, composed of β-d-fructofuranosyl units, in which fructose molecules are linked with sucrose by β(2-1), β(2-6) or both (Verma et al., 2021; Yoshida, 2021). Fructans are prebiotics and also support membrane stability during cold or drought stress (Brzozowski et al., 2023). Fructans are mainly supplied by cereals and their processed products, making them an essential component of our daily dietary intake. The data in Table 2.6 indicates that rye is the richest source of fructans, whereas oats provide relatively low levels of fructans among the cereals discussed here. The percentage of fructans in rye cereal varies between 4.6 and 6.6% (Ikram et al., 2023).

2.4.3 B-glucan

β-glucan is a type of DF that helps in regulating blood cholesterol levels (Schmidt, 2022). β-glucans are polysaccharides, consisting of glucose monomers linked together by β-(1,3) and β-(1,4) glycosidic bonds (Lante et al., 2023). Barley and oats are known to have the highest β-glucan levels among cereal grains, typically ranging from 3.2 to 4.6% and 3.9 to 7.1%, respectively. Rye also contains β-glucan in minute quantities, typically 1.5–2.5% (Andersson et al., 2008) (Table 2.6).

2.5 BIOACTIVE COMPOUNDS COMPARISON

Bioactive compounds are supplementary nutritional components commonly found in minute quantities in food (Gani et al., 2012). Bioactive compounds, including phenolic compounds, phytosterols and tocols, are abundantly present in the aleurone layer of rye (Ikram et al., 2023). Specific bioactive compounds are uniquely associated with each type of cereal: alkylresorcinol (AR) in rye, beta-glucans in oats and barley, γ-oryzanol in rice, avenanthramides and saponins in oats (Gani et al., 2012). Bioactive components contribute to antioxidant functionality and play a key role in maintaining a healthy lifestyle as they help counteract the oxidative damage (Kaur et al., 2021). Rye contains a variety of bioactive compounds that include flavonoids, benzoxazinoids (BXs) and alkylresorcinals, phenolic acids, phytosterols and tocols have also been identified (Németh & Tömösközi, 2021). Rye consumption, either as whole grains or whole grain flour, can significantly boost the intake of health-promoting phytochemicals and provide numerous nutritional benefits (Ikram et al., 2022). Nutritional benefits of bioactive compounds of rye are shown in Figure 2.1.

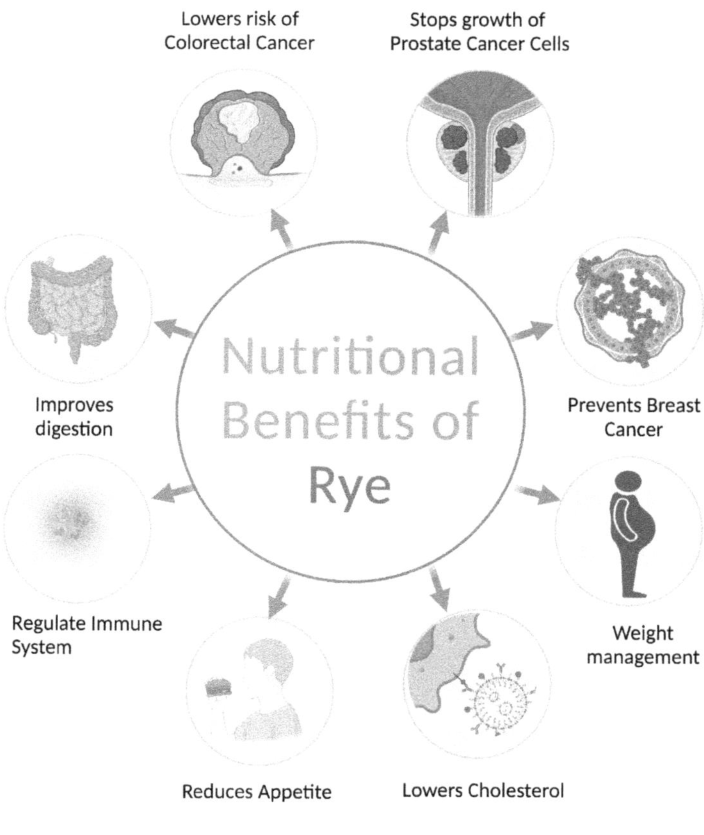

FIGURE 2.1 Nutritional benefits of rye.

2.5.1 Alkylresorcinals

Alkylresorcinals are phenolic lipids of plant origin that are abundantly contained in wheat and rye bran. ARs are associated with lowering the risk and managing long-term illnesses, including cardiovascular disorders, diabetes, cancer and obesity, due to their health-promoting benefits (Tian et al., 2020). Rye is recognized for having the highest content of alkylresorcinals, often reaching levels twice as high as those found in wheat (Sofi et al., 2019). Reports indicated alkylresorcinals concentrations range from 360 to 3200 μg/g in rye, 317 to 1430 μg/g in wheat, 439 to 647 μg/g in triticale and 42 to 51 μg/g in barley, but these compounds are absent in oats, rice, sorghum and millet (Targońska-Karasek et al., 2023; Wang et al., 2024).

2.5.2 Benzoxazinoids

BXs are a group of bioactive natural compounds present in whole grains such as rye, maize and wheat. They are also present in minute quantities in wild barley. In rye and wild barley, the main BX compound is DIBOA (2,4-dihydroxy-2H-1, 4-benzoxazin-3(4H)-one), whereas in wheat and maize, it is DIMBOA (2,4-dihydroxy-7-methoxy-2H-1,4-benzoxazin-3(4H)-one) (Du Fall & Solomon, 2011). These compounds are not found in oats, rice, sorghum or millet.

2.5.3 Phenolic Acids

Rye is rich in phenolic acids, which are the most significant phenolic compounds in this grain. These compounds are primarily concentrated in the bran and aleurone layer (Kulichová et al., 2019). Phenolic acids are further classified as cinammic acids (*p*-coumaric, caffeic, ferulic and sinapic acid) and benzoic acids (*p*-hydroxybenzoic, protocatechuic, gallic, syringic and vanillic acids). The most prevailing phytic acid (PA) in cereals is ferulic acid, accounting for nearly 90% of the total PAs, and is predominantly located in the outer layers of the grains (Ficco et al., 2024). The major PAs present in rye are ferulic acid (85–90%), sinapic acid (9–10%) and *p*-coumaric acid (3–5%) (Kataria et al., 2024). Caffeic, protocatechuic, gallic, syringic and vanillic acids have also been identified in minute quantities in rye grain (Dziki, 2022). Compared to rye grain, oats contain higher levels of ferulic and *p*-coumaric acids (Table 2.7)

2.5.4 Flavonoids

Flavonoids, a diverse group of polyphenolic compounds, are classified into six primary subclasses (flavones, flavans, flavanols, flavonols, flavanones and anthocyanins) (Dziki, 2022). Flavonoids are predominantly found in the outer layers (pericarp) of all grains. Sorghum is known to have the most extensive range of flavonoids identified (Zhao et al., 2022). Flavonoids are present only in trace amounts in most cereals, but barley stands out as it contains detectable levels of catechin and specific di- and tri-procyanidins (Sofi et al., 2019). In barley, the flavonoid levels have been documented to vary between 62 and 300.8 mg/g (Raj et al., 2023). In rye, the overall

TABLE 2.7
Comparison of PAs Composition Across a Range of Cereal Grains (µg/g)

Phenolic Acids (µg/g)	Rye	Wheat	Barley	Oats	Rice	Corn	Sorghum	Millet	Triticale
Ferulic	218	254	323	1035.9	68.2	94.2	66.1	233.4	46.7
p-Coumaric	29.8	14.4	74.9	607.2	22.8	340.5	43.6	46	139.8
Caffeic	12.3	0.53	5.6	9.1	0.9	15	32.1	4.6	9.9
Sinapic	51.6	46.5	43.5	107.1	24.1	66.1	8.2	46.7	83
p-Hydroxybenzoic	17.1	4.8	6.4	8	5.1	8.2	36.2	3	7.1
Gallic	7.7	6.4	6.5	1.7	5.5	55.4	28.0	68.6	333.7
Vanillic	12.9	4.1	5.9	20.5	4.3	10.3	23.2	22.2	446
Syringic	6.3	0.5	6.2	19.7	4.4	45.3	16.9	13.1	173.2
References	Dziki (2022)	Dziki (2022)	Van Hung (2016)	Van Hung (2016)	Obayiuwana (2023)	Khan et al. (2024)	Hithamani and Srinivasan (2014)	Khan et al. (2024)	Irakli et al. (2012)

content of flavonoids in bran and flour was 20.5 and 4.6 mg/100 g (d.w), respectively (Garcia-Castro et al., 2022), whereas the flavonoid content of wheat is higher than that of rye, ranging from 21.6 to 126.1 mg/kg (Dziki, 2022). Reports have disclosed that the flavone content of rye varies between 57 and 137µg/g, depending on the variety of rye. The major flavones are O-glycosides ranging from 50 to 68%, occurring as derivatives of chrysoeriol, tricin and apigenin. Rye generally contains a comparable content of flavones to that of wheat. However, the concentrations of these compounds are lower in rice and barley, typically <20 and 5.5–29 µg/g, respectively (Ravisankar et al., 2020). Anthocyanidins and flavonoid C-glycosides were also identified in wholemeal flour and bran of rye. The highest concentration of anthocyanidins was found in rye bran, approximately 15 mg/kg, whereas wholemeal rye flour contained only 1.8 mg/kg (Pihlava et al., 2018).

2.5.5 Phytosterols

Phytosterols, including sterols and stanols, are plant-based secondary metabolites (Topka et al., 2024). Phytosterols are similar in structure to cholesterol but differ due to the presence of additional methyl or ethyl groups. Phytosterols perform the same function in plants as cholesterol does in humans and animal cells (Pantović et al., 2024). Plant sterols in cereals exist in various forms, such as free sterol, glycosides of sterols, esters of sterols (with fatty acids or phenolic acids) and acylated glycosides of sterols (Meher & Jayadeep, 2024). The main phytosterols found in cereals are sitosterols, campesterols, stigmasterols and avenasteroli. The phytosterols concentration of rye is the highest compared to other cereals, with a mean of 95.5 mg/100 g (wet basis). In contrast, the total sterol content is 76.1 mg/100 g in barley, 69.0 mg/100 g in wheat and 44.7 mg/100 g in oats (wet basis) (Loskutov & Khlestkina, 2021). Sitosterol is the most dominant phytosterol in all cereals (Table 2.8).

TABLE 2.8
Comparison of Phytosterols Including Sitosterol, Campesterol, Stigmasterol, Avenasteroli and Stanoli Across Various Cereals (mg/kg)

Cereals	Rye	Wheat	Barley	Oats	Rice	Millet
Total phytosterols (mg/kg)	707–1134	447–830	720–801	350–611	723	770
Sitosterol	358–607	288–486	437–484	237–323	375	371
Campesterol	128–210	108–150	150–192	32–51	146	112
Stigmasterol	22–37	15–24	24–36	11–38	104	18
Avenasteroli	5–42	22	56–69	15–212	20	87
Stanoli	122–220	151–171	17–19	–	32	
Reference	Pantović et al. (2024)	Pantović et al. (2024)	Raj et al. (2023)	Raguindin et al. (2021)	Liu et al. (2023)	Pantović et al. (2024)

2.5.6 Tocols

Tocols, which are a collective term used for tocopherols and tocotrienols, are antioxidative agents that occur naturally in cereals and are valued for their vitamin E functionality (Kowalska, 2024). Tocopherols are primarily concentrated in the embryo of grain, whereas tocotrienols are located in the outer layers (pericarp) and endosperm (Loskutov & Khlestkina, 2021). Tocols exist in eight different forms, i.e., (α, β, γ, δ)-tocopherol and (α, β, γ, δ)-tocotrienol (Tiwari & Cummins, 2009). Vegetable oils are major sources of tocols, although significant amounts of these components are also found in a variety of cereals (i.e., rye, wheat, barley, oats and triticale), ranging from 10.7 to 74.7 mg/kg (Gani et al., 2012). Barley is known to have the highest content of tocols, typically fluctuating between 30.5 and 80.6 mg/kg, followed by wheat (35–46 mg/kg), oats (20–30 mg/kg) and rye (17–39 mg/kg) (Zielinski, 2008). Rye contains lower concentrations of tocopherols and tocotrienols, usually 13.7 and 14.1, respectively. The main vitamer of rye is α-tocopherol, constituting about 11.4 µg/g. β-tocopherol (2.2 µg/g), α-tocotrienols (5.9 µg/g) and β-tocotrienol (8.1 µg/g) are also found in rye. The primary tocol in barley is α-tocotrienol (17.09 µg/g), while minute quantities of α-tocopherol (3.84 µg/g), β-tocotrienol (1.32 µg/g) and γ-tocotrienol (4.73 µg/g) are also present (Zielinski et al., 2001). The predominant tocol in wheat is β-tocotrienol (16.12 µg/g), followed by α-tocopherols (5.58 µg/g), β-tocopherols (3.07 µg/g) and α-tocotrienols (2.53 µg/g) (Tremmel-Bede et al., 2022). On the other hand, the concentrations of α-tocopherol, β-tocopherol, α-tocotrienols and β-tocotrienol in oats are 7.3–12, 0.8–2.4, 20–34 and 2.4–6.1 µg/g, respectively (Pöysä et al., 2024).

2.5.7 Lignans

Lignans are bioactive polyphenolic compounds (Rodríguez-García et al., 2019). These compounds are phytoestrogens found in fruits, vegetables and cereals. Rye grain contains the highest levels of lignans (11.4–67.0 mg/kg) among the cereals (Dziki, 2022). Wheat grain is also a rich source of lignans compared to other cereals (3.4–22.7 µg/g), while barley contains minute quantities of lignins (4.25 mg/kg) (Bleidere et al., 2022; Zanella et al., 2017; Žilić, 2016). Common lignans found in the human diet are secoisolariciresinol, lariciresinol, pinoresinol, syringaresinol and matairesinol (Bach Knudsen et al., 2020). Seven dietary lignans have been identified in rye grain. Secoisolariciresinol is the most abundant rye lignan, making up approximately 80% of lignan content. Pinoresinol and medioresinol each contribute around 10%, while the rest 10% is other dietary lignans in trace amounts (Feng, 2019).

2.6 CONCLUSION

Rye was discovered in ancient times, around 10,000 BCE, by early ancestors in the Middle East. Since then, it has been considered a healthy cereal, often regarded as more effective than wheat and barley. Rye is a widely adaptable crop grown in temperate climates and is recognized for its ability to thrive in challenging growing environments. Rye is a nutrient-dense cereal with a unique composition compared to other major cereals (wheat, barley, oats, triticale). Starch and protein levels of rye are similar to those of wheat, barley and triticale. However, the main nutritional benefit of rye is its high DF content. Rye and its products are major DF sources worldwide. Among cereals, rye exhibits the highest levels of arabinoxylans and fructans, whereas oats and barley are particularly rich in beta-glucans. In addition to that, rye also provides a wide array of bioactive compounds, including alkylresorcinals, BXs, phenolic acids, tocols, phytosterols, flavonoids and lignans. The comparison of these nutrients highlights rye's unique benefits, particularly in promoting cardiovascular health, improving digestion, preventing diabetes, managing weight and providing antioxidant effects. Considering these factors, incorporating rye into the diet may offer valuable nutritional and health benefits, positioning it as a significant cereal for maintaining health. Future research could aim to focus on gaining deeper insights into the nutritional benefits of rye cereal and explore its potential in reducing disease risk.

REFERENCES

Allai, F. M., Azad, Z. R. A. A., Gul, K., & Dar, B. N. (2022). Wholegrains: A review on the amino acid profile, mineral content, physicochemical, bioactive composition and health benefits. *International Journal of Food Science & Technology*, *57*(4), 1849–1865.

Andersson, A. A. M., Rüegg, N., & Åman, P. (2008). Molecular weight distribution and content of water-extractable β-glucan in rye crisp bread. *Journal of Cereal Science*, *47*(3), 399–406.

Avelar, Z., Pereira, R. N., Vicente, A. A., & Rodrigues, R. M. (2024). Protein quality of cereals: Technological and functional perspectives. *Journal of Cereal Science*, *117*, 103922.

Bach Knudsen, K. E., Nørskov, N., Bolvig, A. K., Hedemann, M. S., & Lærke, H. N. (2020). Lignans. *Dietary Polyphenols: Their Metabolism and Health Effects*, 365–406.

Bhatt, D., Fairos, M., & Mazumdar, A. (2022). Millets: Nutritional composition, production and significance: A review. *Journal of Pharmaceutical Innovation*, *11*, 1577–82.

Biel, W., Kazimierska, K., & Bashutska, U. (2020). Nutritional value of wheat, triticale, barley and oat grains. *Acta scientiarum polonorum. Zootechnica*, *19*(2), 19–28.

Bleidere, M., Zute, S., Gailāne, N., Kantāne, I., Jākobsone, I., & Bartkevičs, V. (2022, April). Lignans in oat and barley grains depending on genotype and farming practice. *Proceedings of the Latvian Academy of Sciences. Section B. Natural, Exact, and Applied Sciences*, *76*(1), 124–130.

Bondia-Pons, I., Aura, A. M., Vuorela, S., Kolehmainen, M., Mykkänen, H., & Poutanen, K. (2009). Rye phenolics in nutrition and health. *Journal of Cereal Science*, *49*(3), 323–336.

Brzozowski, L. J., Szuleta, E., Phillips, T. D., Van Sanford, D. A., & Clark, A. J. (2023). Breeding cereal rye (*Secale cereale*) for quality traits. *Crop Science*, *63*(4), 1964–1987.

Camerlengo, F., & Kiszonas, A. M. (2023). Genetic factors influencing triticale quality for food. *Journal of Cereal Science*, *113*, 103744.

Carcea, M. (2021). Value of wholegrain rice in a healthy human nutrition. *Agriculture*, *11*(8), 720.

Chen, B. R., Wang, C. Y., Ping, W. A. N. G., Zhu, Z. X., Ning, X. U., Shi, G. S., … & Huang, R. D. (2019). Genome-wide association study for starch content and constitution in sorghum (*Sorghum bicolor* (L.) Moench). *Journal of Integrative Agriculture*, *18*(11), 2446–2456.

Chen, H., Chen, Z., Fu, Y., Liu, J., Lin, S., Zhang, Q., … & Qin, W. (2019). Structure, antioxidant, and hypoglycemic activities of arabinoxylans extracted by multiple methods from triticale. *Antioxidants*, *8*(12), 584.

Cooke, R. (2017). ProQuest ebook central. *The Charleston Advisor, 19*(2), 39–43.
Deleu, L. J., Lemmens, E., Redant, L., & Delcour, J. A. (2020). The major constituents of rye (*Secale cereale* L.) flour and their role in the production of rye bread, a food product to which a multitude of health aspects are ascribed. *Cereal Chemistry, 97*(4), 739–754.
Drakos, A., Malindretou, K., Mandala, I., & Evageliou, V. (2017). Protein isolation from jet milled rye flours differing in particle size. *Food and Bioproducts Processing, 104*, 13–18.
Du Fall, L. A., & Solomon, P. S. (2011). Role of cereal secondary metabolites involved in mediating the outcome of plant-pathogen interactions. *Metabolites, 1*(1), 64–78.
Dziki, D. (2022). Rye flour and rye bran: New perspectives for use. *Processes, 10*(2), 293.
Farooq, M., Jang, Y. H., Kim, E. G., Park, J. R., Eom, G. H., Zhao, D. D., & Kim, K. M. (2023). Evaluation of amino acid profiles of rice genotypes under different salt stress conditions. *Plants, 12*(6), 1315.
Feng, G. (2019). Rye. In *Bioactive factors and processing technology for cereal foods* (pp. 151–169). Springer.
Ficco, D. B. M., Petroni, K., Mistura, L., & D'Addezio, L. (2024). Polyphenols in cereals: State of the art of available information and its potential use in epidemiological studies. *Nutrients, 16*(13), 2155.
Gani, A., Wani, S. M., Masoodi, F. A., & Hameed, G. (2012). Whole-grain cereal bioactive compounds and their health benefits: A review. *Journal of Food Processing & Technology, 3*(3), 146–56.
Garcia-Castro, A., Román-Gutiérrez, A. D., Castaneda-Ovando, A., Carino-Cortes, R., Acevedo-Sandoval, O. A., Lopez-Perea, P., & Guzmán-Ortiz, F. A. (2022). Cereals as a source of bioactive compounds with anti-hypertensive activity and their intake in times of COVID-19. *Foods, 11*(20), 3231.
Garg, M., Sharma, A., Vats, S., Tiwari, V., Kumari, A., Mishra, V., & Krishania, M. (2021). Vitamins in cereals: A critical review of content, health effects, processing losses, bioaccessibility, fortification, and biofortification strategies for their improvement. *Frontiers in Nutrition, 8*, 586815.
Garutti, M., Nevola, G., Mazzeo, R., Cucciniello, L., Totaro, F., Bertuzzi, C. A., ... & Puglisi, F. (2022). The impact of cereal grain composition on the health and disease outcomes. *Frontiers in Nutrition, 9*, 888974.
Geng, L., Li, M., Zhang, G., & Ye, L. (2022). Barley: A potential cereal for producing healthy and functional foods. *Food Quality and Safety, 6*, fyac012.
Gomand, S. V., Verwimp, T., Goesaert, H., & Delcour, J. A. (2011). Structural and physicochemical characterisation of rye starch. *Carbohydrate Research, 346*(17), 2727–2735.
Hithamani, G., & Srinivasan, K. (2014). Bioaccessibility of polyphenols from wheat (*Triticum aestivum*), sorghum (*Sorghum bicolor*), green gram (*Vigna radiata*), and chickpea (*Cicer arietinum*) as influenced by domestic food processing. *Journal of Agricultural and Food Chemistry, 62*(46), 11170–11179.
Huang, M., Xiao, Z., Hu, L., Chen, J., & Cao, F. (2024). Amylopectin content rather than amylose or protein content is critical to determining the starch digestion rate in high-amylose rice. *Food Chemistry Advances, 5*, 100758.
Ikram, A., Saeed, F., Afzaal, M., Abdullah, M., Niaz, B., Asif Khan, M., ... & Siddeeg, A. (2022). Comparative study of biochemical properties, anti-nutritional profile, and antioxidant activity of newly developed rye variants. *International Journal of Food Properties, 25*(1), 608–616.
Ikram, A., Saeed, F., Noor, R. A., Imran, A., Afzaal, M., Rasheed, A., ... & Kinki, A. B. (2023). A comprehensive review on biochemical and technological properties of rye (*Secale cereale* L. *International Journal of Food Properties, 26*(1), 2212–2228.
Irakli, M. N., Samanidou, V. F., Biliaderis, C. G., & Papadoyannis, I. N. (2012). Development and validation of an HPLC-method for determination of free and bound phenolic acids in cereals after solid-phase extraction. *Food Chemistry, 134*(3), 1624–1632.

Izydorczyk, M. S. (2021). Arabinoxylans. In: *Handbook of hydrocolloids* (pp. 399–461). Woodhead Publishing.

Jaeger, A., Zannini, E., Sahin, A. W., & Arendt, E. K. (2021). Barley protein properties, extraction and applications, with a focus on brewers' spent grain protein. *Foods, 10*(6), 1389.

Jayaprakash, G., Bains, A., Chawla, P., Fogarasi, M., & Fogarasi, S. (2022). A narrative review on rice proteins: Current scenario and food industrial application. *Polymers, 14*(15), 3003.

Jiang, X. L., Tian, J. C., Zhi, H., & Zhang, W. D. (2008). Protein content and amino acid composition in grains of wheat-related species. *Agricultural Sciences in China, 7*(3), 272–279.

Joye, I. J. (2020). Dietary fibre from whole grains and their benefits on metabolic health. *Nutrients, 12*(10), 3045.

Kataria, D., Kaur, J., Singh, J., Rasane, P., Bhadariya, V., & Kaur, S. (2024). Rye. In *Cereals and nutraceuticals* (pp. 107–119). Springer Nature Singapore.

Kaur, P., Sandhu, K. S., Purewal, S. S., Kaur, M., & Singh, S. K. (2021). Rye: A wonder crop with industrially important macromolecules and health benefits. *Food Research International, 150*, 110769.

Khalid, A., Hameed, A., & Tahir, M. F. (2023). Wheat quality: A review on chemical composition, nutritional attributes, grain anatomy, types, classification, and function of seed storage proteins in bread making quality. *Frontiers in Nutrition, 10*, 1053196.

Khan, A., Khan, N. A., Bean, S. R., Chen, J., Xin, Z., & Jiao, Y. (2023). Variations in total protein and amino acids in the sequenced sorghum mutant library. *Plants, 12*(8), 1662.

Khan, J., Gul, P., Rashid, M. T., Li, Q., & Liu, K. (2024). Composition of whole grain dietary fiber and phenolics and their impact on markers of inflammation. *Nutrients, 16*(7), 1047.

Klempová, T., Slaný, O., Šišmiš, M., Marcinčák, S., & Čertík, M. (2020). Dual production of polyunsaturated fatty acids and beta-carotene with *Mucor wosnessenskii* by the process of solid-state fermentation using agro-industrial waste. *Journal of Biotechnology, 311*, 1–11.11.

Kowalska, A. (2024). The vitamin E: Overview of history of discovery, mechanism of action, role and deficiency. *Prospects in Pharmaceutical Sciences, 22*(3), 76–80.

Kulichová, K., Sokol, J., Nemeček, P., Maliarová, M., Maliar, T., Havrlentová, M., & Kraic, J. (2019). Phenolic compounds and biological activities of rye (*Secale cereale* L.) grains. *Open Chemistry, 17*(1), 988–999.

Kumar, L., Sehrawat, R., & Kong, Y. (2021). Oat proteins: A perspective on functional properties. *LWT, 152*, 112307.

Lante, A., Canazza, E., & Tessari, P. (2023). Beta-glucans of cereals: Functional and technological properties. *Nutrients, 15*(9), 2124.

Leszczyńska, D., Wirkijowska, A., Gasiński, A., Średnicka-Tober, D., Trafiałek, J., & Kazimierczak, R. (2023). Oat and oat processed products—Technology, composition, nutritional value, and health. *Applied Sciences, 13*(20), 11267.

Li, Y., Cheng, W., Qiu, X., Sun, Y., Xia, X., Yang, L., ... & Qian, H. (2023). Effects of β-amylase hydrolysis on the structural, physicochemical and storage properties of wheat starch. *Journal of Cereal Science, 109*, 103605.

Liu, J., Jin, S., Song, H., Huang, K., Li, S., Guan, X., & Wang, Y. (2020). Effect of extrusion pretreatment on extraction, quality and antioxidant capacity of oat (*Avena sativa* L.) bran oil. *Journal of Cereal Science, 95*, 102972.

Liu, Z., Liu, X., Ma, Z., & Guan, T. (2023). Phytosterols in rice bran and their health benefits. *Frontiers in Nutrition, 10*, 1287405.

Long, L., Lin, Q., Wang, J., & Ding, S. (2024). Microbial α-L-arabinofuranosidases: diversity, properties, and biotechnological applications. *World Journal of Microbiology and Biotechnology, 40*(3), 84.

Loskutov, I. G., & Khlestkina, E. K. (2021). Wheat, barley, and oat breeding for health benefit components in grain. *Plants, 10*(1), 86.

Mahajan, P., Bera, M. B., Panesar, P. S., & Chauhan, A. (2021). Millet starch: A review. International. *Journal of Biological Macromolecules, 180*, 61–79.

Manzoor, M. F., Ali, A., Ain, H. B. U., Kausar, S., Khalil, A. A., Aadil, R. M., & Zeng, X. A. (2024). Bioaccessibility mechanisms, fortification strategies, processing impact on bioavailability, and therapeutic potentials of minerals in cereals. *Future Foods, 10*, 100425.

Maqbool, M. A., Beshir Issa, A., & Khokhar, E. S. (2021). Quality protein maize (QPM): Importance, genetics, timeline of different events, breeding strategies and varietal adoption. *Plant Breeding, 140*(3), 375–399.

Meher, T., & Jayadeep, A. (2024). Effect of bioprocessing through germination and hydrothermal treatment on nutrients, nutraceuticals, and antioxidant properties in red and black rice. *Journal of Food Measurement and Characterization, 18*(4), 2935–2951.

Meriles, S. P., Penci, M. C., Curet, S., Boillereaux, L., & Ribotta, P. D. (2022). Effect of microwave and hot air treatment on enzyme activity, oil fraction quality and antioxidant activity of wheat germ. *Food Chemistry, 386*, 132760.

Németh, R., & Tömösközi, S. (2021). Rye: Current state and future trends in research and applications. *Acta Alimentaria, 50*(4), 620–640.

Obayiuwana, O. A. (2023). *Phenolic acid content and fibre in Nigerian wholegrains; their metabolism, and potential cardiovascular benefits* [Doctoral dissertation, University of Roehampton}.

Pantović, J. G., Gordanić, S., & Cvijanović, V. (2024). Role and significance of phytosterol from cereal seed. In *Book of proceedings* (p. 158).

Pihlava, J. M., Hellström, J., Kurtelius, T., & Mattila, P. (2018). Flavonoids, anthocyanins, phenolamides, benzoxazinoids, lignans and alkylresorcinols in rye (*Secale cereale*) and some rye products. *Journal of Cereal Science, 79*, 183–192.

Poutanen, K., Katina, K., & Heiniö, R. L. (2014). Rye. In *Bakery products science and technology* (pp. 75–87). Wiley Online Library.

Poutanen, K. S., Kårlund, A. O., Gómez-Gallego, C., Johansson, D. P., Scheers, N. M., Marklinder, I. M., ... & Landberg, R. (2022). Grains – A major source of sustainable protein for health. *Nutrition Reviews, 80*(6), 1648–1663.

Povilaitis, D., & Venskutonis, P. R. (2015). Optimization of supercritical carbon dioxide extraction of rye bran using response surface methodology and evaluation of extract properties. *The Journal of Supercritical Fluids, 100*, 194–200.

Pöysä, M., Pihlava, J. M., Fedotov, A., Lampi, A. M., Hietaniemi, V., & Piironen, V. (2024). Variations in tocopherol, tocotrienol, avenanthramide and saponin content in oats and the influence of milling and baking processes. *Journal of Cereal Science, 117*, 103902.

Purewal, S. S., Salar, R. K., Bhatti, M. S., Sandhu, K. S., Singh, S. K., & Kaur, P. (2020). Solid-state fermentation of pearl millet with *Aspergillus oryzae* and *Rhizopus azygosporus*: Effects on bioactive profile and DNA damage protection activity. *Journal of Food Measurement and Characterization, 14*, 150–162.

Raguindin, P. F., Itodo, O. A., Stoyanov, J., Dejanovic, G. M., Gamba, M., Asllanaj, E., ... & Kern, H. (2021). A systematic review of phytochemicals in oat and buckwheat. *Food Chemistry, 338*, 127982.

Raj, R., Shams, R., Pandey, V. K., Dash, K. K., Singh, P., & Bashir, O. (2023). Barley phytochemicals and health promoting benefits: A comprehensive review. *Journal of Agriculture and Food Research, 14*, 100677.

Rajnincová, D., Spaleková, A., Gálová, Z., & Romanavá, K. (2019). The protein profile of cereals, pseudocereals and legumes. *Journal of Food Science and Technology, 7*, 49–53.

Rakha, A., Åman, P., & Andersson, R. (2011). Dietary fiber in triticale grain: Variation in content, composition, and molecular weight distribution of extractable components. *Journal of Cereal Science, 54*(3), 324–331.

Ravisankar, S., Queiroz, V. A., & Awika, J. M. (2020). Rye flavonoids – Structural profile of the flavones in diverse varieties and effect of fermentation and heat on their structure and antioxidant properties. *Food Chemistry, 324*, 126871.

Reddappa, S. B., Chhabra, R., Talukder, Z. A., Muthusamy, V., Zunjare, R. U., & Hossain, F. (2022). Development and validation of rapid and cost-effective protocol for estimation of amylose and amylopectin in maize kernels. *3 Biotech*, *12*(3), 62.

Rodehutscord, M., Rückert, C., Maurer, H. P., Schenkel, H., Schipprack, W., Bach Knudsen, K. E., ... & Mosenthin, R. (2016). Variation in chemical composition and physical characteristics of cereal grains from different genotypes. *Archives of Animal Nutrition*, *70*(2), 87–107.

Rodríguez-García, C., Sánchez-Quesada, C., Toledo, E., Delgado-Rodríguez, M., & Gaforio, J. J. (2019). Naturally lignan-rich foods: A dietary tool for health promotion? *Molecules*, *24*(5), 917.

Sá, A. G., & House, J. D. (2024). Protein quality of cereals: Digestibility determination and processing impacts. *Journal of Cereal Science*, *117*, 103892.

Schmidt, M. (2022). Cereal beta-glucans: An underutilized health endorsing food ingredient. *Critical Reviews in Food Science and Nutrition*, *62*(12), 3281–3300.

Selle, P. H., McInerney, B. V., McQuade, L. R., Khoddami, A., Chrystal, P. V., Hughes, R. J., & Liu, S. Y. (2020). Composition and characterisation of kafirin, the dominant protein fraction in grain sorghum. *Animal Production Science*, *60*(9), 1163–1172.

Shewry, P. R. (2023). Cereal grain proteins. In *ICC handbook of 21st century cereal science and technology* (pp. 63–70). Academic Press.

Shewry, P. R., & Hey, S. J. (2015). The contribution of wheat to human diet and health. *Food and Energy Security*, *4*(3), 178–202.

Siminiuc, R., & Coşciug, L. (2021). Effect of dehulling and hydrothermal treatment on the amino acid content of soriz (*Sorghum oryzoidum*). *Food and Nutrition Sciences*, *12*(12), 1232–1242.

Sluková, M., Jurkaninová, L., Švec, I., & Skřivan, P. (2021). Rye – The nutritional and technological evaluation in Czech cereal technology – A review: Grain and flours. *Czech Journal of Food Sciences*, *39*(1).

Sofi, S. A., Nazir, A., & Ashraf, U. (2019). Cereal bioactive compounds: A review. *International Journal of Agriculture, Environment and Biotechnology*, *12*(2), 107–113.

Šramková, Z., Gregová, E., & Šturdík, E. (2009). Chemical composition and nutritional quality of wheat grain. *Acta chimica slovaca*, *2*(1), 115–138.

Swati, P., Rasane, P., Kaur, J., Kaur, S., Ercisli, S., Assouguem, A., ... & Singh, J. (2024). The nutritional, phytochemical composition, and utilisation of different parts of maize: A comparative analysis. *Open Agriculture*, *9*(1), 20220358.

Tanwar, R., Panghal, A., Chaudhary, G., Kumari, A., & Chhikara, N. (2023). Nutritional, phytochemical and functional potential of sorghum: A review. *Food Chemistry Advances*, 100501.

Targońska-Karasek, M., Kwiatek, M., Groszyk, J., Walczewski, J., Kowalczyk, M., Pawelec, S., ... & Rucińska, A. (2023). Characteristic of the gene candidate SecARS encoding alkylresorcinol synthase in *Secale*. *Molecular Biology Reports*, *50*(10), 8373–8383.

Tian, S., Sun, Y., Chen, Z., & Zhao, R. (2020). Bioavailability and bioactivity of alkylresorcinols from different cereal products. *Journal of Food Quality*, *2020*(1), 5781356.

Tiwari, U., & Cummins, E. (2009). Nutritional importance and effect of processing on tocols in cereals. *Trends in Food Science & Technology*, *20*(11–12), 511–520.

Tomičić, Z., Pezo, L., Spasevski, N., Lazarević, J., Čabarkapa, I., & Tomičić, R. (2022). Diversity of amino acids composition in cereals. *Food & Feed Research*, *49*(1), 11–22. https://doi.org/10.5937/ffr0-34322

Topka, P., Rudzińska, M., Poliński, S., Szydłowska-Czerniak, A., & Tańska, M. (2024). Enhancing antioxidant activity and nutritional profile of dark chocolate through enrichment with plant sterols: A study on phytosterol concentrations and functional properties. *Foods*, *13*(22), 3578.

Tremmel-Bede, K., Szentmiklóssy, M., Lampi, A. M., Piironen, V., Shewry, P. R., Vida, G., … & Rakszegi, M. (2022). Variation in the content and composition of tocols in a wheat population. *Foods*, *11*(9), 1343.

Van Hung, P. (2016). Phenolic compounds of cereals and their antioxidant capacity. *Critical Reviews in Food Science and Nutrition*, *56*(1), 25–35.

Verma, D. K., Patel, A. R., Thakur, M., Singh, S., Tripathy, S., Srivastav, P. P., … & Aguilar, C. N. (2021). A review of the composition and toxicology of fructans, and their applications in foods and health. *Journal of Food Composition and Analysis*, *99*, 103884.

Wang, Y., Fan, M., Qian, H., Ying, H., Li, Y., & Wang, L. (2024). Whole grains-derived functional ingredients against hyperglycemia: Targeting hepatic glucose metabolism. *Critical Reviews in Food Science and Nutrition*, *64*(20), 7268–7289.

Xu, D., Hao, J., Wang, Z., Liang, D., Wang, J., Ma, Y., & Zhang, M. (2021). Physicochemical properties, fatty acid compositions, bioactive compounds, antioxidant activity and thermal behavior of rice bran oil obtained with aqueous enzymatic extraction. *LWT*, *149*, 111817.

Yoshida, M. (2021). Fructan structure and metabolism in overwintering plants. *Plants*, *10*(5), 933.

Zanella, I., Biasiotto, G., Holm, F., & di Lorenzo, D. (2017). Cereal lignans, natural compounds of interest for human health? *Natural Product Communications*, *12*(1). https://doi.org/10.1177/1934578X1701200139

Zannini, E., Bravo Núñez, Á., Sahin, A. W., & Arendt, E. K. (2022). Arabinoxylans as functional food ingredients: A review. *Foods*, *11*(7), 1026.

Zhang, K., Dong, R., Hu, X., Ren, C., & Li, Y. (2021). Oat-based foods: Chemical constituents, glycemic index, and the effect of processing. *Foods*, *10*(6), 1304.

Zhao, Y., Zhai, G., Li, X., Tao, H., Li, L., He, Y., … & Zhu, Y. (2022). Metabolomics reveals nutritional diversity among six coarse cereals and antioxidant activity analysis of grain sorghum and sweet sorghum. *Antioxidants*, *11*(10), 1984.

Zieliński, H. (2008). Tocotrienols: Distribution and sources cereals – Role in human health. In *Tocotrienols: Vitamin E beyond tocopherols* (pp. 23–42). CRC Press.

Zielinski, H., Ciska, E., & Kozlowska, H. A. L. I. N. A. (2001). The cereal grains: Focus on vitamin E. *Czech Journal of Food Sciences*, *19*(5), 182–188.

Žilić, S. (2016). Phenolic compounds of wheat. Their content, antioxidant capacity and bioaccessibility. *MOJ Food Processing & Technology*, *2*(3), 00037.

3 Rye

An In-Depth Exploration into Its Physical, Chemical and Nutritional Properties

Gholamreza Abdi, Ubaida Akbar, B. S. Adithya, Md Sabir Ahmed Mondol, Maryam Abbasi Tarighat and Charalampos Proestos

3.1 INTRODUCTION

Cereal grains have become a focal point for both the food and pharmaceutical sectors due to their potential in reducing many chronic diseases. Presently, there is a growing emphasis on diets enriched with targeted nutrients and bioactive compounds known for their health benefits (Purewal et al., 2022; Salazar-López et al., 2020). Researchers worldwide are examining the health-promoting properties of diverse cereal grains. Important grains under study for their nutritional composition and health benefits include wheat (Gupta et al., 2021; Singh, 2023), rice (Kowsalya et al., 2022), maize (Meena et al., 2022; Rouf Shah et al., 2016) and millets (Kaur et al., 2023; Tripathi et al., 2021).

Rye, a resilient member of the Poaceae family, has been cultivated for centuries across Europe, where it remains a staple in many traditional foods and beverages (Targonska-Karasek et al., 2020). Originating from the Triticeae tribe, rye shares close botanical ties with wheat and barley, but it stands out for its ability to flourish in regions with challenging climates and soil conditions (Kataria et al., 2024). Unlike most other cereals, rye can grow in sandy and low-fertility soils, demonstrating an extraordinary hardiness that makes it suitable for areas with harsh winters and cooler climates (Gruner & Miedaner, 2021; Korzun et al., 2021). This adaptability has established rye as a valuable crop across northern and eastern Europe, where countries like Germany, Poland and Russia lead in its production. In fact, Europe supplies approximately 85% of the world's rye, underscoring its regional significance (Nemeth & Tomoskozi, 2021). Rye makes up only about 13% of global grain trade, as it is mainly produced for local food needs and on-farm use (Wrigley & Bushuk, 2017). Its global cereal production may be relatively modest, and it holds a unique position in food culture and nutrition. Moreover, grain composition is notably rich in dietary fiber, primarily arabinoxylans, which

DOI: 10.1201/9781003585817-3

are beneficial for digestive health and blood sugar regulation (Jonsson et al., 2018; Kaur et al., 2021).

The starch content, accounting for more than half of the grain's weight, shares structural similarities with wheat but brings different functional properties to baked goods, contributing to the dense, flavorful texture of rye bread (Deleu et al., 2020; Ikram et al., 2023). Protein, though less concentrated than in some other cereals, remains nutritionally valuable, with albumins and prolamins being the predominant types (Dziki, 2022; El-Mahis et al., 2023). Additionally, it also contains important minerals like phosphorus, potassium and magnesium, adding to its nutritional profile (Kaur et al., 2024).

The chemical composition of rye, influenced by factors such as genetic traits, type of soil and climate, contributes to its growing appeal in health-conscious food markets (Ghafoor et al., 2024). As interest in functional foods and whole grains increases, researchers are examining its bioactive compounds, including phenolic acids and alkylresorcinols, for their potential antioxidant, anti-inflammatory and gut health benefits. This renewed scientific attention positions it as a promising ingredient for innovative product development, from nutrient-dense breads and snacks to fermented beverages like rye-based whiskies and vodkas (Kataria et al., 2024).

3.2 BOTANICAL AND MORPHOLOGICAL CHARACTERISTICS OF RYE

The term 'rye' has its origins in the Germanic and Old Norse languages, indicating its integration into English culture from those areas (Wrigley & Bushuk, 2017). It is commonly referred to as 'ryecorn' or 'cereal rye' to differentiate it from ryegrass, which belongs to the *Lolium* genus (Pinnamaneni et al., 2022). Botanically, rye is part of the Gramineae family (also known as Poaceae), within the Pooideae subfamily and Triticeae tribe, and is primarily classified as *Secale cereale* (L.), with a smaller quantity of *Secale fragile* cultivated in southwestern Asia (Tang et al., 2011). It is believed to have originated from *Secale montanum*, a wild species perennial grass native to southern Europe (Daskalova & Spetsov, 2020).

3.3 GENETIC COMPOSITION OF RYE

Cultivated rye is typically diploid, featuring seven pairs of chromosomes (1R-7R), which correspond to the chromosome pairs of wheat (1A-7A, 1B-7B, 1D-7D) (Mikhailova & Doležel, 2021). Although tetraploid varieties of rye have been developed, they are not commonly cultivated due to their vulnerability to ergot and lower resistance to cold temperatures (Wrigley & Bushuk, 2017). Rye is categorized as either winter or spring, depending on the planting season, with winter varieties being well-suited for colder climates. Genetic research has identified differences between spring and winter types, often correlating with their geographical origins, underscoring their unique genetic adaptations to different environmental conditions (Hackauf et al., 2021).

3.4 PHYSICAL COMPOSITION OF RYE

Rye is a cereal grain recognized for its distinctive physical properties, which plays an important role in its effectiveness and versatility in milling, baking, and broader food industry applications (Kataria et al., 2024). An in-depth understanding of the physical composition of this grain—covering factors like density, particle size distribution, structural characteristics and grain hardness—is crucial for optimizing its processing efficiency, enhancing its nutritional profile and tailoring its use across different food products (Dziki et al., 2023; El-Mahis et al., 2023).

3.5 PHYSICAL PROPERTIES

3.5.1 Density and Bulk Density

Density, a critical physical property, influences the handling, storage and processing efficiency of rye. Specifically, it refers to the mass of the grain per unit volume, which in rye is often measured using the pycnometer method for accurate specific density (Eadkhong & Danworaphong, 2021). Bulk density, on the other hand, accounts for the packing of grains in a given volume, including air spaces between kernels, providing insight into the weight of rye when stored or transported (Chiputula et al., 2021).

Density and bulk density directly impact rye processing and quality. A higher bulk density usually signifies grains with lower moisture content and fewer impurities, which are often markers of superior quality. In the milling process, a uniform density is beneficial as it allows for consistent milling performance, leading to higher flour yields and better quality (Drakos et al., 2017). Additionally, bulk density has implications for storage, as grains with lower bulk density require more space and may be more prone to spoilage and pest issues. By understanding these density characteristics, processors can optimize storage, milling and transportation conditions to ensure product quality and minimize losses (Bhardwaj & Sharma, 2020).

3.5.2 Particle Size Distribution

The distribution of particle sizes within milled rye is another fundamental property impacting its baking and processing characteristics. During milling, the resulting particle size distribution can affect flour texture, hydration properties and end-product quality. Fine particles yield a smoother flour, while coarser particles are preferred in applications requiring a distinct texture, such as artisanal breads and crackers (Dziki, 2022). Two main methods analyze particle size distribution in rye: sieve analysis and laser diffraction. Sieve analysis separates particles by passing them through stacked sieves of varying mesh sizes, giving a breakdown of particle proportions. Laser diffraction offers precise size profiling by examining light scattering from particles (Silventoinen et al., 2021).

3.6 PHYSICAL STRUCTURE

3.6.1 Kernel Structure

Rye kernels have a distinct layered structure comprising three primary components: the hull (or bran), the starchy endosperm and the germ (Kaur et al., 2021). The endosperm makes up approximately 80–85% of the kernel's weight, while the embryo accounts for 2–3%, and the outer layers constitute about 10–15%. The embryo is located at the basal end and features thin-walled cells containing a large nucleus, protein bodies and fat globules known as spherosomes (Poutanen et al., 2014). The germ consists of the embryo proper and the scutellum, which acts as a shield-like structure against the endosperm. During germination, the scutellum plays a vital role by digesting and absorbing nutrients from the endosperm, which are then transported to support the growth of the embryonic axis. The hull serves as a protective layer, rich in dietary fiber, promoting digestive health while adding a coarser texture to whole-grain products (Ratnavathi, 2019). The endosperm is primarily composed of starches and proteins, providing energy and forming the bulk of flour essential for dough structure and development in baking (Dziki, 2022; Nemeth & Tomoskozi, 2021). Despite its smaller size, the germ is nutritionally rich, contributing fats, vitamins and minerals that enhance the overall quality of whole rye flour (Feng, 2019).

3.6.2 Grain Hardness

Grain hardness greatly affects milling and flour quality. Softer rye grains require less energy to grind and produce finer flour, while harder grains lead to coarser flour. Rye is generally softer than wheat or barley, making it easier to mill, though hardness can vary by variety. By choosing the right hardness, millers can control energy use, maximize flour yield and achieve consistent texture in the final product (Kselíkova et al., 2020).

3.7 CHEMICAL COMPOSITION OF RYE

The chemical composition of rye shares common traits with other cereal grains, viz, wheat, barley and triticale. However, it is particularly noted for its elevated fiber content, particularly in the form of pentosans, found in both the whole grain and endosperm. This high fiber content is regarded as one of the key nutritional advantages of rye along with its significant levels of bioactive compounds (Nemeth & Tomoskozi, 2021; Poutanen et al., 2014). The chemical constituents of rye can be broadly categorized into macronutrients and micronutrients, each playing a vital role in human health.

3.7.1 Macronutrients

3.7.1.1 Proteins

Rye is a valuable cereal grain with a protein content that generally ranges from 8 to 15%, though these levels can fluctuate based on environmental factors such as soil composition, growth conditions and crop variety (Laidig et al., 2017).

Interestingly, studies suggest that genetic factors may exert a stronger influence on the quality and composition of rye protein than environmental conditions alone (Kunkulberga et al., 2017; Li et al., 2021). Rye proteins can be divided into several types, with albumins as the primary proteins, making up 29–40% of the total protein content. Globulins follow at 8–11%, while storage proteins, such as the prolamins (or secalins) and glutelins, account for 17–19% and 9–15%, respectively (Figure 3.1). Furthermore, nearly a quarter of rye's protein, roughly 21–30%, is unextractable when using the Osborne fractionation method, indicating a complex protein profile (Redant et al., 2017). Rye is also recognized for its amino acid profile, which is slightly superior to that of wheat due to its higher lysine content, an essential amino acid often limited in cereal grains. This makes rye particularly advantageous in nutritional contexts, as lysine supports protein synthesis and various metabolic functions (Rani et al., 2021). However, rye remains deficient in some essential amino acids, such as tryptophan and isoleucine, meaning it does not serve as a complete protein source. Among its amino acids, rye contains substantial amounts of glutamic and aspartic acids, which are critical for

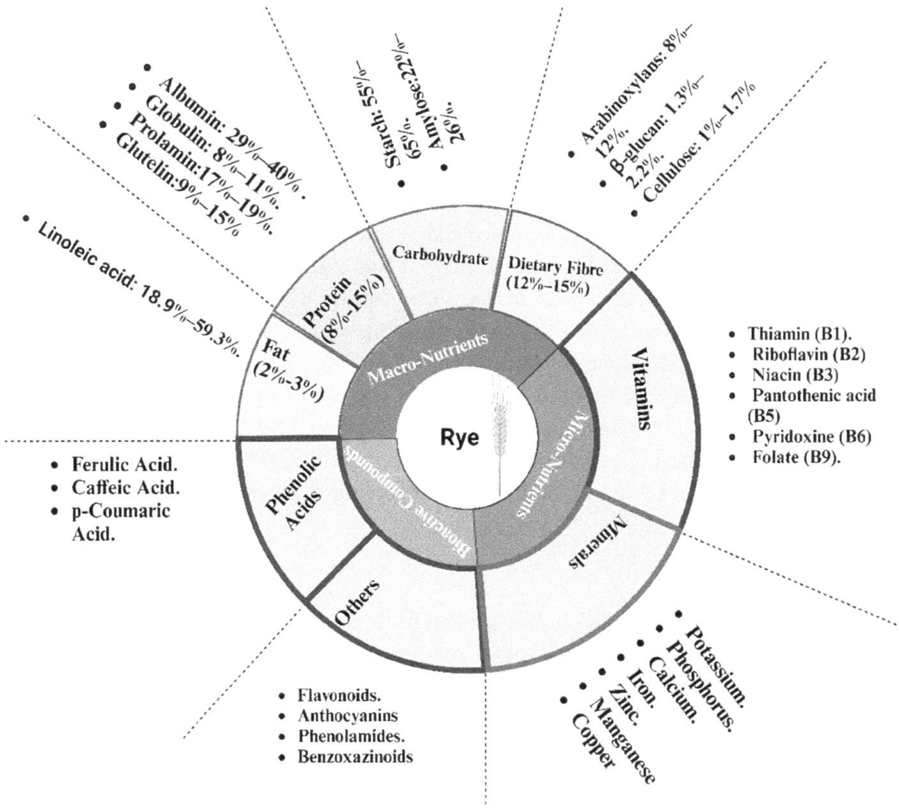

FIGURE 3.1 Chemical composition (major macronutrients, micronutrients and bioactive compounds) present in rye.

metabolic processes. As a result, rye can enhance the nutritional value of other protein sources when combined in a balanced diet (Nemeth & Tomoskozi, 2021; Redant et al., 2017).

3.7.1.2 Carbohydrates

Carbohydrates serve as the primary macronutrient in rye with starch being the most prominent component, similar to other cereal grains. The starch content in rye generally ranges from 55 to 65%, slightly lower than in wheat (63–72%) but higher than in barley (50–64%) (Poutanen et al., 2014). Rye starch, like that of wheat and barley, consists of large A-type granules, which make up 60–90% of the starch and are typically 23–40 µm in diameter, as well as smaller B-type granules, comprising 10–40% and measuring less than 10 µm (Grossmann & Koehler, 2016). The amylose content in rye starches typically falls between 22 and 26%, though some studies have reported higher values, such as 30.1% (Nemeth & Tomoskozi, 2021). Additionally, rye contains other carbohydrate types, including non-starch polysaccharides (NSPs) and simple sugars, which contribute to its nutritional profile (Figure 3.1). Notably, rye starches digest more slowly, resulting in a lower glycemic index, which makes rye a suitable carbohydrate source for individuals managing blood sugar levels (Knudsen & Laerke, 2010).

3.7.1.3 Dietary Fiber

Rye is rich in dietary fiber, with levels ranging from 12 to 15% depending on variety and processing (Andersson et al., 2009). Its fiber profile comprises primarily arabinoxylans (8–12%), β-glucan (1.3–2.2%) and cellulose (1–1.7%), with fructans, including fructooligosaccharides, contributing an additional 4.6–6.6% as shown in Figure 3.1 (Ikram et al., 2023). Arabinoxylan is the predominant fiber in rye, featuring a backbone of (1→4)-β-D-xylopyranosyl units with α-L-arabinofuranosyl side chains. This polysaccharide varies across grain tissues, with high water solubility in the endosperm but limited solubility in the aleurone and pericarp/testa layers (Cyran, 2015). Ferulic acid esterification on arabinoxylan enables cross-linking, enhancing rye's texture and stability in food applications (Ikram et al., 2023). Soluble fibers, including β-glucan, contribute to viscosity and support blood glucose management and cholesterol reduction. Additionally, soluble dietary fibers and fructans provide fermentable substrates for gut microbiota, fostering digestive health (Dahl et al., 2017; Jonsson et al., 2018).

3.7.1.4 Fats

With a lipid content of around 2–3%, similar to that of wheat, rye has a notable fatty acid profile despite its relatively low-fat concentration. Linoleic acid, a polyunsaturated fatty acid, is the predominant component, ranging from 18.9 to 59.3% of total fatty acids (Bagcı et al., 2019; Nemeth & Tomoskozi, 2021). Other essential fatty acids include oleic acid (monounsaturated) and smaller amounts of saturated fats, such as palmitic and stearic acids (Figure 3.1). The unsaturated fats present offer health benefits, contributing to cholesterol management and cardiovascular health (Sacks et al., 2017).

3.7.2 Micronutrients

Micronutrients, required in minimal amounts, are essential for critical biochemical functions, including energy metabolism, immune support and hormone regulation (Elmadfa & Meyer, 2019) Unlike macronutrients, which provide energy, micronutrients facilitate these processes, helping lower risks of diseases like heart issues and diabetes. A diet abundant in diverse micronutrients promotes better health and strengthens the body's defense against chronic conditions (Kaur et al., 2021; Kumar et al., 2024).

3.7.2.1 Vitamins

Rye is naturally enriched with a range of essential vitamins that provide numerous health benefits and support various bodily functions (El-Mahis et al., 2023; Kaur et al., 2021). A standout feature of its nutritional profile is its high content of B vitamins, including thiamin (B1), riboflavin (B2), niacin (B3), pantothenic acid (B5), pyridoxine (B6) and folate (B9) (Figure 3.1). These vitamins are crucial for energy production, nervous system health and cellular repair, contributing to overall wellbeing (Dziki, 2022; Nemeth & Tomoskozi, 2021; Wrigley & Bushuk, 2017).

Thiamin (B1) plays a fundamental role in carbohydrate metabolism, converting glucose into energy to fuel both physical activity and brain function (Tardy et al., 2020; Wolak et al., 2017). Riboflavin, also known as vitamin B2, supports fat and protein breakdown and functions as an antioxidant, safeguarding cells against oxidative stress (Olfat et al., 2022; Suwannasom et al., 2020). Niacin (vitamin B3) aids in DNA repair, supports energy metabolism and helps regulate cholesterol levels, which contributes to heart health (Doroftei et al., 2020; Tiwari et al., 2024). Pantothenic acid (vitamin B5) is crucial for the synthesis of coenzyme A, a key molecule in fatty acid synthesis and energy production (Miallot et al., 2023; Zhao et al., 2023). Pyridoxine, or vitamin B6, is necessary for amino acid metabolism and the synthesis of neurotransmitters, supporting brain function, mood regulation and immune health (Mikkelsen & Apostolopoulos, 2019; Paez-Hurtado et al., 2023). Folate (vitamin B9), vital for DNA synthesis and cell division, is particularly important during pregnancy, as it promotes healthy fetal growth and aids in the prevention of neural tube defects (Tola, 2024; Yingngam, 2024).

Alongside B vitamins, rye is also a source of tocopherol, commonly known as vitamin E, which offers powerful antioxidant properties (El-Mahis et al., 2023; Feng, 2019). Vitamin E defends cells from oxidative damage, boosts immune health and contributes to healthy skin by protecting cell membranes from free radicals (Chu et al., 2023; Khadim & Al-Fartusie, 2021). Another significant nutrient in rye is choline, often grouped with B vitamins due to its similar role in supporting cellular function (Mikkelsen & Apostolopoulos, 2018; Nemeth & Tomoskozi, 2021). Choline is essential for brain health and cognitive function as it aids in the production of acetylcholine, a neurotransmitter involved in memory and muscle control (Kansakar et al., 2023). Furthermore, choline supports lipid metabolism and liver function, helping prevent excess fat buildup in the liver (Mazidi et al., 2019; Moretti et al., 2020).

3.7.2.2 Minerals

Rye is well-known for its adaptability across a range of culinary applications, but also for its impressive mineral profile, which significantly enhances its nutritional value (Kaur et al., 2024). Among the key minerals found in rye, potassium positions out as the most abundant and contributes significantly to various essential bodily functions. This essential mineral is crucial for sustaining fluid balance, managing blood pressure and ensuring the proper functioning of muscles and nerves (Nemeth & Tomoskozi, 2021; Sharma et al., 2020). Phosphorus, another important component (Figure 3.1), works synergistically with calcium to strengthen bones and teeth while also playing an essential role in energy metabolism and the formation of cellular structures, including DNA and RNA (Hayann et al., 2024; Müller et al., 2019). Additionally, magnesium is present in substantial amounts in rye and is involved in over 300 biochemical reactions within the body, including those related to muscle contraction, protein synthesis and energy production (Al-Fartusie & Mohssan, 2017). Calcium, while present in smaller quantities compared to other minerals, remains essential for maintaining strong bones and supporting several physiological processes, such as blood clotting and muscle function (Ciosek et al., 2021; Godswill et al., 2020).

In addition to these major minerals, rye is a rich source of trace elements like iron, zinc, manganese and copper, which are crucial for various bodily functions (Bagcı et al., 2019; Nemeth & Tomoskozi, 2021). Iron is essential for the formation of haemoglobin, the protein responsible for oxygen transport in the bloodstream, and is vital for energy production and metabolic processes (Yiannikourides & Latunde-Dada, 2019). Zinc is fundamental to immune function, wound healing and DNA synthesis, rendering it essential for overall health and development (Chasapis et al., 2020; Jarosz et al., 2017). Manganese contributes to bone formation and antioxidant defense mechanisms, helping to combat oxidative stress in the body (Mezzaroba et al., 2019; Wroblewski et al., 2024). Copper, while required in smaller amounts, is important for iron metabolism, the formation of red blood cells and maintaining healthy connective tissues (Collins, 2017; Jomova et al., 2022). The diverse mineral content of rye (Figure 3.1) not only underscores its status as a nutrient-dense food but also highlights its potential in addressing dietary deficiencies and promoting overall health. By including rye in their diet, individuals can benefit from its rich mineral composition, supporting various physiological processes and lowering the likelihood of chronic illnesses (Adepoju et al., 2024; Matías et al., 2024).

3.7.3 Bioactive Compounds

S. cereale (rye) is distinguished by its rich composition of bioactive compounds that greatly enhance its health-promoting qualities. This grain is abundant in phenolic compounds and specialized alkaloids, which contribute substantially to its antioxidant potential. Compounds such as alkylresorcinols, ferulic acid, catechol, sinapic acid, vanillin and vanillic acid play a vital role in countering oxidative stress, thus protecting cells from damage and supporting immune function. These bioactive components as mention in Figure 3.1 act as powerful free radical scavengers, potentially reducing age-related degeneration and bolstering long-term health (Kaur et al., 2021; Kulichova et al., 2019).

3.7.3.1 Phenolic Compounds

A range of phenolic compounds can be found in rye, including phenolic acids (Table 3.1), flavonoids and lignans, all of which contribute substantially to its health-promoting effects (Dynkowska, 2019). Phenolic acids in rye, such as ferulic acid and caffeic acid, are the most abundant and are well-regarded for antioxidant properties (Pihlava et al., 2018; Verhulst et al., 2023). These acids act as scavengers of free radicals and have anti-inflammatory, anti-cancer and cardioprotective effects. They have been shown to support cardiovascular health by improving blood vessel

TABLE 3.1
Phenolic Compounds in Rye: Sources, Composition and Health Benefits

Phenolic Compound Group	Primary Location in Rye	Notable Compounds	Approximate Composition in Total Phenolics (%)	Health Benefits
p-Hydroxybenzoic Acid Derivatives	Aleurone layer and bran	Vanillic acid and syringic acid	–	Contributes to antioxidant defenses, supports heart health
p-Coumaric Acid Derivatives	Aleurone layer and bran	Ferulic acid and caffeic acid	–	Anti-inflammatory effects, boosts antioxidant capacity, offers antimicrobial properties
Free Phenolic Acids	Throughout rye grain (minor)	Caffeic acid and gallic acid	<3	Enhances antioxidant properties, releases in colon fermentation, promoting gut health
Ferulic Acid	Throughout rye grain	4-Hydroxy-3-methoxycinnamic acid	85–90	Strong antioxidant, combats inflammation, supports cardiovascular health and protects against certain cancers
Sinapic Acid	Common in rye and other cereals	3,5-Dimethoxy-4-hydroxycinnamic acid	9–10	Anti-cancer, anti-inflammatory, neuroprotective, blood sugar regulating, antioxidant properties
p-Coumaric Acid	Widespread in plant-based foods	4-Hydroxycinnamic acid	3–5	Anti-inflammatory, offers antimicrobial properties, potential anti-cancer effects
Other Minor Phenolic Compounds	Throughout rye grain	Protocatechuic acid, vanillic acid and syringic acid	–	General antioxidant properties, may contribute to overall health

Source: Iftikhar et al. (2020), Ferreira et al. (2019), Kulichova et al. (2019), Dynkowska (2019), Pihlava et al. (2015), Adom & Liu (2002), Bondia-Pons et al. (2009), Uraji et al. (2013), and Chen (2016).

function, enhancing lipid metabolism and reducing inflammation, which in turn can lower the risk of cardiovascular disease (Griffiths et al., 2016; Saleem et al., 2022). Additionally, phenolic acids can inhibit tumor cell growth, which highlights their potential role in cancer prevention (Bonta, 2020).

Flavonoids, though less studied in rye than in other grains, play a significant role in defending the body against oxidative stress (Dziki, 2022; Lee et al., 2015). In rye, these compounds typically occur as glycosidic forms, meaning they are bonded to sugar molecules. This structure enhances their stability and bioavailability in the human body (Philava et al., 2015; Ravisankar et al., 2020). Flavonoids work by neutralizing free radicals, thereby reducing oxidative damage to cells. They have also been associated with anti-aging properties, as oxidative stress is a known factor in cellular aging and various age-related diseases (Forni et al., 2019). The presence of these flavonoids in rye adds an extra layer of functional value to the grain, as they may provide protection against a range of degenerative health issues (Dong et al., 2024).

Lignans, a group of polyphenols also found in high amounts in rye, have unique hormonal and antioxidant activities (El-Mahis et al., 2023; Soleymani et al., 2020). These compounds are phytoestrogens, meaning they can mimic estrogen in the body, which may be beneficial for hormone balance, especially in postmenopausal women (Dynkowska, 2019; Kaur et al., 2021). Research has shown that lignans have protective effects against conditions like breast cancer and cardiovascular disease. Rye is one of the richest sources of lignans among cereals, further supporting its role as a functional food with potential long-term health benefits (Jang et al., 2022; Witkowska et al., 2018).

The antioxidant properties of phenolic compounds in rye are enhanced by their metal ion-chelating abilities, which add to their effectiveness in combating oxidative stress (Kaur et al., 2021; Kulichova et al., 2019). By binding to metal ions, these compounds prevent the catalysis of reactions that produce free radicals, thus reducing oxidative damage (Warechowska et al., 2019). Processing methods, including sourdough fermentation, can also influence the bioavailability of these compounds, sometimes making them more accessible for absorption in the human digestive system (Dapcevic-Hadnadev et al., 2022). This factor further underscores the health potential of rye, as these bioactive compounds can be preserved or even enhanced through appropriate food processing techniques, making rye a valuable dietary component for promoting health and preventing chronic diseases (Wang & Wang, 2024).

3.7.3.2 Benzoxazinoids

Benzoxazinoids are nitrogen-rich secondary metabolites found in various cereals, including maize, wheat, rye and triticale. These compounds can be categorized into several groups: hydroxamic acids like DIBOA (2,4-dihydroxy-1,4-benzoxazin-3-one) and DIMBOA (2,4-dihydroxy-7-methoxy-1,4-benzoxazin-3-one); lactams such as HBOA (2-hydroxy-1,4-benzoxazin-3-one) and HMBOA (2-hydroxy-7-methoxy-1,4-benzoxazin-3-one); and benzoxazolinones like BOA (benzoxazolin-2-one) and MBOA (6-methoxy-benzoxazolin-2-one). While the protective roles of benzoxazinoids against insects and pathogens have been explored for some time, their occurrence in rye grains was only recently documented. The predominant form of benzoxazinoids in rye grains is the dihexoside of DIBOA (Hanhineva et al., 2014; Pihlava et al., 2018; Tanwir et al., 2017).

3.7.3.3 Arabinoxylans

Arabinoxylans are heteropolymers consisting of arabinose and xylose sugars, forming a key component of the NSPs found in cereal grains. These polysaccharides play an essential role in maintaining the structural integrity of cell walls, influencing cell interactions and acting as receptors that contribute to various regulatory functions (Wang et al., 2020) As important constituents of the human diet, arabinoxylans have an important impact on metabolic health and are known to affect the physical and chemical properties of cereal products (Zannini et al., 2022).

In rye, arabinoxylans account for about 64% of the total dietary fiber polysaccharides, with their concentration in rye grains ranging from 6.5 to 11.5% on a dry matter basis. Within the endosperm, levels of arabinoxylans are lower, typically between 4.0 and 4.7%. The distribution of soluble dietary fiber, including arabinoxylans, is relatively uniform across cereal grains (Dynkowska, 2020; Hansen et al., 2004). For instance, rye wholemeal flour contains soluble arabinoxylans at levels of 2.1–2.5%, while endosperm flour has a similar concentration of 2.5–2.8% (Cyran & Cygankiewicz, 2004). However, the insoluble fraction of arabinoxylans is significantly lower in endosperm flour compared to wholemeal flour, highlighting the importance of whole grains in achieving optimal dietary fiber intake (Dynkowska, 2020).

3.8 ANTIOXIDANT ACTIVITY

The antioxidant activity of rye is a fundamental aspect of its bioactive profile, contributing significantly to its health benefits (Kaur et al., 2021). When compared to other cereal grains, rye demonstrates a notably higher antioxidant capacity, largely due to its rich content of phenolic compounds and other phytochemicals (Kulichova et al., 2019). Research indicates that rye grain exhibits greater antioxidant activity than both wheat and corn, positioning it as a superior option for individuals aiming to boost their dietary intake of antioxidants (El-Mahis et al., 2023; Korycinska et al., 2009). The unique profile of phytochemicals found in rye works in synergy, enhancing its health-promoting properties. Furthermore, the remarkable antioxidant properties of rye make it an important ingredient in traditional baked goods, as rye flour and rye-based bread show better antioxidative characteristics compared to wheat-based products. The antioxidant compounds in rye seeds play a vital role in detoxification processes, aiding in seed survival against oxidative stress and cellular damage. Moreover, the distribution and profile of these antioxidants are influenced by various genetic and environmental factors, underscoring the importance of understanding rye's antioxidant potential for promoting health (Bailly, 2004; Garcia-Caparros et al., 2021).

3.9 ANTINUTRITIONAL COMPONENTS

Rye is a highly nutritious grain, yet it contains certain antinutritional factors that can interfere with nutrient absorption, digestion and overall bioavailability. Among the primary antinutritional compounds in rye is phytic acid, which acts as a mineral binder (Ram et al., 2020). This compound can sequester essential minerals like calcium, magnesium, iron and zinc, forming complexes that the digestive system finds challenging to break down and absorb (Feizollahi et al., 2021). Without proper processing, a diet high in rye may lead to deficiencies in these minerals, particularly when rye is a staple food (Wrigley & Bushuk, 2017). Fortunately, fermentation methods, such as those used in sourdough bread, can break down phytic acid, thus enhancing the bioavailability of minerals (Wang & Wang, 2024).

Another significant antinutritional component in rye is alkylresorcinols, a group of phenolic lipids concentrated in the outer layers of the grain. While these compounds possess antioxidant properties, they can reduce the digestibility of proteins and lipids, limiting the energy and protein absorption from rye-based foods (Feng, 2019; Grabinski et al., 2021). Additionally, rye contains enzyme inhibitors, especially protease inhibitors, which interfere with digestive enzymes like trypsin and chymotrypsin (Piasecka-Kwiatkowska et al., 2012). These inhibitors can reduce protein digestibility and amino acid availability, which may cause mild digestive discomfort (Joye, 2019). However, cooking and fermentation can deactivate these inhibitors, thereby improving the protein content's accessibility and digestibility (Cirkovic Velickovic & Stanic-Vucinic, 2018; Sa et al., 2020).

The complex polysaccharides in rye, known as pentosans, are another noteworthy factor (Poutanen et al., 2014; Stępniewska et al., 2024). Pentosans have a high water-binding capacity, which can be beneficial for increasing stool bulk and supporting digestive health but may also affect nutrient absorption and make rye dough challenging to handle due to its stickiness (David et al., 2019; Dynkowska, 2020). Enzymatic treatments can be applied to manage pentosan content, improving the dough's workability and texture in baking (Naderi et al., 2023).

Dietary fiber, particularly its NSPs like arabinoxylans, also poses a dual role. While these fibers aid in digestion and promote gut health, they can reduce the digestibility of other nutrients when consumed in large quantities (Bederska-Lojewska et al., 2017). Rye also contains small amounts of tannins, polyphenolic compounds that bind with proteins and minerals, reducing their absorption and impacting protein quality by limiting essential amino acid availability (El-Mahis et al., 2023).

To counter these antinutritional effects, several processing methods are widely used. Fermentation, as in sourdough preparation, effectively reduces phytic acid and tannins, improving mineral bioavailability (Wang & Wang, 2024). Soaking and sprouting methods can break down enzyme inhibitors and reduce phytic acid, while cooking and baking help inactivate enzyme inhibitors, enhancing protein absorption. By applying these processing techniques, antinutritional effects can be minimized, unlocking its full nutritional potential and making it more digestible and beneficial for consumption (Elliott et al., 2022; Kumar et al., 2022).

3.10 NUTRITIONAL ASPECTS OF RYE

Rye (*S. cereale*) is increasingly valued for its distinctive taste and flexibility in culinary applications, as well as for its notable nutritional qualities. With its high fiber levels, essential minerals and beneficial bioactive components, rye is an excellent choice for enhancing a healthy diet.

3.11 POTENTIAL HEALTH BENEFITS AND RISKS OF RYE CONSUMPTION

Rye, distinguished for its fiber-rich profile, offers a variety of health benefits due to its unique composition, particularly when consumed as whole grain. The European Food Safety Authority (EFSA) has affirmed that rye fiber aids in promoting digestive health by supporting regular bowel function, an effect especially noted when rye is eaten in sufficient quantities (EFSA Panel on Dietetic Products, Nutrition and Allergies (NDA), 2014).

Numerous studies underscore potential of rye to benefit cardiovascular health, aid in managing type 2 diabetes and potentially reduce the risk of colorectal cancer. The high dietary fiber content of rye, combined with its bioactive compounds, plays an essential role in these health benefits. Fiber helps to regulate blood glucose levels by slowing digestion, reducing blood sugar spikes and improving glycemic control, which is predominantly beneficial for individuals with or at risk for type 2 diabetes (Arens, 2015; Jonsson et al., 2018). Moreover, it also appears to have favorable effects on blood cholesterol, especially LDL cholesterol, which contributes to improved heart health and may reduce the incidence of cardiovascular diseases. However, research findings on some health outcomes remain inconsistent. The type of rye product, whether whole grain, refined or mixed with other grains, can influence the fiber and nutrient profiles, potentially altering health outcomes (Nemeth & Tomoskozi, 2021).

Despite various benefits of rye, some individuals may experience adverse effects. Rye contains gluten-related proteins, making it unsuitable for individuals with celiac disease who require a gluten-free diet (Malalgoda & Simsek, 2017). Moreover, rye is high in fructan, a type of FODMAP (Fermentable Oligosaccharides, Disaccharides, Monosaccharides and Polyols) that can be challenging for people with irritable bowel syndrome (IBS) to digest. Wholemeal rye products or those high in bran are particularly rich in fructans, which can accumulate up to 6.6 g per 100 g in the bran fraction and around 3 g per 100 g in refined light rye flours (Pejcz et al., 2020). In individuals with IBS, undigested FODMAPs ferment in the colon, leading to gas production, bloating, abdominal pain and altered bowel habits (Nemeth & Tomoskozi, 2021).

3.12 PROCESSING AND APPLICATIONS OF RYE

Rye undergoes particular processing methods to convert the grain into different types of flour, and its applications in cooking range from classic to modern recipes, extending to nutraceutical products.

3.12.1 Milling and Flour Production

3.12.1.1 Milling Process

The milling of rye grains is a distinctive process, shaped by the grain's specific structural and compositional properties. Compared to wheat, rye kernels contain more water-soluble fibers and lower gluten content, making the milling process unique and tailored to preserve these features (Rosell & Koehler, 2023; Wrigley, 2019).

According to research, rye grains undergo a series of initial processing steps that include cleaning and conditioning (Papageorgiou & Skendi, 2018). Conditioning involves adjusting the moisture content of the grain to prepare it for grinding, which is a critical step because it affects the separation of the bran and endosperm layers and influences the quality of the final flour product (Dziki et al., 2023). Studies have shown that conditioning rye to an optimal moisture level aids in achieving a consistent texture and improves the flour yield (Meeus et al., 2020).

Once conditioned, rye kernels can be milled using either traditional stone mills or modern roller mills. Stone milling is typically chosen for whole-grain rye products, as it grinds the grains more gently, maintaining the nutrient density, fiber and natural bioactive compounds found within the bran and germ layers (Carcea et al., 2019). This traditional approach produces a coarser flour, which is popular in health-oriented baking due to its high fiber, vitamin and mineral content. Stone milling is also associated with a more robust flavor profile, making it well-suited for artisanal and rustic breads (Wrigley & Bushuk, 2017).

In contrast, roller mills are generally employed for creating refined rye flour, particularly white rye flour, which has less bran and a lighter texture (Aprodu & Banu, 2017). Roller milling provides a faster, more efficient grinding method, especially useful in large-scale production settings. While refined rye flour contains fewer nutrients than whole grain, it still retains more fiber and nutrients than comparable refined wheat flours, appealing to consumers who seek both refinement and nutritional value (Gao et al., 2020). The roller milling process divides rye into several fractions, which vary in phytochemical and nutrient content based on the specific milling stream from which they are derived.

The internal structure of rye affects its milling in other ways. Unlike wheat, where the separation of bran and endosperm is less intensive, rye's compact kernel layers require a more forceful milling approach with fewer stages. During milling, the outer layers of rye remain strongly attached to the endosperm, making it challenging to achieve complete separation (Dziki, 2022).Consequently, rye flour typically has higher bran and germ content than refined wheat flour, naturally increasing its dietary fiber and micronutrient levels, including B vitamins and minerals (El-Mahis et al., 2023).

A notable feature of rye milling is the susceptibility of its starch granules to mechanical damage (Wang & Zheng, 2020). Rye starch granules are more vulnerable than those in wheat, meaning they can be easily fractured during grinding. This mechanical damage has implications for the flour's functional properties, particularly in fermentation and baking (Dziki, 2022). Damaged starch absorbs water more readily, which can influence the consistency and stability of dough and impact the final texture (Eliasson, 2017).

Producing whole grain rye flour is simpler than refined flour, but the coarseness of bran and flour streams significantly affects wholemeal rye flour properties (Michalska et al., 2007). Various grinding methods are employed to finely grind the outer grain layers, improving flour characteristics. Wholemeal rye flour typically contains around 2% ash and 15–21% fiber, with almost twice the phenolic content compared to white rye flour (Kołodziejczyk et al., 2018). A dry fractionation technique was proposed to produce fiber-rich rye flour, using roller and ball mills to generate coarse and fine fractions, with the fine fraction containing more soluble fiber (Kołodziejczyk et al., 2018). The milling technique, grinding conditions and extraction parameters greatly affect the chemical composition and baking properties of rye flour (Dziki, 2022; Gómez et al., 2009). Bread made from wholemeal rye flour has a denser, darker crumb and smaller loaf volume than refined flour bread. Additionally, bioactive compounds like phytates, minerals and lignans in rye's outer layers offer significant health benefits (Dziki, 2022).

Ultrafine grinding (UFG), also known as micronization, is an efficient technique for reducing the particle size of cereal grains and other foods, which significantly affects their functional properties and suitability for processing (Chen et al., 2018; Hemery et al., 2011). This process can be fine-tuned by selecting the appropriate mill type and settings (Dhiman & Prabhakar, 2021). UFG methods typically include dry milling, favored for processing cereal grains and wet milling. Common UFG techniques in cereal processing include jet milling, ball milling and impact milling (Protonotariou et al., 2021; Van Craeyveld et al., 2009; Yan et al., 2020). Kołodziejczyk et al. (2018) demonstrated that combining ball milling and roller milling for high-fiber rye flour yields fine and coarse fractions with distinct dietary fiber contents. Similarly, jet milling, which accelerates particles via pressurized gas for size reduction through collision, has been effective for rye, though it minimally affects phenolic extraction despite increased starch granule damage (Drakos et al., 2017). Air classification and UFG was used for rye bran, achieving protein-rich fractions with enhanced emulsion stability and a notable increase in soluble fiber content (Silventoinen et al., 2021).

3.13 CURRENT DEVELOPMENTS IN THE UTILIZATION OF RYE BRAN AND WHOLEMEAL RYE FLOUR

The utilization of rye bran and wholemeal rye flour has gained significant attention in recent years, driven by their nutritional benefits and versatile applications in food production, packaging and bioengineering. The increasing utilization of rye bran and wholemeal rye flour is gaining traction across various sectors, particularly due to their notable nutritional benefits and versatility. One significant application is the incorporation of wholemeal rye flour into wheat wafer batter, which has been shown to enhance the porosity and fracturability of the wafers while concurrently decreasing their hardness (Nasabi et al., 2021). In the domain of snack production, wholemeal rye flour has been effectively combined with wholemeal milk powder, leveraging extrusion-based 3D printing techniques. This innovative method not only augments the shape stability of the resulting snacks but also enriches them with dietary fiber, presenting a healthier alternative for consumers (Lille et al., 2020). In addition, rye flour has been utilized as a 10% component in biocomposite materials for packaging purposes. This application has resulted in decreased water permeability and increased stiffness, enhancing the overall functionality of the packaging (Beigmohammadi et al., 2020).

The use of dephytated rye bran as a substitute for 20% of wheat flour in pasta production has yielded promising outcomes, leading to an increase in the levels of minerals, fiber and total phenolic compounds, thereby enriching the nutritional profile of the pasta (Levent et al., 2020). Furthermore, research indicates that partial replacement of wheat flour with enzyme-treated dry rye bran in bread production significantly elevates the levels of free phenolic acids, particularly ferulic acid, which contributes positively to health benefits (Koistinen et al., 2016). Another interesting application involves the fermentation and acidification of rye bran as an additive in rye flour extrudates, resulting in improvements in texture, such as reduced hardness and enhanced crispness, along with an increase in soluble dietary fiber content (Nikinmaa et al., 2020). In the realm of snack production, the incorporation of rye bran as a partial replacement for 20 and 40% of corn grits has led to products containing approximately 9.04% dietary fiber, while maintaining acceptable sensory attributes, particularly with the lower substitution level (Makowska et al., 2015). Moreover, the use of rye bran in meat processing has been explored, specifically as a partial fat replacement in Frankfurter-like sausages. This application revealed that while the sole addition of rye bran produced undesirable stale flavors, combining it with collagen resulted in a more favorable flavor profile for the sausages (Hjelm et al., 2019). Additionally, rye bran has been employed in the production of fermentable sugars, yielding hydrolyzates that contain 0.72 grams of reducing sugars per gram of bran, predominantly in the form of glucose. This highlights its potential role in bioethanol production (Demirel et al., 2022).

The extraction of water-extractable arabinoxylans from rye bran for application in frozen steamed bread has demonstrated promising results, such as increased bread volume and reduced crumb hardness (Wang et al., 2016). Furthermore, rye bran is also recognized for its capability to produce enzymes necessary for the hydrolysis of waste materials, facilitating the conversion of lignocellulosic substrates into fermentable sugars for ethanol production (Mihajlovski et al., 2021). Alkylresorcinols extracted from rye bran have shown effectiveness in protecting apples from fungal diseases, reducing infestations during harvest and storage without adversely affecting the quality of the fruit (Tahir et al., 2019). Henceforth, the multifaceted applications of rye bran and wholemeal rye flour not only enhance the nutritional quality of various food products but also contribute to sustainable practices in food processing and packaging. Their adaptability across different contexts underscores their significance in the contemporary food industry, paving the way for innovative solutions that benefit both consumers and producers alike (Dziki, 2022).

3.14 CONCLUSION

The physical and chemical composition of rye highlight its value as a versatile and nutritionally rich grain with extensive applications across food, health and industrial sectors. Physical characteristics, viz, dense grain structure and high fiber content, enhance the functionality and textural quality of various food products, while its chemical composition of complex carbohydrates, proteins and diverse dietary fibers, particularly arabinoxylans and β-glucans, supports digestive health and blood sugar management. Additionally, rye is a potent source of bioactive compounds, including phenolic acids, alkylresorcinols and lignans, which impart significant antioxidant, anti-inflammatory and potential anti-cancer benefits, making it ideal for the development of functional foods aimed at promoting health and wellness. Its resilience to processing, from fermentation to extrusion, ensures that rye-based products maintain their nutritional integrity, catering to the rising consumer demand for sustainable, health-focused ingredients. By integrating rye into health-promoting and environmentally conscious food products, the food industry can leverage this remarkable grain to meet the evolving demands for nutrient-rich, functional and resilient food solutions.

REFERENCES

Adepoju, M., Verheecke-Vaessen, C., Pillai, L. R., Phillips, H., & Cervini, C. (2024). Unlocking the potential of teff for sustainable, gluten-free diets and unravelling its production challenges to address global food and nutrition security: A review. *Foods, 13*(21), 3394. https://doi.org/10.3390/foods13213394

Adom, K. K., & Liu, R. H. (2002). Antioxidant activity of grains. *Journal of Agricultural and Food Chemistry, 50*(21), 6182–6187. https://doi.org/10.1021/jf0205099

Al-Fartusie, F. S., & Mohssan, S. N. (2017). Essential trace elements and their vital roles in human body. *Indian Journal of Advances in Chemical Science, 5*(3), 127–136. https://doi.org/10.22607/IJACS.2017.503003

Andersson, R., Fransson, G., Tietjen, M., & Åman, P. (2009). Content and molecular-weight distribution of dietary fiber components in whole-grain rye flour and bread. *Journal of Agricultural and Food Chemistry, 57*(5), 2004–2008. https://doi.org/10.1021/jf801280f

Aprodu, I., & Banu, I. (2017). Milling, functional and thermo-mechanical properties of wheat, rye, triticale, barley and oat. *Journal of Cereal Science, 77*, 42–48. https://doi.org/10.1016/j.jcs.2017.07.009

Arens, U. (2015). Authorised EU health claim for rye fibre. In *Foods, nutrients and food ingredients with authorised EU health claims* (Vol. 2, pp. 129–137). Woodhead Publishing. https://doi.org/10.1016/B978-1-78242-382-9.00007-4

Bagcı, A., Gecgel, U., Dursun, N., Ozcan, M. M., Tamkoç, A., Ozer, İ., & Ozcan, M. M. (2019). The oil yields, mineral contents and fatty acid compositions of some rye (*Secale cereale*) grains. *Iranian Journal of Chemistry and Chemical Engineering (IJCCE), 38*(5), 285–292.

Bailly, C. (2004). Active oxygen species and antioxidants in seed biology. *Seed Science Research, 14*(2), 93–107. https://doi.org/10.1079/SSR2004159

Bederska-Lojewska, D., Swiatkiewicz, S., Arczewska-Włosek, A., & Schwarz, T. (2017). Rye non-starch polysaccharides: Their impact on poultry intestinal physiology, nutrients digestibility and performance indices – A review. *Annals of Animal Science, 17*(2), 351–369. https://doi.org/10.1515/aoas-2016-0090

Beigmohammadi, F., Barzoki, Z. M., & Shabanian, M. (2020). Rye flour and cellulose reinforced starch biocomposite: A green approach to improve water vapor permeability and mechanical properties. *Starch-Stärke, 72*(5–6), 1900169. https://doi.org/10.1002/star.201900169

Bhardwaj, S., & Sharma, R. (2020). The challenges of grain storage: A review. *International Journal of Farm Sciences, 10*(2), 18–22. https://doi.org/10.5958/2250-0499.2020.00028.2

Bondia-Pons, I., Aura, A. M., Vuorela, S., Kolehmainen, M., Mykkänen, H., & Poutanen, K. (2009). Rye phenolics in nutrition and health. *Journal of Cereal Science, 49*(3), 323–336. https://doi.org/10.1016/j.jcs.2009.01.007

Bonta, R. K. (2020). Dietary phenolic acids and flavonoids as potential anti-cancer agents: Current state of the art and future perspectives. *Anti-Cancer Agents in Medicinal Chemistry (Formerly Current Medicinal Chemistry-Anti-Cancer Agents)*, *20*(1), 29–48. https://doi.org/10.2174/1871520619666191019112712

Carcea, M., Turfani, V., Narducci, V., Melloni, S., Galli, V., & Tullio, V. (2019). Stone milling versus roller milling in soft wheat: Influence on products composition. *Foods*, *9*(1), 3. https://doi.org/10.3390/foods9010003

Chasapis, C. T., Ntoupa, P. S. A., Spiliopoulou, C. A., & Stefanidou, M. E. (2020). Recent aspects of the effects of zinc on human health. *Archives of Toxicology*, *94*, 1443–1460. https://doi.org/10.1007/s00204-020-02702-9

Chen, C. (2016). Sinapic acid and its derivatives as medicine in oxidative stress-induced diseases and aging. *Oxidative Medicine and Cellular Longevity*, *2016*(1), 3571614. https://doi.org/10.1155/2016/3571614

Chen, T., Zhang, M., Bhandari, B., & Yang, Z. (2018). Micronization and nanosizing of particles for an enhanced quality of food: A review. *Critical Reviews in Food Science and Nutrition*, *58*(6), 993–1001. https://doi.org/10.1080/10408398.2016.1236238

Chiputula, J., Ajayi, E., Bucklin, R., & Blount, A. R. (2021). Effects of moisture content and compaction pressure on bulk density of rye. *Applied Engineering in Agriculture*, *37*(3), 491–494. https://doi.org/10.13031/aea.14385

Chu, C. C., Chew, S. C., Liew, W. C., & Nyam, K. L. (2023). Review article vitamin E: A multi-functional ingredient for health enhancement and food preservation. *Journal of Food Measurement and Characterization*, *17*(6), 6144–6156. https://doi.org/10.1007/s11694-023-02042-z

Ciosek, Ż., Kot, K., Kosik-Bogacka, D., Łanocha-Arendarczyk, N., & Rotter, I. (2021). The effects of calcium, magnesium, phosphorus, fluoride, and lead on bone tissue. *Biomolecules*, *11*(4), 506. https://doi.org/10.3390/biom11040506

Cirkovic Velickovic, T. D., & Stanic-Vucinic, D. J. (2018). The role of dietary phenolic compounds in protein digestion and processing technologies to improve their antinutritive properties. *Comprehensive Reviews in Food Science and Food Safety*, *17*(1), 82–103. https://doi.org/10.1111/1541-4337.12320

Collins, J. F. (2017). Copper: Basic physiological and nutritional aspects. In *Molecular, genetic, and nutritional aspects of major and trace minerals* (pp. 69–83). Academic Press. https://doi.org/10.1016/B978-0-12-802168-2.00007-5

Cyran, M., & Cygankiewicz, A. (2004). Variability in the content of water-extractable and water-unextractable non-starch polysaccharides in rye flour and their relationship to baking quality parameters. *Cereal Research Communications*, *32*, 143–150. https://doi.org/10.1007/BF03543292

Cyran, M. R. (2015). Dietary fiber arabinoxylans in processed rye: Milling-and breadmaking-induced changes. In *Processing and impact on active components in food* (pp. 319–328). Academic Press. https://doi.org/10.1016/B978-0-12-404699-3.00038-X

Dahl, W. J., Agro, N. C., Eliasson, Å. M., Mialki, K. L., Olivera, J. D., Rusch, C. T., & Young, C. N. (2017). Health benefits of fiber fermentation. *Journal of the American College of Nutrition*, *36*(2), 127–136. https://doi.org/10.1080/07315724.2016.1188737

Dapcevic-Hadnadev, T., Stupar, A., Stevanović, D., Skrobot, D., Maravic, N., Tomic, J., & Hadnadev, M. (2022). Ancient wheat varieties and sourdough fermentation as a tool to increase bioaccessibility of phenolics and antioxidant capacity of bread. *Foods*, *11*(24), 3985. https://doi.org/10.3390/foods11243985

Daskalova, N., & Spetsov, P. (2020). Taxonomic relationships and genetic variability of wild *Secale* L. species as a source for valued traits in rye, wheat and triticale breeding. *Cytology and Genetics*, *54*(1), 71–81. https://doi.org/10.3103/S0095452720010041

David, I., Velciov, A., Corina, M., Bujancă, G., & Danci, M. (2019). The influence of pentosanases enzymes on rye flour. *International Multidisciplinary Scientific GeoConference: SGEM*, *19*(6.1), 985–992.

Deleu, L. J., Lemmens, E., Redant, L., & Delcour, J. A. (2020). The major constituents of rye (*Secale cereale* L.) flour and their role in the production of rye bread, a food product to which a multitude of health aspects are ascribed. *Cereal Chemistry*, *97*(4), 739–754. https://doi.org/10.1002/cche.10306

Demirel, F., Germec, M., & Turhan, I. (2022). Fermentable sugars production from wheat bran and rye bran: Response surface model optimization of dilute sulfuric acid hydrolysis. *Environmental Technology*, *43*(24), 3779–3800. https://doi.org/10.1080/09593330.2021.1934563

Dhiman, A., & Prabhakar, P. K. (2021). Micronization in food processing: A comprehensive review of mechanistic approach, physicochemical, functional properties and self-stability of micronized food materials. *Journal of Food Engineering*, *292*, 110248. https://doi.org/10.1016/j.jfoodeng.2020.110248

Dong, L., Li, Y., Chen, Q., Liu, Y., Wu, Z., Pan, D., Zhao, Y., & Liu, L. (2024). Cereal polyphenols inhibition mechanisms on advanced glycation end products and regulation on type 2 diabetes. *Critical Reviews in Food Science and Nutrition*, *64*(26), 9495–9513. https://doi.org/10.1080/10408398.2023.2213768

Doroftei, B., Ilie, O.-D., Cojocariu, R.-O., Ciobica, A., Maftei, R., Grab, D., Anton, E., McKenna, J., Dhunna, N., & Simionescu, G. (2020). Minireview exploring the biological cycle of vitamin B3 and its influence on oxidative stress: Further molecular and clinical aspects. *Molecules*, *25*(15), 3323. https://doi.org/10.3390/molecules25153323

Drakos, A., Kyriakakis, G., Evageliou, V., Protonotariou, S., Mandala, I., & Ritzoulis, C. (2017). Influence of jet milling and particle size on the composition, physicochemical and mechanical properties of barley and rye flours. *Food Chemistry*, *215*, 326–332. https://doi.org/10.1016/j.foodchem.2016.07.169

Drakos, A., Malindretou, K., Mandala, I., & Evageliou, V. (2017). Protein isolation from jet milled rye flours differing in particle size. *Food and Bioproducts Processing*, *104*, 13–18. https://doi.org/10.1016/j.fbp.2017.04.004

Dynkowska, W. (2020). Rye (*Secale cereale* L.) arabinoxylans: Molecular structure, physicochemicals properties and the resulting pro-health effects. *Plant Breeding and Seed Science*, *81*, 19–38. https://doi.org/10.37317/pbss-2020-0002

Dynkowska, W. M. (2019). Rye (*Secale cereale* L.) phenolic compounds as health-related factors. *Plant Breeding and Seed Science*, *79*, 9–24. https://doi.org/10.37317/pbss-2019-0002

Dziki, D. (2022). Rye flour and rye bran: New perspectives for use. *Processes*, *10*(2), 293. https://doi.org/10.3390/pr10020293

Dziki, D. (2023). The latest innovations in wheat flour milling: A review. *Agricultural Engineering*, *27*(1), 147–162. https://doi.org/10.2478/agriceng-2023-0011

Dziki, D., Hassoon, W. H., Kramek, A., & Krajewska, A. (2023). Grinding characteristics of new varieties of winter triticale grain. *Processes*, *11*(5), 1477. https://doi.org/10.3390/pr11051477

Eadkhong, T., & Danworaphong, S. (2021). On the sensitivity of an air pycnometer. *Measurement*, *169*, 108511. https://doi.org/10.1016/j.measurement.2020.108511

EFSA Panel on Dietetic Products, Nutrition and Allergies (NDA). (2014). Scientific opinion on the substantiation of a health claim related to high fibre sourdough rye bread and reduction of post-prandial glycaemic responses pursuant to Article 13 (5) of Regulation (EC) No 1924/2006. *EFSA Journal*, *12*(10), 3837. https://doi.org/10.2903/j.efsa.2014.3837

Eliasson, A. C. (2017). Starch: Physicochemical and functional aspects. In *Carbohydrates in food* (pp. 501–600). CRC Press. https://doi.org/10.1201/9781315372822

Elliott, H., Woods, P., Green, B. D., & Nugent, A. P. (2022). Can sprouting reduce phytate and improve the nutritional composition and nutrient bioaccessibility in cereals and legumes? *Nutrition Bulletin*, *47*(2), 138–156. https://doi.org/10.1111/nbu.12549

Elmadfa, I., & Meyer, A. L. (2019). The role of the status of selected micronutrients in shaping the immune function. *Endocrine, Metabolic & Immune Disorders-Drug Targets*

(Formerly Current Drug Targets-Immune, Endocrine & Metabolic Disorders), *19*(8), 1100–1115. https://doi.org/10.2174/1871530319666190529101816

El-Mahis, A., Baky, M. H., & Farag, M. A. (2023). How does rye compare to other cereals? A comprehensive review of its potential nutritional value and better opportunities for its processing as a food-based cereal. *Food Reviews International*, *39*(7), 4288–4311. https://doi.org/10.1080/87559129.2021.2023817

Feizollahi, E., Mirmahdi, R. S., Zoghi, A., Zijlstra, R. T., Roopesh, M. S., & Vasanthan, T. (2021). Review of the beneficial and anti-nutritional qualities of phytic acid, and procedures for removing it from food products. *Food Research International*, *143*, 110284. https://doi.org/10.1016/j.foodres.2021.110284

Feng, G. (2019). Rye. In *Bioactive factors and processing technology for cereal foods* (pp. 151–169). https://doi.org/10.1007/978-981-13-6167-8_9

Ferreira, P. S., Victorelli, F. D., Fonseca-Santos, B., & Chorilli, M. (2019). A review of analytical methods for p-coumaric acid in plant-based products, beverages, and biological matrices. *Critical Reviews in Analytical Chemistry*, *49*(1), 21–31. https://doi.org/10.1080/10408347.2018.1459173

Forni, C., Facchiano, F., Bartoli, M., Pieretti, S., Facchiano, A., D'Arcangelo, D., Norelli, S., Valle, G., Nisini, R., Beninati, S., Tabolacci, C., & Jadeja, R. N. (2019). Beneficial role of phytochemicals on oxidative stress and age-related diseases. *BioMed Research International*, *2019*(1), 8748253. https://doi.org/10.1155/2019/8748253

Gao, W., Chen, F., Wang, X., & Meng, Q. (2020). Recent advances in processing food powders by using superfine grinding techniques: A review. *Comprehensive Reviews in Food Science and Food Safety*, *19*(4), 2222–2255. https://doi.org/10.1111/1541-4337.12580

Garcia-Caparros, P., De Filippis, L., Gul, A., Hasanuzzaman, M., Ozturk, M., Altay, V., & Lao, M. T. (2021). Oxidative stress and antioxidant metabolism under adverse environmental conditions: A review. *The Botanical Review*, *87*, 421–466. https://doi.org/10.1007/s12229-020-09231-1

Ghafoor, A. Z., Karim, H., Studnicki, M., Raza, A., Javed, H. H., & Asghar, M. A. (2024). Climate change and rye (*Secale cereale* L.) production: Challenges, opportunities and adaptations. *Journal of Agronomy and Crop Science*, *210*(4), e12725. https://doi.org/10.1111/jac.12725

Godswill, A. G., Somtochukwu, I. V., Ikechukwu, A. O., & Kate, E. C. (2020). Health benefits of micronutrients (vitamins and minerals) and their associated deficiency diseases: A systematic review. *International Journal of Food Sciences*, *3*(1), 1–32. https://doi.org/10.47604/ijf.1024

Gómez, M., Pardo, J., Oliete, B., & Caballero, P. A. (2009). Effect of the milling process on quality characteristics of rye flour. *Journal of the Science of Food and Agriculture*, *89*(3), 470–476. https://doi.org/10.1002/jsfa.3475

Grabinski, J., Sułek, A., Wyzinska, M., Stuper-Szablewska, K., Cacak-Pietrzak, G., Nierobca, A., & Dziki, D. (2021). Impact of genotype, weather conditions and production technology on the quantitative profile of anti-nutritive compounds in rye grains. *Agronomy*, *11*(1), 151. https://doi.org/10.3390/agronomy11010151

Griffiths, K., Aggarwal, B. B., Singh, R. B., Buttar, H. S., Wilson, D., & De Meester, F. (2016). Food antioxidants and their anti-inflammatory properties: A potential role in cardiovascular diseases and cancer prevention. *Diseases*, *4*(3), 28. https://doi.org/10.3390/diseases4030028

Grossmann, I., & Koehler, P. (2016). Fractionation-reconstitution studies to determine the functional properties of rye flour constituents. *Journal of Cereal Science*, *70*, 1–8. https://doi.org/10.1016/j.jcs.2016.05.006

Gruner, P., & Miedaner, T. (2021). Perennial rye: Genetics of perenniality and limited fertility. *Plants*, *10*(6), 1210. https://doi.org/10.3390/plants10061210

Gupta, R., Meghwal, M., & Prabhakar, P. K. (2021). Bioactive compounds of pigmented wheat (*Triticum aestivum*): Potential benefits in human health. *Trends in Food Science & Technology*, *110*, 240–252. https://doi.org/10.1016/j.tifs.2021.02.003

Hackauf, B., Rabanus-Wallace, M. T., Korzun, V. (2021). Bridging the genotype–phenotype gap for precision breeding in rye. In M. T. Rabanus-Wallace & N. Stein (Eds.), *The rye genome. Compendium of plant genomes*. Springer. https://doi.org/10.1007/978-3-030-83383-1_9

Hanhineva, K., Keski-Rahkonen, P., Lappi, J., Katina, K., Pekkinen, J., Savolainen, O., Timonen, O., Paananen, J., Mykkänen, H., & Poutanen, K. (2014). The postprandial plasma rye fingerprint includes benzoxazinoid-derived phenylacetamide sulfates. *The Journal of Nutrition*, *144*(7), 1016–1022. https://doi.org/10.3945/jn.113.187237

Hansen, H. B., Møller, B., Andersen, S. B., Jørgensen, J. R., & Hansen, Å. (2004). Grain characteristics, chemical composition, and functional properties of rye (*Secale cereale* L.) as influenced by genotype and harvest year. *Journal of Agricultural and Food Chemistry*, *52*(8), 2282–2291. https://doi.org/10.1021/jf0307191

Hayann, L., Ciancaglini, P., Ramos, A. P., & Napierala, D. (2024). Calcium and phosphate and their role in matrix vesicles: A biological view. In *Mineralizing vesicles* (pp. 151–173). Academic Press. https://doi.org/10.1016/B978-0-323-99158-2.00010-3

Hemery, Y., Holopainen, U., Lampi, A.-M., Lehtinen, P., Nurmi, T., Piironen, V., Edelmann, M., & Rouau, X. (2011). Potential of dry fractionation of wheat bran for the development of food ingredients, part II: Electrostatic separation of particles. *Journal of Cereal Science*, *53*(1), 9–18. https://doi.org/10.1016/j.jcs.2010.06.014

Hjelm, L., Mielby, L. A., Gregersen, S., Eggers, N., & Bertram, H. C. (2019). Partial substitution of fat with rye bran fibre in Frankfurter sausages – Bridging technological and sensory attributes through inclusion of collagenous protein. *LWT*, *101*, 607–617. https://doi.org/10.1016/j.lwt.2018.11.055

Iftikhar, M., Zhang, H., Iftikhar, A., Raza, A., Khan, M., Sui, M., & Wang, J. (2020). Comparative assessment of functional properties, free and bound phenolic profile, antioxidant activity, and in vitro bioaccessibility of rye bran and its insoluble dietary fiber. *Journal of Food Biochemistry*, *44*(10), e13388. https://doi.org/10.1111/jfbc.13388

Ikram, A., Saeed, F., Noor, R. A., Imran, A., Afzaal, M., Rasheed, A., Islam, F., Iqbal, A., Zahoor, T., Naz, S., Waheed, W., Shahid, M. Q., Khan, A. W., & Kinki, A. B. (2023). A comprehensive review on biochemical and technological properties of rye (*Secale cereale* L.). *International Journal of Food Properties*, *26*(1), 2212–2228. https://doi.org/10.1080/10942912.2023.2244697

Izydorczyk, M. S. (2021). Arabinoxylans. In *Handbook of hydrocolloids* (pp. 399–461). Woodhead Publishing. https://doi.org/10.1016/B978-0-12-820104-6.00016-4

Jang, W. Y., Kim, M. Y., & Cho, J. Y. (2022). Antioxidant, anti-inflammatory, anti-menopausal, and anti-cancer effects of lignans and their metabolites. *International Journal of Molecular Sciences*, *23*(24), 15482. https://doi.org/10.3390/ijms232415482

Jarosz, M., Olbert, M., Wyszogrodzka, G., Młyniec, K., & Librowski, T. (2017). Antioxidant and anti-inflammatory effects of zinc. Zinc-dependent NF-κB signaling. *Inflammopharmacology*, *25*, 11–24. https://doi.org/10.1007/s10787-017-0309-4

Jomova, K., Makova, M., Alomar, S. Y., Alwasel, S. H., Nepovimova, E., Kuca, K., Rhodes, C. J., & Valko, M. (2022). Essential metals in health and disease. *Chemico-Biological Interactions*, *367*, 110173. https://doi.org/10.1016/j.cbi.2022.110173

Jonsson, K., Andersson, R., Knudsen, K. E. B., Hallmans, G., Hanhineva, K., Katina, K., Kolehmainen, M., Kyrø, C., Langton, Maud., Nordlund, E., Lærke, H. N., Olsen, A., Poutanen, K., Tjønneland, A., & Landberg, R. (2018). Rye and health – Where do we stand and where do we go? *Trends in Food Science & Technology*, *79*, 78–87. https://doi.org/10.1016/j.tifs.2018.06.018

Joye, I. (2019). Protein digestibility of cereal products. *Foods*, *8*(6), 199. https://doi.org/10.3390/foods8060199

Kansakar, U., Trimarco, V., Mone, P., Varzideh, F., Lombardi, A., & Santulli, G. (2023). Choline supplements: An update. *Frontiers in Endocrinology*, *14*, 1148166. https://doi.org/10.3389/fendo.2023.1148166

Karlund, A., Paukkonen, I., Gómez-Gallego, C., & Kolehmainen, M. (2021, August). Intestinal exposure to food-derived protease inhibitors: Digestion physiology-and gut health-related effects. In *Healthcare* (Vol. 9, No. 8, p. 1002). MDPI. https://doi.org/10.3390/healthcare9081002

Kataria, D., Kaur, J., Singh, J., Rasane, P., Bhadariya, V., & Kaur, S. (2024). Rye. In *Cereals and nutraceuticals* (pp. 107–119). Springer Nature Singapore. https://doi.org/10.1007/978-981-97-2542-7_5

Kaur, B., Singh, A., Suri, S., Usman, M., & Dutta, D. (2023). Minor millets: A review on nutritional composition, starch extraction/modification, product formulation, and health benefits. *Journal of the Science of Food and Agriculture*, *103*(10), 4742–4754. https://doi.org/10.1002/jsfa.12493

Kaur, P., Sandhu, K. S., Bangar, S. P., Purewal, S. S., Kaur, M., Ilyas, R. A., Asyraf, M. R. M., & Razman, M. R. (2021). Unraveling the bioactive profile, antioxidant and DNA damage protection potential of rye (*Secale cereale*) flour. *Antioxidants*, *10*(8), 1214. https://doi.org/10.3390/antiox10081214

Kaur, P., Sandhu, K. S., Purewal, S. S., Kaur, M., & Singh, S. K. (2021). Rye: A wonder crop with industrially important macromolecules and health benefits. *Food Research International*, *150*, 110769. https://doi.org/10.1016/j.foodres.2021.110769

Kaur, S., Kapoor, H. K., Romana, C. K., Sharma, H., & Rani, R. (2024). Nutritional characteristics and phenolic compounds in pigmented oats and rye. In *Pigmented grains* (pp. 135–155). Academic Press. https://doi.org/10.1016/B978-0-443-15906-0.00006-1

Khadim, R. M., & Al-Fartusie, F. S. (2021, March). Antioxidant vitamins and their effect on immune system. *Journal of Physics: Conference Series*, *1853*(1), 012065). IOP Publishing. https://doi.org/10.1088/1742-6596/1853/1/012065

Knudsen, K. E. B., & Laerke, H. N. (2010). Rye arabinoxylans: Molecular structure, physicochemical properties and physiological effects in the gastrointestinal tract. *Cereal Chemistry*, *87*(4), 353–362. https://doi.org/10.1094/CCHEM-87-4-0353

Koistinen, V. M., Katina, K., Nordlund, E., Poutanen, K., & Hanhineva, K. (2016). Changes in the phytochemical profile of rye bran induced by enzymatic bioprocessing and sourdough fermentation. *Food Research International*, *89*, 1106–1115. https://doi.org/10.1016/j.foodres.2016.06.027

Kołodziejczyk, P., Makowska, A., Pospieszna, B., Michniewicz, J., & Paschke, H. (2018). Chemical and nutritional characteristics of high-fibre rye milling fractions. *Acta Scientiarum Polonorum Technologia Alimentaria*, *17*(2), 149–157. https://www.food.actapol.net/volume17/issue2/6_2_2018.pdf

Korycinska, M., Czelna, K., Jaromin, A., & Kozubek, A. (2009). Antioxidant activity of rye bran alkylresorcinols and extracts from whole-grain cereal products. *Food Chemistry*, *116*(4), 1013–1018. https://doi.org/10.1016/j.foodchem.2009.03.056

Korzun, V., Ponomareva, M. L., Sorrells, M. E. (2021). Economic and academic importance of rye. In M. T. Rabanus-Wallace & N. Stein (Eds.), *The rye genome. Compendium of plant genomes*. Springer. https://doi.org/10.1007/978-3-030-83383-1_1

Kowsalya, P., Sharanyakanth, P. S., & Mahendran, R. (2022). Traditional rice varieties: A comprehensive review on its nutritional, medicinal, therapeutic and health benefit potential. *Journal of Food Composition and Analysis*, *114*, 104742. https://doi.org/10.1016/j.jfca.2022.104742

Kselíkova, V., Vyhnanek, T., Hanack, P., & Martinek, P. (2020). *Grain hardness in triticale: A physical and molecular evaluation*. https://doi.org/10.17221/96/2019-CJGPB

Kulichova, K., Sokol, J., Nemecek, P., Maliarova, M., Maliar, T., Havrlentova, M., & Kraic, J. (2019). Phenolic compounds and biological activities of rye (*Secale cereale* L.) grains. *Open Chemistry*, *17*(1), 988–999. https://doi.org/10.1515/chem-2019-0103

Kumar, M., Kumar, D., Sharma, A., Bhadauria, S., Thakur, A., & Bhatia, A. (2024). Micronutrients throughout the life cycle: Needs and functions in health and disease. *Current Nutrition & Food Science*, *20*(1), 62–84. https://doi.org/10.2174/1573401319666230420094603

Kumar, Y., Basu, S., Goswami, D., Devi, M., Shivhare, U. S., & Vishwakarma, R. K. (2022). Anti-nutritional compounds in pulses: Implications and alleviation methods. *Legume Science*, *4*(2), e111. https://doi.org/10.1002/leg3.111

Kunkulberga, D., Linina, A., Kronberga, A., Kokare, A., & Lenenkova, I. (2017). *Grain quality of winter rye cultivars grown in Latvia*. https://doi.org/10.22616/foodbalt.2017.015

Laidig, F., Piepho, H. P., Rentel, D., Drobek, T., Meyer, U., & Huesken, A. (2017). Breeding progress, variation, and correlation of grain and quality traits in winter rye hybrid and population varieties and national on-farm progress in Germany over 26 years. *Theoretical and Applied Genetics*, *130*, 981–998. https://doi.org/10.1007/s00122-017-2865-9

Lee, Y. M., Han, S. I., Song, B. C., & Yeum, K. J. (2015). Bioactives in commonly consumed cereal grains: Implications for oxidative stress and inflammation. *Journal of Medicinal Food*, *18*(11), 1179–1186. https://doi.org/10.1089/jmf.2014.3394

Levent, H., Koyuncu, M., Bilgicli, N., Adıgüzel, E., & Dedeoglu, M. (2020). Improvement of chemical properties of noodle and pasta using dephytinized cereal brans. *LWT*, *128*, 109470. https://doi.org/10.1016/j.lwt.2020.109470

Li, G., Wang, L., Yang, J., He, H., Jin, H., Li, X., Ren, T., Ren, Z., Li, F., Han, X., Zhao, X., Dong, L., Li, Y., Song, Z., Yan, Z., Zheng, N., Shi, C., Wang, Z., Yang, S., … Wang, D. (2021). A high-quality genome assembly highlights rye genomic characteristics and agronomically important genes. *Nature Genetics*, *53*(4), 574–584. https://doi.org/10.1038/s41588-021-00808-z

Lille, M., Kortekangas, A., Heiniö, R. L., & Sozer, N. (2020). Structural and textural characteristics of 3D-printed protein-and dietary fibre-rich snacks made of milk powder and wholegrain rye flour. *Foods*, *9*(11), 1527. https://doi.org/10.3390/foods9111527

Makowska, A., Polcyn, A., Chudy, S., & Michniewicz, J. (2015). Application of oat, wheat and rye bran to modify nutritional properties, physical and sensory characteristics of extruded corn snacks. *Acta Scientiarum Polonorum Technologia Alimentaria*, *14*(4), 375–386.

Malalgoda, M., & Simsek, S. (2017). Celiac disease and cereal proteins. *Food Hydrocolloids*, *68*, 108–113. https://doi.org/10.1016/j.foodhyd.2016.09.024

Matías, J., Rodríguez, M. J., Carrillo-Vico, A., Casals, J., Fondevilla, S., Haros, C. M., Pedroche, J., Aparicio, N., Fernández-García, N., Aguiló-Aguayo, I., Soler-Rivas, C., Caballero, P. A., Morte, A., Rico, D., Reguera, M., & NutriCrop Spanish Consortium/Group. (2024). From "farm to fork": Exploring the potential of nutrient-rich and stress-resilient emergent crops for sustainable and healthy food in the Mediterranean region in the face of climate change challenges. *Plants*, *13*(14), 1914. https://doi.org/10.3390/plants13141914

Mazidi, M., Katsiki, N., Mikhailidis, D. P., & Banach, M. (2019). Adiposity may moderate the link between choline intake and non-alcoholic fatty liver disease. *Journal of the American College of Nutrition*, *38*(7), 633–639. https://doi.org/10.1080/07315724.2018.1507011

Meena, K. K., Taneja, N. K., Jain, D., Ojha, A., Saravanan, C., & Mudgil, D. (2022). Bioactive components and health benefits of maize-based fermented foods: A review. *Biointerface Research in Applied Chemistry*, *13*(4), 338. https://doi.org/10.33263/BRIAC134.338

Meeus, Y., Janssen, F., Wouters, A. G., Delcour, J. A., & Moldenaers, P. (2020). Linear and non-linear rheology of bread doughs made from blends of wheat (*Triticum aestivum* L.) and rye (*Secale cereale* L.) flour. *Food and Bioprocess Technology*, *13*, 159–171. https://doi.org/10.1007/s11947-019-02393-w

Mezzaroba, L., Alfieri, D. F., Simão, A. N. C., & Reiche, E. M. V. (2019). The role of zinc, copper, manganese and iron in neurodegenerative diseases. *Neurotoxicology*, *74*, 230–241. https://doi.org/10.1016/j.neuro.2019.07.007

Miallot, R., Millet, V., Galland, F., & Naquet, P. (2023). The vitamin B5/coenzyme A axis: A target for immunomodulation? *European Journal of Immunology*, *53*(10), 2350435. https://doi.org/10.1002/eji.202350435

Michalska, A., Ceglinska, A., Amarowicz, R., Piskula, M. K., Szawara-Nowak, D., & Zielinski, H. (2007). Antioxidant contents and antioxidative properties of traditional rye breads. *Journal of Agricultural and Food Chemistry*, *55*(3), 734–740. https://doi.org/10.1021/jf062425w

Mihajlovski, K., Buntic, A., Milic, M., Rajilic-Stojanovic, M., & Dimitrijevic-Brankovic, S. (2021). From agricultural waste to biofuel: Enzymatic potential of a bacterial isolate Streptomyces fulvissimus CKS7 for bioethanol production. *Waste and Biomass Valorization*, *12*, 165–174. https://doi.org/10.1007/s12649-020-00960-3

Mikhailova, E., & Doležel, J. (2021). Rye cytogenetics and chromosome genomics. In M. T. Rabanus-Wallace & N. Stein (Eds.), *The rye genome. Compendium of plant genomes*. Springer. https://doi.org/10.1007/978-3-030-83383-1_3

Mikkelsen, K., & Apostolopoulos, V. (2018). B vitamins and ageing. *Biochemistry and Cell Biology of Ageing: Part I Biomedical Science*, 451–470. https://doi.org/10.1007/978-981-13-2835-0_15

Mikkelsen, K., & Apostolopoulos, V. (2019). Vitamin B1, B2, B3, B5, and B6 and the immune system. *Nutrition and Immunity*, 115–125. https://doi.org/10.1007/978-3-030-16073-9_7

Moretti, A., Paoletta, M., Liguori, S., Bertone, M., Toro, G., & Iolascon, G. (2020). Choline: An essential nutrient for skeletal muscle. *Nutrients*, *12*(7), 2144. https://doi.org/10.3390/nu12072144

Müller, W. E., Schröder, H. C., & Wang, X. (2019). Inorganic polyphosphates as storage for and generator of metabolic energy in the extracellular matrix. *Chemical Reviews*, *119*(24), 12337–12374. https://doi.org/10.1021/acs.chemrev.9b00460

Naderi, B., Nasabi, M., Akbari, M., Zargaraan, A., & Amini, M. (2023). Effect of enzymatic treatment on rheological properties of wafer batter and textural properties of wafer sheet. *Journal of Food Measurement and Characterization*, *17*(3), 2213–2223. https://doi.org/10.1007/s11694-022-01760-0

Nasabi, M., Naderi, B., Akbari, M., Aktar, T., Kieliszek, M., & Amini, M. (2021). Physical, structural and sensory properties of wafer batter and wafer sheets influenced by various sources of grains. *LWT*, *149*, 111826.

Nemeth, R., & Tomoskozi, S. (2021). Rye: Current state and future trends in research and applications. *Acta Alimentaria*, *50*(4), 620–640. https://doi.org/10.1556/066.2021.00162

Nikinmaa, M., Kajala, I., Liu, X., Nordlund, E., & Sozer, N. (2020). The role of rye bran acidification and in situ dextran formation on structure and texture of high fibre extrudates. *Food Research International*, *137*, 109438. https://doi.org/10.1016/j.foodres.2020.109438

Olfat, N., Ashoori, M., & Saedisomeolia, A. (2022). Riboflavin is an antioxidant: A review update. *British Journal of Nutrition*, *128*(10), 1887–1895. https://doi.org/10.1017/S0007114521005031

Paez-Hurtado, A. M., Calderon-Ospina, C. A., & Nava-Mesa, M. O. (2023). Mechanisms of action of vitamin B1 (thiamine), B6 (pyridoxine), and B12 (cobalamin) in pain: A narrative review. *Nutritional Neuroscience*, *26*(3), 235–253. https://doi.org/10.1080/1028415X.2022.2034242

Papageorgiou, M., & Skendi, A. (2018). Introduction to cereal processing and by-products. In *Sustainable recovery and reutilization of cereal processing by-products* (pp. 1–25). Woodhead Publishing. https://doi.org/10.1016/B978-0-08-102162-0.00001-0

Piasecka-Kwiatkowska, D., Warchalewski, J. R., Zielinska-Dawidziak, M., & Michalak, M. (2012). Digestive enzyme inhibitors from grains as potential components of nutraceuticals. *Journal of Nutritional Science and Vitaminology*, *58*(3), 217–220. https://doi.org/10.3177/jnsv.58.217

Pihlava, J. M., Hellström, J., Kurtelius, T., & Mattila, P. (2018). Flavonoids, anthocyanins, phenolamides, benzoxazinoids, lignans and alkylresorcinols in rye (*Secale cereale*) and some rye products. *Journal of Cereal Science*, *79*, 183–192. https://doi.org/10.1016/j.jcs.2017.09.009

Pihlava, J. M., Nordlund, E., Heiniö, R. L., Hietaniemi, V., Lehtinen, P., & Poutanen, K. (2015). Phenolic compounds in wholegrain rye and its fractions. *Journal of Food Composition and Analysis*, *38*, 89–97. https://doi.org/10.1016/j.jfca.2014.10.004

Pinnamaneni, S. R., Mubvumba, P., Anapalli, S. S., & Reddy, K. N. (2022). Cereal rye (*Secale cereale* L.) cover crop improves soil physico-chemical properties with no influence on soybean (*Glycine max* L.) root growth parameters. *Frontiers in Soil Science*, *2*, 970380. https://doi.org/10.3389/fsoil.2022.970380

Poutanen, K., Katina, K., & Heinio, R. L. (2014). Rye. In *Bakery products science and technology* (pp. 75–87). https://doi.org/10.1002/9781118792001.ch4

Protonotariou, S., Ritzoulis, C., & Mandala, I. (2021). Jet milling conditions impact on wheat flour particle size. *Journal of Food Engineering*, *294*, 110418. https://doi.org/10.1016/j.jfoodeng.2020.110418

Purewal, S. S., Kaur, P., Sandhu, K. S., Bangar, S. P., Siroha, A. K., Singh, S. K., Punia, S., Kaur, M., Salar, R. K., & Markandey, D. K. (2022). Nutritional profile of maize and effect of processing methods. *In Maize* (pp. 77–100). CRC Press.

Ram, S., Narwal, S., Gupta, O. P., Pandey, V., & Singh, G. P. (2020). Anti-nutritional factors and bioavailability: Approaches, challenges, and opportunities. In *Wheat and barley grain biofortification* (pp. 101–128). https://doi.org/10.1016/B978-0-12-818444-8.00004-3

Rani, M., Singh, G., Siddiqi, R. A., Gill, B. S., Sogi, D. S., & Bhat, M. A. (2021). Comparative quality evaluation of physicochemical, technological, and protein profiling of wheat, rye, and barley cereals. *Frontiers in Nutrition*, *8*, 694679. https://doi.org/10.3389/fnut.2021.694679

Ratnavathi, C. V. (2019). Grain structure, quality, and nutrition. In *Breeding sorghum for diverse end uses* (pp. 193–207). Woodhead Publishing. https://doi.org/10.1016/B978-0-08-101879-8.00012-7

Ravisankar, S., Queiroz, V. A., & Awika, J. M. (2020). Rye flavonoids – Structural profile of the flavones in diverse varieties and effect of fermentation and heat on their structure and antioxidant properties. *Food Chemistry*, *324*, 126871. https://doi.org/10.1016/j.foodchem.2020.126871

Redant, L., Buggenhout, J., Brijs, K., & Delcour, J. A. (2017). Extractability and chromatographic separation of rye (*Secale cereale* L.) flour proteins. *Journal of Cereal Science*, *73*, 68–75. https://doi.org/10.1016/j.jcs.2016.11.010

Rosell, C. M., & Koehler, P. (2023). Chemistry of cereal grains. In *Handbook on sourdough biotechnology* (pp. 25–66). Springer International Publishing. https://doi.org/10.1007/978-3-031-23084-4_3

Rouf Shah, T., Prasad, K., & Kumar, P. (2016). Maize—A potential source of human nutrition and health: A review. *Cogent Food & Agriculture*, *2*(1), 1166995. https://doi.org/10.1080/23311932.2016.1166995

Sa, A. G. A., Moreno, Y. M. F., & Carciofi, B. A. M. (2020). Food processing for the improvement of plant proteins digestibility. *Critical Reviews in Food Science and Nutrition*, *60*(20), 3367–3386. https://doi.org/10.1080/10408398.2019.1688249

Sacks, F. M., Lichtenstein, A. H., Wu, J. H. Y., Appel, L. J., Creager, M. A., Kris-Etherton, PM., Miller, M., Rimm, E. B., Rudel, L. L., Robinson, J. G., Stone, N. J., Van Horn, L. V., & American Heart Association. (2017). Dietary fats and cardiovascular disease: A presidential advisory from the American heart association. *Circulation*, *136*(3), e1–e23. https://doi.org/10.1161/CIR.0000000000000510

Salazar-López, N. J., Ovando-Martínez, M., & Domínguez-Avila, J. A. (2020). Cereal/grain by-products. In *Food wastes and by-products: Nutraceutical and health potential* (pp. 1–34). https://doi.org/10.1002/9781119534167.ch1

Afnan, Saleem, A., Akhtar, M. F., Sharif, A., Akhtar, B., Siddique, R., Ashraf, G. M., Alghamdi, B. S., & Alharthy, S. A. (2022). Anticancer, cardio-protective and anti-inflammatory potential of natural-sources-derived phenolic acids. *Molecules*, *27*(21), 7286. https://doi.org/10.3390/molecules27217286

Sharma, K., Tayade, A., Singh, J., & Walia, S. (2020). Bioavailability of nutrients and safety measurements. In *Functional foods and nutraceuticals: Bioactive components, formulations and innovations* (pp. 543–593). https://doi.org/10.1007/978-3-030-42319-3_25

Silventoinen, P., Kortekangas, A., Ercili-Cura, D., & Nordlund, E. (2021). Impact of ultra-fine milling and air classification on biochemical and techno-functional characteristics of wheat and rye bran. *Food Research International*, *139*, 109971. https://doi.org/10.1016/j.foodres.2020.109971

Singh, A. P. (2023). Nutritional composition, bioactive compounds, and phytochemicals of wheat grains. In *Wheat science* (pp. 125–181). CRC Press.

Soleymani, S., Habtemariam, S., Rahimi, R., & Nabavi, S. M. (2020). The what and who of dietary lignans in human health: Special focus on prooxidant and antioxidant effects. *Trends in Food Science & Technology*, *106*, 382–390. https://doi.org/10.1016/j.tifs.2020.10.015

Stępniewska, S., Cacak-Pietrzak, G., Szafrańska, A., Ostrowska-Ligęza, E., Salamon, A., & Kowalska, H. (2024). Assessment of the baking properties of rye flour based on the polysaccharide content and properties. *Applied Sciences*, *14*(7), 2772. https://doi.org/10.3390/app14072772

Suwannasom, N., Kao, I., Pruß, A., Georgieva, R., & Bäumler, H. (2020). Riboflavin: The health benefits of a forgotten natural vitamin. *International Journal of Molecular Sciences*, *21*(3), 950. https://doi.org/10.3390/ijms21030950

Tahir, I., Dey, E., & Nybom, H. (2019). Application of alkylresorcinols in an organic apple orchard for protection against storage diseases. *European Journal of Horticultural Science*, *84*, 142–151. https://doi.org/10.17660/eJHS.2019/84.3.4

Tang, Z. X., Ross, K., Ren, Z. L., Yang, Z. J., Zhang, H. Y., Chikmawati, T., Miftahudin, & Gustafson, J. P. (2011). Secale. In C. Kole (Ed.), *Wild crop relatives: Genomic and breeding resources* (pp. 263–295). Springer. https://doi.org/10.1007/978-3-642-14228-4_8

Tanwir, F., Dionisio, G., Adhikari, K. B., Fomsgaard, I. S., & Gregersen, P. L. (2017). Biosynthesis and chemical transformation of benzoxazinoids in rye during seed germination and the identification of a rye Bx6-like gene. *Phytochemistry*, *140*, 95–107. https://doi.org/10.1016/j.phytochem.2017.04.020

Tardy, A. L., Pouteau, E., Marquez, D., Yilmaz, C., & Scholey, A. (2020). Vitamins and minerals for energy, fatigue and cognition: A narrative review of the biochemical and clinical evidence. *Nutrients*, *12*(1), 228. https://doi.org/10.3390/nu12010228

Targonska-Karasek, M., Boczkowska, M., Podyma, W., Pasnik, M., Niedzielski, M., Rucinska, A., & Rakoczy-Trojanowska, M. (2020). Investigation of obsolete diversity of rye (*Secale cereale* L.) using multiplexed SSR fingerprinting and evaluation of agronomic traits. *Journal of Applied Genetics*, *61*, 513–529. https://doi.org/10.1007/s13353-020-00579-z

Tiwari, N., Mehrotra, N., Tripathi, P., & Sharma, S. (2024). Vitamins and their impact on human health. In *Clinical applications of biomolecules in disease diagnosis: A comprehensive guide to biochemistry and metabolism* (pp. 129–153). Springer Nature Singapore. https://doi.org/10.1007/978-981-97-4723-8_6

Tola, F. S. (2024). The concept of folic acid supplementation and its role in prevention of neural tube defect among pregnant women: PRISMA. *Medicine, 103*(19), e38154. https://doi.org/10.1097/MD.0000000000038154

Tripathi, M. K., Mohapatra, D., Jadam, R. S., Pandey, S., Singh, V., Kumar, V., & Kumar, A. (2021). Nutritional composition of millets. In *Millets and millet technology* (pp. 101–119). https://doi.org/10.1007/978-981-16-0676-2_5

Uraji, M., Kimura, M., Inoue, Y., Kawakami, K., Kumagai, Y., Harazono, K., & Hatanaka, T. (2013). Enzymatic production of ferulic acid from defatted rice bran by using a combination of bacterial enzymes. *Applied Biochemistry and Biotechnology, 171*(5), 1085–1093. https://doi.org/10.1007/s12010-013-0190-6

Van Craeyveld, V., Holopainen, U., Selinheimo, E., Poutanen, K., Delcour, J. A., & Courtin, C. M. (2009). Extensive dry ball milling of wheat and rye bran leads to in situ production of arabinoxylan oligosaccharides through nanoscale fragmentation. *Journal of Agricultural and Food Chemistry, 57*(18), 8467–8473. https://doi.org/10.1021/jf901870r

Verhulst, E. P., Brunton, N. P., & Rai, D. K. (2023). Polyphenols in agricultural grassland crops and their health-promoting activities—A review. *Foods, 12*(22), 4122. https://doi.org/10.3390/foods12224122

Wang, J., Bai, J., Fan, M., Li, T., Li, Y., Qian, H., Wang, L., & Rao, Z. (2020). Cereal-derived arabinoxylans: Structural features and structure–activity correlations. *Trends in Food Science & Technology, 96*, 157–165. https://doi.org/10.1016/j.tifs.2019.12.016

Wang, P., Tao, H., Jin, Z., & Xu, X. (2016). Impact of water extractable arabinoxylan from rye bran on the frozen steamed bread dough quality. *Food Chemistry, 200*, 117–124. https://doi.org/10.1016/j.foodchem.2016.01.027

Wang, Q., Li, L., & Zheng, X. (2020). A review of milling damaged starch: Generation, measurement, functionality and its effect on starch-based food systems. *Food Chemistry, 315*, 126267. https://doi.org/10.1016/j.foodchem.2020.126267

Wang, Z., & Wang, L. (2024). Impact of sourdough fermentation on nutrient transformations in cereal-based foods: Mechanisms, practical applications, and health implications. *Grain & Oil Science and Technology*. https://doi.org/10.1016/j.gaost.2024.03.001

Warechowska, M., Warechowski, J., Tyburski, J., Siemianowska, E., Nawrocka, A., Miś, A., & Skrajda-Brdak, M. (2019). Evaluation of physicochemical properties, antioxidant potential and baking quality of grain and flour of primitive rye (*Secale cereale* var. Multicaule). *Journal of Food Science and Technology, 56*, 3422–3430. https://doi.org/10.1007/s13197-019-03827-1

Witkowska, A. M., Waskiewicz, A., Zujko, M. E., Szczesniewska, D., Stepaniak, U., Pająk, A., & Drygas, W. (2018). Are total and individual dietary lignans related to cardiovascular disease and its risk factors in postmenopausal women? A nationwide study. *Nutrients, 10*(7), 865. https://doi.org/10.3390/nu10070865

Wolak, N., Zawrotniak, M., Gogol, M., Kozik, A., & Rapala-Kozik, M. (2017). Vitamins B1, B2, B3 and B9 – Occurrence, biosynthesis pathways and functions in human nutrition. *Mini Reviews in Medicinal Chemistry, 17*(12), 1075–1111. https://doi.org/10.2174/1389557516666160725095729

Wrigley, C., & Bushuk, W. (2017). Rye: Grain-quality characteristics and management of quality requirements. In *Cereal grains* (pp. 153–178). Woodhead Publishing. https://doi.org/10.1016/B978-0-08-100719-8.00007-3

Wrigley, C. W. (2019). Rye. In *Whole grains* (pp. 173–196). CRC Press. https://doi.org/10.1201/9781351104760

Wroblewski, M., Wroblewska, W., & Sobiesiak, M. (2024). The role of selected elements in oxidative stress protection: Key to healthy fertility and reproduction. *International Journal of Molecular Sciences*, *25*(17), 9409. https://doi.org/10.3390/ijms25179409

Yan, X., Liu, C., Huang, A., Chen, R., Chen, J., & Luo, S. (2020). The nutritional components and physicochemical properties of brown rice flour ground by a novel low temperature impact mill. *Journal of Cereal Science*, *92*, 102927. https://doi.org/10.1016/j.jcs.2020.102927

Yiannikourides, A., & Latunde-Dada, G. O. (2019). A short review of iron metabolism and pathophysiology of iron disorders. *Medicines*, *6*(3), 85. https://doi.org/10.3390/medicines6030085

Yingngam, B. (2024). Folic acid deficiency. In *Causes and management of nutritional deficiency disorders* (pp. 116–164). IGI Global. https://doi.org/10.4018/979-8-3693-2947-4.ch006

Zannini, E., Bravo Núñez, Á., Sahin, A. W., & Arendt, E. K. (2022). Arabinoxylans as functional food ingredients: A review. *Foods*, *11*(7), 1026. https://doi.org/10.3390/foods11071026

Zhao, K., Tang, H., Zhang, B., Zou, S., Liu, Z., & Zheng, Y. (2023). Microbial production of vitamin B5: Current status and prospects. *Critical Reviews in Biotechnology*, *43*(8), 1172–1192. https://doi.org/10.1080/07388551.2022.2104690

4 Processing Techniques of Rye Cereals

Tehseena Sabir, Muhammad Nouman Ijaz, Samra Yasmin, Aanchal Sharma, Ghulam Mustafa Kamal and Muhammad Khalid

4.1 INTRODUCTION

Rye, scientifically called as *Secale cereale*, belongs to the Poaceae family, also known as the grass family. This family includes many important cereal grains like wheat, barley and oats. It is a versatile and nutritious grain that adds a unique taste to the meals and famous for its robust flavor and extensive nutritional benefits. It has been cultivated for centuries, for its utilization in bread making, whiskey industry and animal feed. Ripened rye plant and rye grains are shown in Figure 4.1. Rye is rich in fiber, vitamins and minerals, making it a healthy addition to any diet. It's particularly popular in Northern and Eastern Europe, where it's used to make dense, flavorful rye bread.

FIGURE 4.1 Rye grains and rye plant.

DOI: 10.1201/9781003585817-4

4.2 CLASSIFICATION OF RYE (*Secale cereale*)

Taxonomic classification of rye is given below:

Scientific name:	*Secale cereale*
Family:	Poaceae (grass family)
Genus:	*Secale*
Species:	*S. cereale*
Binomial name:	*S. cereale*
Order:	Poales
Class:	Lilioposida
Kingdom:	Plantae
2n = 14 (rye has 14 chromosomes)	

4.2.1 Origin

Rye has one of the more recent origins than most other cereal grains. It was not discovered in any Egyptian ruins or Swiss lake houses (Lorenz, 2000). Rye, recognized scientifically as *S. cereale*, is a grass broadly cultivated for its grain. It is believed to have originated in southwestern Asia around 6500 BCE, spreading westward throughout the Balkan Peninsula and into Europe. The grain was a staple meal in Scandinavia in the course of the Middle Ages and stays quintessential to the location's delicacies, especially in the shape of rye crispbread. Today, Europe produces approximately half of the arena's rye, which is also used for making bread, beer, rye whiskey, and animal fodder. Despite its lesser recognition in comparison to wheat, rye flourishes in harsh climates and bad soil, making it an important crop in certain areas (Sencer & Hawkes, 1980).

4.2.2 Evolutionary History

Rye, a grain that has come to be a staple in lots of diets around the world, has a wealthy record that dates lower back lots of years. It is assumed to have originated from a wild species known as *Secale montanum*. This ancestral shape of rye turned into possibly first cultivated in the location this is now contemporary-day Turkey, as early as four, 000–5000 years in the past. Over time, rye unfolds throughout Europe and Asia, turning into a crucial crop because of its resilience in cold climates and terrible soil conditions.

The domestication of rye turned into a vast development inside the records of agriculture. It provided early farmers with a reliable supply of meals that could withstand harsh situations wherein different crops would possibly fail. Rye's capacity to develop in marginal soils made it specifically valuable in mountainous regions and contributed to its spread throughout Europe, mainly in Germany, Poland and Russia, where it has become a nutritional mainstay.

Rye's adaptability isn't always its best top-notch feature; it also possesses a completely unique nutritional profile. It is wealthy in dietary fiber, that's beneficial for digestive fitness, and carries loads of nutrients and minerals, which includes magnesium, phosphorus and B complex nutrients. These dietary blessings, in conjunction with its different taste, have made rye a famous choice for bread-making and different culinary makes use of.

In addition to its culinary makes use of, rye has played a function inside the production of alcoholic liquids. It is a key aspect in rye whiskey and a few sorts of beer, where it imparts a spicy, hearty flavor that is especially prized by using connoisseurs. The use of rye in these beverages has an extended lifestyle, especially in Eastern Europe and North America.

The cultivation of rye has also had environmental implications. As a cover crop, rye can enhance soil health by using preventing erosion, suppressing weeds and enhancing soil fertility via its huge root gadget. These environmental advantages, alongside its economic importance, have ensured that rye remains a valued crop in modern agriculture.

4.3 NUTRITIONAL COMPOSITION

Rye is a distinctly nutritious grain this is frequently considered a healthier opportunity to wheat. A 169 g serving of rye grain includes about 571 calories, with a macronutrient distribution of 2.8 g of fats, 17 g of protein and 128 g of carbohydrates. Of the carbohydrates, 26 g is dietary fiber, that's 93% of each day price, and 1.7 g is sugar. Toasting and extrusion caused the greatest modifications in its composition. After toasting the grain, the sugar content increased 5.81 times, but the number of soluble components and protein degradability fell by 17.92% and 11.65%, respectively (Fattakhova et al., 2020). Rye is also rich in nutrients and minerals, offering an excellent source of B nutrients which include thiamin, riboflavin, niacin, nutrition B6, folate and pantothenic acid. It additionally consists of nutrition E and minerals like iron, zinc, magnesium, phosphorus and selenium, which can vary relying on the soil content. The grain's excessive fiber content is useful for digestive health, and its nutrient density makes it a treasured addition to a balanced weight loss plan. Rye bread, crafted from rye grain, shares lots of those dietary advantages and can be a more fit option as compared to everyday wheat bread, presenting a terrific source of B vitamins and minerals like copper, manganese and selenium consistent with slice (Kaur et al., 2021).

4.4 HEALTH BENEFITS OF RYE

Rye, a grain much like wheat and barley, has been identified for its several health benefits. Rye cultivation has a 20% lower greenhouse gas emissions and an 8% lower carbon footprint than wheat production, according to a new partial life cycle assessment and mixed model analysis (Hackauf et al., 2022). It is mainly noted for helping in weight loss due to its high fiber content, which can assist suppress urge for food and manage weight benefit. The fiber in rye also contributes to preventing gallstones by using lowering bile acid concentrations. Moreover, rye can enhance metabolism, which is important for preserving a healthy weight and universal properly being. For those managing diabetes, rye may be beneficial because it facilitates alter blood sugar degrees. Additionally, its nutrient-wealthy profile supports coronary heart health and has anti-inflammatory properties which could benefit conditions like bronchial asthma. Rye's composition consists of important vitamins and minerals inclusive of magnesium, iron and B vitamins, which are critical for diverse body capabilities, such as hormonal balance and skeletal health. Incorporating rye into one's food plan may be a simple yet effective way to beautify nutritional intake and promote a healthier way of life.

Rye foods are gaining attention for their potential benefits in managing insulin metabolism, which is a critical factor in diabetes prevention and management. Studies have indicated that rye-based foods, compared to wheat bread, can induce a lower insulin response in the postprandial phase, which refers to the period following a meal. This effect, often called the 'rye factor', may be attributed to the slower uptake of glucose in the intestine when consuming rye products. Furthermore, the consumption of whole grain rye has been consistently associated with a lower risk of developing type-2 diabetes, cardiovascular disease, and other non-communicable diseases. The unique composition of rye, which includes a variety of dietary fibers and bioactive components, is thought to play a role in its health benefits. These components may contribute to improved glycemic control and a steadier increase in blood sugar levels, which is particularly beneficial for individuals with diabetes, as it helps avoid sudden spikes in blood glucose. While more research is needed to fully understand the mechanisms behind the 'rye factor' and its impact on health, current evidence supports the inclusion of rye-based foods in a balanced diet for their positive effects on insulin and glucose responses (Jonsson et al., 2018).

Rye is a nutritious grain this is excessive in dietary fiber, that's useful for digestive fitness and can assist sell feelings of fullness. A slice of rye bread generally incorporates about 1.9 g of fiber, contributing to each day dietary desires and assisting a healthful gut microbiome. The fiber in rye additionally enables with blood sugar manipulate, making it an excellent desire for those coping with their glucose stages. Additionally, rye incorporates important vitamins consisting of B nutrients, iron and selenium, making it a healthy addition to a balanced food plan.(Rakha et al., 2010).

Rye is indeed a treasured source of protein, which is important for muscle growth and renovation. An unmarried cup of rye grain includes approximately 17 g of protein, contributing substantially to the everyday protein requirements. This protein content, along with a wealthy profile of amino acids, makes rye an exceptional preference for the ones trying to help their muscle fitness. Additionally, rye is packed with dietary fiber and different critical vitamins, similarly enhancing its nutritional profile and advantages for average fitness (Németh & Tömösközi, 2021).

Rye, an entire grain, is a wealthy source of sinapic acid, which has been diagnosed for its numerous fitness benefits. Studies advocate that sinapic acid reveals residences that may contribute to most cancers' prevention, discount of inflammation, regulation of blood sugar stages, and safety toward mutations. Additionally, it has been located to provide neuroprotective advantages and act as an amazing antioxidant, similarly supporting the overall fitness-promoting profile of rye as part of a balanced weight-reduction plan (Dziki, 2022). Phenolic acid have antioxidant characteristics and can protect against degenerative illnesses when oxygen reactive species are present (Dykes & Rooney, 2007).

Rye, a grain recognized for its wealthy, hearty flavor, is indeed a supply of gluten. This makes it improper for people with celiac ailment or gluten sensitivity, as their bodies react adversely to gluten proteins, leading to damage within the small gut.

Fortunately, there are numerous gluten-unfastened options to be had that permit the ones affected to enjoy similar ingredients without the fitness risks, together with breads and pastas made from rice, corn or other gluten-unfastened grains. It's important for people with those situations to cautiously study food labels to keep away from gluten-containing elements (Biesiekierski, 2017).

Rye bran is indeed a powerhouse of vitamins, rich in nutritional fiber and a bunch of bioactive compounds, which include plant lignans. These additives are vital to what's known as the dietary fiber complicated, which plays a critical position in human health. Studies have proven that lignans, like enterolactone, can serve as biomarkers for the consumption of lignan-rich foods. While the evidence isn't conclusive, a few research shows that rye and whole grains may additionally offer protecting blessings in opposition to sure sicknesses, inclusive of cardiovascular conditions and diabetes (Bondia-Pons et al., 2009).

4.5 OTHER USES OF RYE

As an inexperienced plant, it serves multiple purposes: it provides nutritious pasture for farm animals, acts as effective inexperienced manure improving soil great and performs an essential role in crop rotations. Its substantial root device now not only helps prevent soil erosion but additionally captures nitrogen, improving soil fertility. Furthermore, rye's straw is a precious byproduct, used in numerous packages from thatching roofs to creating paper products. Its adaptability and utility make it a staple in sustainable farming practices (Bushuk, 2001). It is certainly a popular cereal for bread production, valued for its unique taste and dietary advantages. However, wheat remains the maximum widely used grain for breadmaking because of its gluten content material, which gives the dough its elasticity and sponginess, making it ideal for a huge type of bread kinds. Wheat's adaptability to distinct soils and climates additionally contributes to its prevalence in bread manufacturing globally. While rye is a great crop, especially in certain cultures and areas, wheat's versatility and the ideal characteristics it imparts to bread dough make sure its function as the primary cereal for bread production (Deleu et al., 2020).

Rye flour is indeed a flexible aspect, normally used to make a variety of breads, consisting of the dense and darkish 'black' bread that may be a staple in Eastern Europe and elements of Asia. This type of bread is known for its one-of-a-kind taste, that's richer and more severe than bread made with wheat flour. It's additionally valued for its nutritional advantages, as rye is wealthy in fiber and crucial vitamins. In many cultures, black bread is greater than just a meals object; it's an image of sustenance and is deeply embedded in the location's culinary traditions. The technique of creating black bread can vary; however, it frequently involves a protracted fermentation length, which contributes to its precise taste and texture (Bushuk, 2001).

Whole grain rye porridge has been studied for its effects on satiety, and research indicates that it is able to indeed have prolonged satiating houses when ate up for breakfast. Having a look at comparing whole grain rye porridge to subtle wheat bread determined that the rye porridge ended in a notably prolonged feeling of fullness, decreased hunger and a reduced choice to eat for up to 8 hours after consumption. However, this increased satiety did no longer cause a lower in food consumption later inside the day. This might be because of the complicated network of mental and physiological mechanisms that regulate brief-time period urge for food, along with elements like cognitive influences, mouthfeel and gastric distension. Despite the extended satiety, individuals did no longer file a faded consumption at subsequent meals, indicating that whilst complete grain rye porridge can help maintain a feeling of fullness, it does no longer necessarily impact common each day calorie intake. This data is treasured for the ones looking to manage their starvation levels at some stage in the day without always reducing their overall meals intake (Isaksson et al., 2008).

Rye flour, regarded for its strong taste and nutritional benefits, is a versatile component in baking. It's frequently used to create numerous baked goods, along with cakes, cookies, and pastries. When utilized in desserts, rye flour can impart a rich, deep flavor and a moist, dense texture, making it a five for spice cakes and fruitcakes. In cookies, rye provides a unique earthy taste and can contribute to a crisper texture. For pastries, rye flour can be blended with different flours to achieve a delicate balance of flavor and shape, frequently ensuing in a flakier and more tender crust. Bakers cost rye flour for its high fiber content material and rich nutrient profile, which incorporates crucial nutrients and minerals. While it consists of gluten, its lower gluten content material in comparison to wheat flour makes it a suitable choice for gentle baked items. However, because of its one-of-a-kind gluten shape, it's essential to modify recipes thus whilst substituting rye for all-motive flour (das Graças Costa & de Souza, 2023).

In conventional medicine, in particular in Eastern Europe, rye extracts and flour poultices have been implemented to deal with skin situations and decrease inflammation. The excessive fiber content of rye additionally made it a popular treatment for digestive issues. While current science continues to explore the entire range of rye's medicinal properties, its use in folks' treatments reflects a protracted-standing perception in its health blessings. As with any traditional treatment, it's continually endorsed to seek advice from healthcare specialists earlier than use.

The health benefits and other useful products of rye cereal are summarized in Figure 4.2. The long-term use of rye as a cover crop has shown enormous benefits for soil fitness and water first-rate. Studies have observed that it is able to increase soil water storage capacity by way of as much as 22% in the top layers of the soil, which is critical for plant boom and resilience toward drought. Moreover, the deep root machine of rye enables to enhance soil structure and save you compaction, making an allowance for higher water infiltration and nutrient retention. This no longer most effective enhances the soil's bodily residences however also reduces the risk of sediment and nutrient runoff, contributing to cleaner waterways and a reduction inside the environmental impact of farming practices (Sharma et al., 2018).

Rye extracts are certainly a treasured element in skin care, in particular because of their rich antioxidant properties. These extracts, derived from *S. cereale* or rye seeds, had been shown to beautify skin firmness and radiance. For example, a selected rye extract known as Coheliss has been suggested to easy the skin's surface and improve complexion radiance after regular utility over a time period. It works by means of stimulating the skin's mechanical resistance to gravity, that's important in retaining youthful and colorful skin. Additionally, rye extracts are regarded to stimulate the synthesis of proteins which are important for the skin's mechanical shape, thereby

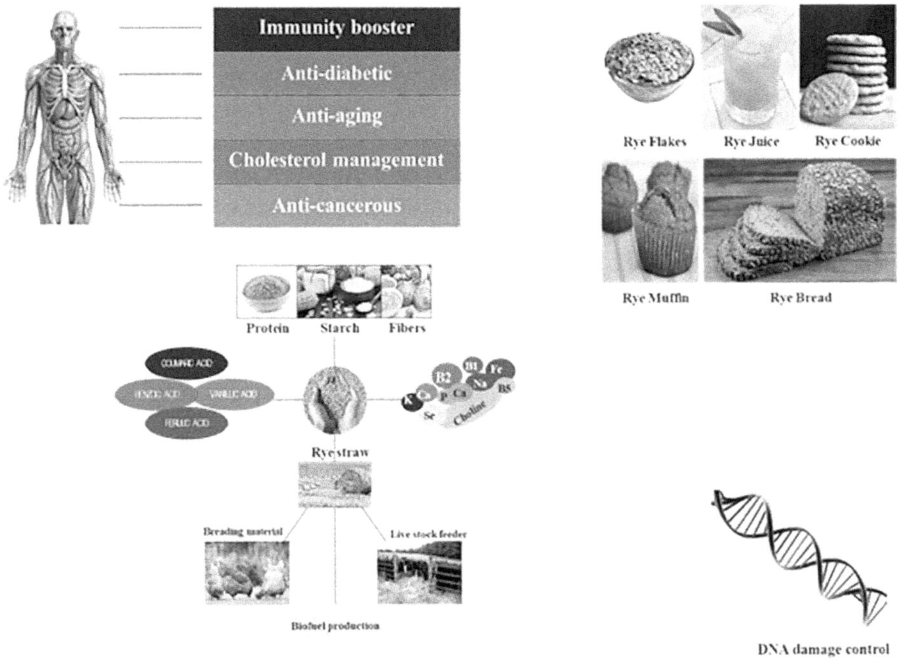

FIGURE 4.2 Health benefits and other useful products of rye cereal.

lowering the advent of wrinkles and imparting a lifting impact. This makes rye extract a sought-after aspect in anti-growing old and pores and skin-smoothing merchandise, offering a natural way to beautify pores and skin health and appearance.

4.6 GENETIC INFORMATION

To further boost rye's agronomic potential, we created a chromosome-scale annotated assembly of the 7.9-gigabase rye genome and comprehensively evaluated its quality using a variety of molecular genetic tools (Rabanus-Wallace et al., 2021). As we appearance to the destiny, rye's position may emerge as even extra huge. With issues about climate change and meals protection, plants like rye which can thrive in less-than-ideal situations might be essential. Researchers are exploring the genetic diversity of rye to enhance its disorder resistance and yield, making sure that it maintains to feed populations around the world.

The story of rye is a testament to the ingenuity of early agricultural societies and the continuing significance of crop diversity. From its wild origins to its cutting-edge repute as a global commodity, rye's journey through records reflects humanity's enduring courting with the land and our continual version to the demanding situations of feeding a developing population. As we hold to domesticate rye, we carry ahead a lifestyle that has sustained human beings for millennia, and we honor the ancient grain that has evolved alongside human civilization (Schreiber et al., 2021).

Rye, a member of the wheat tribe and closely related to barley and wheat, has a significantly massive genome length of approximately 89 GB. This expansive genome carries a rich variety of genes, predicted to be within the range of 30,000–40,000. The complexity of rye's genome makes it a captivating problem for genetic research, especially in the regions of plant breeding and biotechnology, where know-how its genetic make-up can lead to advances in crop resilience and yield (Korzun et al., 2001). In rye, environmental circumstances influenced lignan content via grain size; smaller grains had considerably less total lignan, syringaresinol, and lariciresinol than bigger grains (Smeds et al., 2009). Rye is a diploid species that is famous for a completely unique reproductive mechanism via its gametophytic self-incompatibility. This guarantees genetic variety via stopping self-pollination and promoting pass-pollination, that's critical for the species' survival and adaptability. As an obligatory cross-pollinator, rye is based on outside factors inclusive of wind for the switch of pollen among plant life, which contributes to its genetic variability and resilience (Tenhola-Roininen et al., 2005).

4.7 CULTIVATION OF RYE (*Secale cereale*)

Rye is a tall, grass-like plant with long, flat leaves and a slim stem. The grain is small, tough and brownish in color. Rye, a hardy grain acknowledged for its resilience and adaptability, is often sown in well-draining soil to make certain premier boom. The timing of planting rye seeds is vital; fall planting lets in the seeds to establish them before the wintry weather, taking benefit of the cool, moist conditions that rye flourishes in. This timing also lets in for a strong root gadget to develop, which is crucial for the plant's survival through the chillier months. On the other hand, early spring planting can also be beneficial, as the soil begins to warm up and the last of the frost has passed, providing a distinct set of conditions which can contribute to the increase of the rye. The choice between fall and spring planting often relies upon on the specific weather and geography of the region, in addition to the intended use of the rye crop, whether or not it be for grain, forage or as a cover crop to enhance soil health. In both cases, the properly draining nature of the soil is a key element, as rye does no longer tolerate waterlogged conditions nicely, which could cause root rot and different illnesses. The life cycle of rye plant is exhibited in Figure 4.3. Proper soil preparation, together with tilling and adding organic rely, can decorate drainage and offer the seeds with the high-quality viable begin. Additionally, rye's potential to germinate at lower temperatures as compared to other grains makes it a versatile preference for farmers and gardeners seeking to maximize their growing season.

Rye certainly holds a considerable region within the hierarchy of bread grains, second simplest to wheat. It is extraordinarily valued for its rich dietary fiber content and an array of bioactive compounds that make contributions to its nutritional profile. Rye's benefits make bigger past its fiber richness; it also offers plenty of vitamins and minerals, together with B vitamins and crucial minerals like magnesium and iron. Its decrease glycemic index as compared to wheat, which makes it a good choice for blood sugar management. Furthermore, the inclusion of rye in one's food regimen has been associated with progressed digestive fitness and ability benefits for heart fitness (Popović et al., 2022).

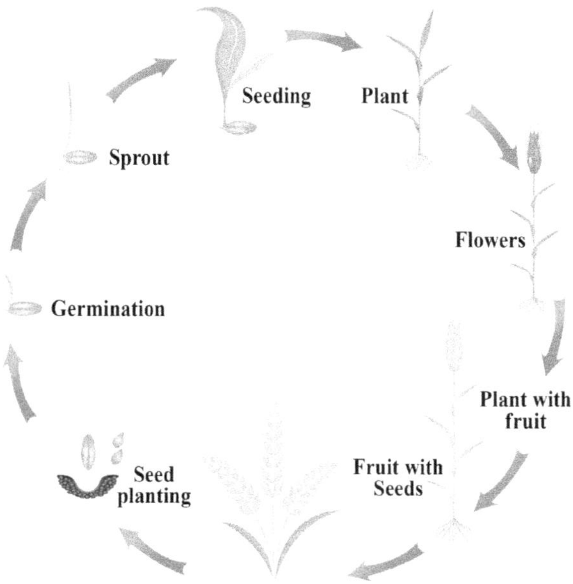

FIGURE 4.3 Rye plant life cycle.

Rye is a resilient and flexible grain that thrives in environments in which other cereals struggle. Its capability to withstand bad soil situations and colder temperatures makes it a really perfect crop for regions in Europe and North America that face agricultural demanding situations. The most important rye belt extends from northern Germany thru Poland, Ukraine and into principal and northern Russia, areas recognized for their harsh climates. Rye's adaptability is not just limited to soil and temperature; it additionally exhibits a strong resistance to diseases that commonly affect different grains, making it a dependable preference for farmers in those areas. Furthermore, rye performs a vast role in crop rotation structures because of its capability to improve soil shape and suppress weeds, enhancing the yield of next plants. Its deep roots help prevent soil erosion, contributing to sustainable farming practices. The grain itself is used in quite a few merchandises, from bread and crispbread, particularly in Scandinavia, to rye whiskey and animal fodder. Despite its importance, rye remains a crop with fantastically low worldwide exchange, with Europe generating approximately half of the sector's supply and little change between countries. This self-sufficiency displays rye's adaptability and the cultural significance it holds within the areas where it's far maximum typically grown. The cultivation of rye additionally has a wealthy history, with archeological evidence suggesting its systematic cultivation started around 13,000 years ago, even though this declare is issue to debate. Today, rye's cultivation is still a testimony to human ingenuity in adapting vegetation to numerous and tough environments (Ikram et al., 2023).

For most beneficial rye growth, it's crucial to apply fertilizers and insecticides appropriately. Nitrogen needs to be top-wearing established stands in early spring, without the need for split applications. Phosphate and potash should be broadcast and included before planting, with precise amounts relying on the expected yield and soil take a look at consequences. Pesticides may be carried out as seed treatments to defend against soil-borne risks and to sell growth. Always observe nearby hints and policies for the usage of fertilizers and insecticides to make certain environmental safety and crop fitness (Quinn et al., 2023). Rye used to be a meter-tall grass. Roofs were constructed from the long straw. Rye, meanwhile, evolved into a cutting-edge crop plant with agronomic and technological benefits (Schlegel, 2013).With the exception of a few perennial wild relatives, rye reproduction is annual. Typically, a fruitful stem produces a single spike. Rye inflorescence is a thin, imperfectly shaped spike that is complex in nature.

Rye grain, with its rich chemical composition, has been diagnosed for its fitness benefits, especially due to its high dietary fiber content material. Dietary fiber is crucial for digestive fitness, and rye's abundance of this nutrient helps promote a wholesome intestine microbiome, which is crucial for basic nicely being. The fiber in rye is predominantly insoluble, which aids in digestion and may help save you constipation. Moreover, the soluble fiber found in rye has been connected to advanced blood sugar control, making it a beneficial preference for individuals with diabetes or those looking to manipulate their blood sugar tiers.

The nutritional profile of rye includes a good supply of crucial nutrients and minerals, consisting of B nutrients, which might be crucial for power metabolism and neurological characteristic. Rye additionally contains vital minerals like magnesium, phosphorus and selenium, contributing to bone fitness and antioxidant defenses. The presence of these nutrients, at the side of the fiber, makes rye a coronary heart-healthy alternative that can help decrease levels of cholesterol and reduce the hazard of cardiovascular ailment.

In addition to its nutritional benefits, rye grain has been associated with weight management. Its excessive fiber content material contributes to a sense of fullness, which could assist control urge for food and help weight loss efforts. Studies have also counseled that the consumption of entire-grain rye merchandise can lead to better satiety compared to sophisticated wheat products, further supporting its function in a healthful food plan. Rye's low glycemic index is any other advantage, as it results in a slower upward thrust in blood sugar degrees after food, offering an extra sustained power launch. This belonging, mixed with its nutrient density, makes rye a wonderful desire for a wholesome and balanced food plan. It's important to notice that while rye offers several fitness blessings, it does contain gluten, so it can no longer be suitable for individuals with celiac disease or gluten sensitivity (Linina et al., 2019).

Utilizing rye as a cover crop offers a mess of environmental blessings, together with nitrogen scavenging, soil erosion reduction and weed suppression, that are instantaneous and glaring upon implementation. These benefits aren't simplest vital for keeping ecological stability but additionally contribute to lengthy-term financial gains by way of maintaining soil resources. Economically, soil conservation translates into decreased soil degradation, which in flip leads to sustained agricultural productivity and profitability. Studies have proven that cowl crops like rye can reduce nitrate losses, sequester carbon and even provide renewable herbal gasoline, thereby supplying a sustainable technique to agriculture that aligns with each environmental stewardship and economic viability. Moreover, the adoption of conservation agriculture practices, which include using cover crops, is associated with decrease hard work and farm-strength inputs, greater strong yields and an improved soil nutrient alternate ability, all of which make a contribution to increased crop production profitability over time (Krueger et al., 2011).

Climate change poses great demanding situations to agriculture, with rye (*S. cereale* L.) being no exception. Research suggests that the implementation of iciness rye as a cowl crop can provide a couple of advantages, doubtlessly mitigating a few negative influences of weather alternate on crop manufacturing. For instance, wintry weather rye can reduce nitrate leaching, sequester carbon and even contribute in producing bioenergy. Moreover, research the use of area-scale laptop models have shown that wintry weather rye cowl plants can result in a giant reduction in nitrate tiers in drainage water, that's vital for retaining water fine and stopping environmental troubles like algal blooms. However, the outcomes of climate alternate, together with elevated temperatures and variable rainfall, can nonetheless lead to decreased yields over time. It's vital to maintain exploring sustainable agricultural practices that may adapt to and fight the outcomes of climate change to make sure meals safety and environmental health (Ghafoor et al., 2024).

Salinity can appreciably affect the increase and yield of rye, as it does with many other vegetation. Research has proven that an increase in salinity above 11.4 dS m^{-1} can lessen rye yield via 10.8%, placing rye in the salt-tolerant category of vegetation. The consequences of salinity on rye are multifaceted, influencing not best the grain yield but additionally the satisfactory and vegetative boom of the plant. For instance, higher salinity stages have been associated with reduced spike weight and man or woman seed weight, as opposed to a lower in the range of spikes. Additionally, salinity can have an effect on the bread exceptional made from rye, despite the fact that the effect is moderate. It's also worth noting that straw yield is touchier to salinity than grain yield. Understanding these consequences is essential for farmers and agronomists to expand powerful strategies for handling salinity in rye cultivation, making sure that the crop stays a feasible and efficient choice even in hard growing conditions (Francois et al., 1989).

Harvesting wintry weather cereal rye not only offsets the prices related to cover crop status quo but also promotes its adoption because of the instant monetary go back from its use as emergency forage. This practice gives you a twin advantage: it provides a valuable feed supply for livestock at the same time as still contributing to soil health and erosion control. Additionally, it is able to act as a safeguard in opposition to forage shortages resulting from unpredictable climate activities, such as drought or alfalfa winterkill (Sadeghpour et al., 2022).

The efficiency of integrate harvesters, mainly inside the threshing method, is essential for minimizing grain loss and maximizing yield. Recent research has shown that with well-tuned threshing gadgets, it is possible to attain remarkably low grain loss costs of approximately 1%, making sure that over 93% of the whole grain makes it into the bunker. Innovations along with multi-threshing gaps in concentric threshing gadgets have in addition advanced performance via calmly dispensing the threshed output mixture and decreasing overall loss rates via a substantial margin. These improvements constitute a soar forward in agricultural generation, offering farmers enhanced harvesting performance and reduced waste (Barac et al., 2012).

Initially, the rye grains are segregated from the panicle, that's the flowering a part of the plant. Following this, the grains are sorted from the straw, the dry stalks of the plant. Finally, winnowing is carried out to separate the chaff, the husks and different impurities, from the grain. These operations are important in traditional agriculture to prepare rye for numerous uses, consisting of baking and brewing (Kulshreshtha et al., 2020).

In the context of grain garage, moisture content material is a vital factor that determines the satisfactory and safety of the stored product. Rye with a 14% moisture content is normally labeled as instantly grade, which is acceptable for instant use or brief-time period storage. However, for long-time period garage, a moisture content of 13% or decrease is suggested to save you the increase of mold and different spoilage organisms. The variety of 10.0%–17.5% moisture content for the take a look at displays the various conditions below which rye is probably stored (Sathya et al., 2008).

4.8 TYPES

There are several varieties of rye, including:

- Winter rye (planted in fall, harvested in summer season)
- Spring rye (planted in spring, harvested in fall)
- Perennial rye (a sort of grass that comes lower back 12 months after year)

4.8.1 Winter Rye

Winter rye certainly performs a giant function in European agriculture, especially within the cooler climates where it prospers. Because it may start growing again early in the spring and is winter hardy, winter rye is a suggested cover crop in cold locations (Krueger et al., 2011). Winter rye germination was studied on the surface of three distinct soils that were placed in three low-temperature incubators at 10, 18 and 25°C and equilibrated to a water potential of -50, 200 and 500 kPa (Wilson et al., 2013). It is valued for its hardiness and potential to develop in negative soils wherein other cereals might not prosper. Winter rye is typically sown in autumn and harvested inside the following summer time, making it a top-notch cover crop that forestalls soil erosion at some point of the wintry weather months. Its deep roots help to improve soil structure and organic content material. Additionally, rye grain is used for a whole lot of merchandise, from bread and whiskey to animal feed, and its straw can be used for bedding, biomass gas and as a creation fabric. The cultivation of iciness rye displays the adaptability of agricultural practices to exclusive environmental situations and the importance of crop variety in meals security and sustainability (Chmielewski & Köhn, 2000). Chromosome 5 controls the development of freezing tolerance and the buildup of antifreeze proteins in winter wheat. Although rye antifreeze proteins may have developed from pathogenesis-related proteins, they still have catalytic properties and may help overwintering plants become more resistant to disease and freezing (Griffith et al., 1997). The apoplast of winter rye leaves, which is where ice develops following freezing, is where the protein was collected (Griffith et al., 1992).

FIGURE 4.4 Winter rye seedlings.

Winter rye certainly dominates the rye manufacturing marketplace, being a hardy crop that may face up to harsh winters and bad soil situations. The new seedlings of winter rye grown up to 5 cm can be seen in Figure 4.4. It is mainly grown in the Northern Hemisphere, with Germany, Poland and Russia main in manufacturing. The versatility of wintry weather rye, from its use as a cowl crop to its role in bread-making, contributes to its incidence. Moreover, it calls for much less water than wheat, making it an extra sustainable alternative in certain areas. Despite being inferior to the major cereal crops like wheat, rice and maize in a number of ways, rye will remain a valuable crop for farmers in many nations due to its resilience to harsh winters, its capacity to thrive in unfavorable soil conditions and consumer demand for baked goods with rye's distinct flavor (Bushuk, 2001).

4.8.2 Spring Rye

Prolific is a real spring rye that grows upright. It is in the early stages of maturity. If it appears in the autumn, *prolific* will only survive mild winter temperatures. It features medium-sized blue green kernels (Briggle, 1959). The leaf weight of the spring rye variety increased almost to the point of earning. The lowest leaves began to discolor and dry faster (Strebeyko & Rosiek, 1964). Rye is indeed a resilient crop, capable of withstanding harsh wintry weather conditions that different cereals cannot bear. It's the most winter-hardy of all cereal grains, with the capacity to germinate and develop at temperatures as low as 33°F. Ploughing the green matter and straw of spring rye is a good source of carbon in the soil when no manure is used, and it is also a practice for controlling global warming by lowering greenhouse gas emissions (Klikocka et al., 2024). This robustness makes it a vital crop in regions like Canada, wherein extreme winters are common. Despite the challenges, winter rye types have been evolved to thrive in these conditions, imparting a lifeline for wildlife and contributing to soil fitness by preventing erosion and scavenging nitrogen. While spring rye cultivars may not in shape the wintry weather types in yield or excellent, they nevertheless play an important role in crop rotation and soil conservation techniques (Bushuk, 2001). Rye spread as a 'hitchhiking' weed in wheat and barley fields (Schreiber et al., 2019).

4.8.3 Perennial Rye

Perennial crops may provide a chance to produce grain in a more environmentally and economically sustainable way (Daly et al., 2022). The quest to develop perennial rye cultivars thru the crossing of *S. cereale* and *Secale strictum* is a charming enterprise that has captured the eye of agricultural scientists globally. These hybridization pursuits to mix the appropriate developments of *S. cereale*, normally called cereal rye, with the perennial nature of *S. strictum*, to create a rye variety which could regrow after seed set, potentially lowering the want for annual replanting. The benefits of this kind of improvement are manifold, inclusive of advanced soil conservation and decreased labor and input costs. The demanding situations, however, are equally vast, particularly in overcoming the reduced fertility regularly related to those crosses due to chromosomal complexities consisting of multivalents. Despite these challenges, progress has been made, as evidenced by using the release of tetraploid sorts like 'Permontra' and 'Sopertra' in Germany, even though these did no longer gain legitimate variety registration. The integration of contemporary genetic tools, inclusive of Quantitative Trait Locus (QTL) mapping and molecular markers, has provided new insights into the genetics of perenniality and fertility in rye, paving the way for more focused breeding techniques. For example, the identity of additively inherited QTLs on particular chromosomes has shed light at the genetic variance influencing perennial traits and fertility, which are crucial for the successful established order of perennial rye cultivars. As research continues, the potential of perennial rye as a genetic useful resource and its software in opportunity breeding techniques offer promising avenues for the improvement of sustainable agricultural structures. The pursuit of perennial rye cultivars represents a sizeable step toward progressive crop answers that would have an enduring impact on each agriculture and the environment (Gruner & Miedaner, 2021).

4.9 RYE CEREAL PROCESSING TECHNIQUES

Grain must be processed in order to be used in culinary products. It entails applying a variety of units, including hydration, mechanical or thermal energy and activation of the grain's natural biological enzyme system (Ikram et al., 2023).

The post-harvest processing of rye is done before its applications in making bread, alcoholic drinks and animal feed, mostly farmed in North America, Northern and Eastern Europe, Russia and China, Additionally, rye can be cultivated for biomass that can be utilized as green manure, feed or to produce bioenergy (Bahrani et al., 2021). Both triticale and rye are minor crops grown only in some northern parts of China (He et al., 2021). Depending on when it is planted, the plant's growth cycle usually lasts six to eight months, and the grains are harvested in late summer or early fall.

4.9.1 Harvesting and Initial Processing

Throughout human civilization, rye has had a significant agronomic, nutritional and social function (Korzun et al., 2021). The second most significant cereal for making bread, after wheat, is rye (*S. cereale* L.). The production of rye in Europe accounts for almost 90% of global production (Dziki, 2022). Rye is usually harvested in late summer or early fall. Additionally, rye is a profitable crop that thrives in peaty or sandy soils. Likewise because of its climate-resilient qualities, this grain may be grown in regions with low temperatures and high elevations (Dziki, 2022). Climate and cultivation conditions, soil quality, rye variety and other factors all affect the chemical makeup of rye grain (Sluková et al., 2021). Although rye grain quality testing is similar to wheat, there are differences in test weight, ergot bodies and grain sensitivity (Ikram et al., 2023). Rye grains, crop harvesting and ripened rye kernels ready for harvesting are shown in Figure 4.5. Rye is harvested using combines that cut, thresh and separate the grain from the chaff, just as other cereal grains. After being

FIGURE 4.5 Rye grains, crop harvesting and ripened rye kernels ready for harvesting.

harvested, rye grains are taken to processing plants where they are first cleaned to get rid of impurities including dust, dirt, stones and any leftover plant matter. Usually, a mix of sieves, air aspirators and gravity separators is used to achieve this.

4.9.2 Drying

Rye grains typically have high moisture content during harvest, frequently exceeding 20%. Rye grain were dried in an oven for 48 hours at 40°C to achieve a moisture level of 4.9%–5.2% before being pin disc milled (Silventoinen et al., 2021). Mechanical dryers or heated air are typically used for this, as they effectively lower moisture content without compromising the structure of the grain. The grain condition and weather patterns determine which of the two conditioning processes is best. The farmer can employ ambient air drying and save energy costs if the harvested grain can be stored at the harvested moisture and temperature conditions for an extended amount of time without experiencing any noticeable spoiling (Sathya et al., 2008).

4.9.3 Milling

Rye is mostly utilized in human nutrition as a raw material in the milling industry, and the conventional milling procedure of rye produces white (light) and particularly dark bread flours (Sluková et al., 2021). Rye grains are ground in a manner similar to that of wheat. Compared to wheat, rye needs a more thorough milling procedure with fewer passes (Ikram et al., 2023). Corrugated iron trucks used rather than smooth rolling rollers for a shorter heating period. Rye flour needs a large surface area to prevent lumps since the albuminous seed of rye is brittle in comparison to the hard albuminous seed of wheat (Ikram et al., 2023). Instead of being ground into fine flour, which keeps more of the bran and germ, rye is frequently ground into coarse meal or whole-grain flour. Prior to going through the milling process, the grains are tempered and wetted. These procedures are meant to extract the endosperm and separate the germs and outer layers (Sluková et al., 2021). The flow chart of rye grains milling process is shown in Figure 4.6. The material is carried higher in the mill chamber by the air stream, and it is disintegrated by revolving steel segments, or 'hammers', arranged in several rows above one another and the mills properly adjusted inner housing. The material travels in the stream of air through the disintegration space until the particles pass through the selected screen, which is attached to the upper side of the mill drum and has an adjustable mesh size (Sluková et al., 2021). The number of rye flours varies from country to country and is often categorized based on the amount of ash present (Németh & Tömösközi, 2021). The components and microstructure of flour can be greatly impacted by the milling procedure. Rye is commonly used as a raw ingredient by the milling industry to create its main products, white and black bread flour (Ikram et al., 2023).

4.9.4 Leavening and Fermentation (for the Production of Bread)

People frequently use rye grain to make crisp bread, which is made by manufacturers using rye grain and no yeast. Low alpha-amylase production is necessary for the

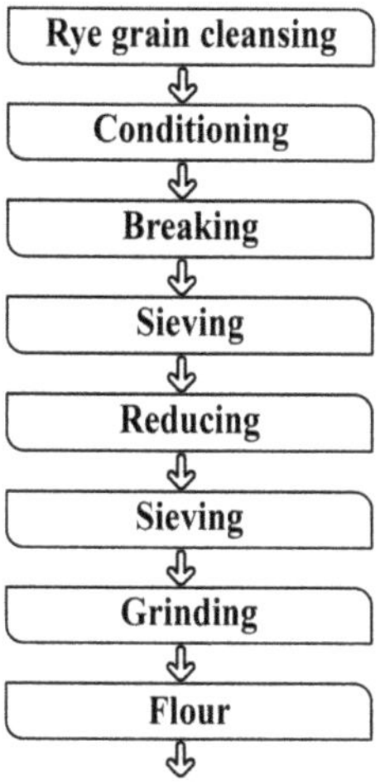

FIGURE 4.6 Flow chart of rye grains milling process.

rye flour used for this purpose due to the manufacturing method (Ikram et al., 2023). The primary usage of rye flour is as a component in sourdough bread. But yeast and a single-phase approach can also be used to make rye bread (Dziki, 2022). Every factor involved in making sourdough and bread, includes essential components like alcohol, heterocyclic compounds, esters, furans, pyrazines, lactones, ketones, bread aldehydes, furans and other alkanes (Ikram et al., 2023). Rye is less elastic than wheat and requires different fermentation techniques because it contains less gliadin and glutenin, two proteins that create gluten. Fermentation can alter the physical properties of grain biopolymers, including their molecular weight, viscosity, absorption and melting points (Ikram et al., 2023). Sourdough starters are used to naturally ferment the majority of rye bread. The first step in making sourdough is usually using sourdough starters. In contrast to spontaneous fermentation throughout the production process, these starters include specific microbes that provide dependability. The ingredients used to make each of the sourdoughs were flour, water and levain. A total of 200 g of the matching flour, 200 mL of water and 20 g of sourdough starter were combined to make the levain, which was then fermented for 16 hours at 26°C. The levain was kept for no more than a week at 5°C (Schmidt & Sciurba, 2021). Commercial yeast may occasionally be added to rye dough to improve the rising process, particularly in industrial bakeries. For many traditional rye bread variations, however, sourdough fermentation is still preferred. Sourdough rye bread is exhibited in Figure 4.7.

Rye bread lessens the need for insulin synthesis more efficiently than wheat bread, even though the two types of bread have comparable amounts of accessible carbs (Ikram et al., 2023). Antioxidant activity of rye bread is higher as compared to wheat bread. Kvass, a high source of vitamins such as B1 (thiamine), PP (niacin), B2 (riboflavin) and B6 (pyridoxine), can be made from whole meal rye bread. This product is used to treat a variety of digestive ailments because it is also a high source of folate and has a favorable effect on metabolism (Ikram et al., 2023).

FIGURE 4.7 Sourdough rye bread.

4.9.5 Processing into Other Food Products

Rye flour is used to make bread as well as a number of other meals, including cookies, crackers, breakfast cereals and snack foods. Instead of fermented bread, rye is utilized to make a variety of biscuits and crackers (Ikram et al., 2023). Additionally, rye flour is used in pancake flour in the United States, for creating gingerbread in France, and as filler for sauces, soups and custard powder. In the US, biscuits and crackers are made with a blend of 90% wheat flour and 10% rye flour (Rosentrater & Evers, 2018). While rye meal is frequently used as a base for soups and stews or as an ingredient in porridge, rye flakes and puffed rye are frequently found in breakfast cereals. Puffed rye is made by heating and pressing rye grains to a high temperature, while rye flakes are made by steaming and flattening them into flakes that can be used in granola, muesli or even baked goods. Together with flour, rye grain and other processed grains, such as those that have been fermented, roasted or hydrothermally treated, are used to make flakes (Ikram et al., 2023). A co-culture symbiotic beverage was created utilizing a blend of mung bean and rye sprouts treated with *Lactobacillus plantarum* (B-28) and *Lactobacillus casei* (B-29), together with inulin and oligofructose as prebiotic. Prebiotics are indigestible food ingredients that specifically promote the growth and activity of particular intestinal bacteria, improving the host's overall health (Shamsaie et al., 2023). Major applications of rye cereal are shown in Figure 4.8.

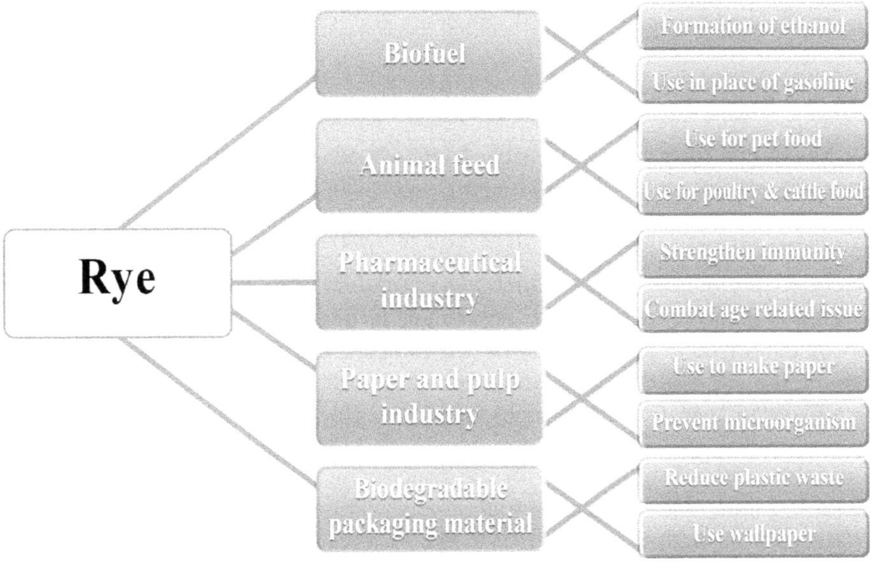

FIGURE 4.8 Major applications of rye cereal.

4.10 INDUSTRIAL APPLICATIONS OF RYE

Additionally, rye grains undergo fermentation and enzymatic hydrolysis, which increases the amount and bioavailability of phytochemicals and produces a variety of physiologically active substances. These differences make it possible to employ rye bran in food, feed, medicine, cosmetics and to shield plants from dangerous organisms (Dziki, 2022). A vital raw material, rye (*S. cereale*) is rich in vitamins, minerals, fatty acids and amino acids (Shamsaie et al., 2023). Beyond the culinary business, rye is a significant raw material for a range of industrial applications due to its distinct qualities and chemical makeup. These applications include materials research, pharmaceutical applications and the manufacture of biofuel.

4.10.1 Animal Feed

Similar to other grains, rye is an important portion of animal feed, especially for pigs, poultry and cattle. Rye can co-produce energy and waste that could be useful as animal feed in bioenergy system (Herbstritt et al., 2022). Since rye is a green plant, it can be utilized as green manure in a variety of crops in addition to providing grass for cattle (Ikram et al., 2023). Rye is frequently used to mixed meals in place of other grains like wheat and corn. It is a desirable feed ingredient because of its high fiber content, energy density and vital elements including vitamins and minerals. However, because rye is less digestible than other grains, its usage in animal feed may be restricted in some areas. The rye plant is occasionally used by pet food manufacturers, generally together with other grains. To prepare pet food, they grind rye into flour and then use offal (Ikram et al., 2023).

4.10.2 Biofuel

Rye has drawn interest as a possible feedstock for the manufacturing of biofuel, especially ethanol, as the need for sources of clean energy grows. The processing and utilization of rye grain in the manufacturing of ethanol and related biofuels are similar, albeit not the same (Ikram et al., 2023). Molted rye grain has greater quantities of amylase, which is merely a byproduct of its use (Ikram et al., 2023). Because of its high starch content, rye can be fermented to produce ethanol, which may be employed in place of gasoline in vehicles. Rye is grown especially for the manufacture of biofuel in several nations, where it is a more environmentally friendly crop than traditional ones like corn. Enzymatic hydrolysis transforms the rye grains' starch into sugars, which are subsequently fermented by yeast to provide ethanol. Flow chart of the processing of biofuel from rye grains is shown in Figure 4.9. To create a high-concentration ethanol solution, the ethanol is extracted through distillation from the fermentation mixture. Ethanol undergoes purification and processing to produce biofuel-grade ethanol. Compared to ethanol made from corn, rye-based ethanol has a number of benefits, including as greater climate adaptation and the possibility for larger yields in areas where corn is less practical.

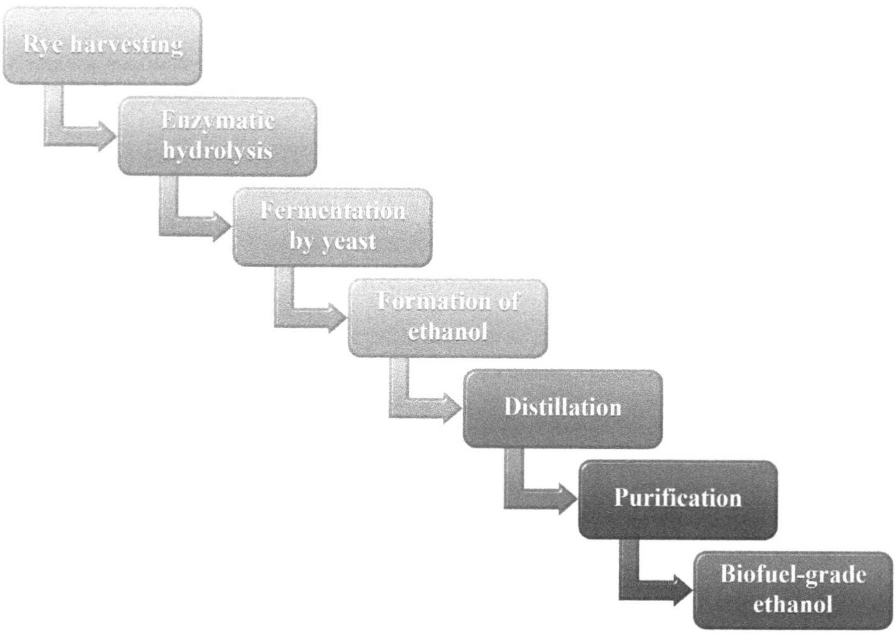

FIGURE 4.9 Flow chart of the processing of biofuel from rye grains.

4.10.3 Pharmaceutical Industry

Rye and wheat are closely related. Because it naturally contains numerous genes that provide protection against a variety of wheat diseases, such as powdery mildew, rust (stripe, leaf and stem) and sharp eyespot, it is considered a crucial genetic resource for wheat improvement.(Li et al., 2023) because of its chemical characteristics, rye is frequently used in the pharmaceutical sector. Numerous bioactive substances found in rye, such as alkaloids, flavonoids and phenolic acids, have been investigated for possible medicinal uses. Alkylresorcinols, ferulic acid, catechol, sinapic acid, vanillin and vanillic acid are among the bioactive substances found in rye grains that have antioxidant potential and may help strengthen immunity and combat age-related problems. Because they participate in processes that reduce the harmful effects caused by free radicals and oxidative stress, bioactive substances with antioxidant qualities are essential for maintaining a healthy lifestyle (Kaur et al., 2021). Rye, for instance, has been utilized in the creation of functional meals and nutritional supplements intended to support digestion and cardiovascular health. Rich in insoluble as well as soluble fiber rye has been demonstrated to improve blood sugar management and cholesterol levels while also being excellent for gastrointestinal health. Certain alkaloids found in rye, such ergot alkaloids, have been used medicinally historically and currently. Despite being harmful in high concentrations, ergot alkaloids have been used historically to cure a number of illnesses, including migraine headaches.

4.10.4 Biodegradable Packaging Material

Utilizing agricultural waste, such rye straw, to create biodegradable packaging materials has gained popularity in recent years. A common byproduct of rye farming is rye straw, which can be used into environmentally beneficial packing materials due to its high cellulose content. After being processed, the straw is turned into pulp, which can be shaped into many kinds of packaging. Packaging made of rye has various environmental advantages, including reduced waste from plastic and the capacity to break down organically in landfills. Plywood, corrugated boxes and wallpaper are all made using rye flour as an adhesive (Wrigley et al., 2004).

4.10.5 Paper and Pulp Industry

Additionally, the cereals contain fibers that might be added to the pulp, the raw material used to make paper (Johansson & Hedberg, 2024). Similar to other cereal straws, rye straw can be utilized to make paper goods for the pulp and paper industries. The main ingredient in paper, cellulose, is abundant in rye straw. Rye straw offers an alternative to conventional wood supplies for papermaking, reducing deforestation and encouraging more environmentally friendly methods. In the production of paper, microbial contamination results in both financial and environmental issues. Bioactive compounds found in fermented rye bran effectively inhibit the growth of almost all microorganisms (Johansson & Hedberg, 2024).

4.11 GLOBAL PRODUCTION OF RYE

Throughout human civilization, rye has had a significant agronomic, nutritional and social role. Rye grain yields have risen during the past 50 years, but not by enough to keep up with the reduction in cropping area (Korzun et al., 2021). Here we discuss about the production and cultivation of rye globally. Rye, a cereal grain this is carefully associated with wheat and barley, has seen a remarkable increase in each cultivation place and production globally. Over a latest 5–12 months period, the average rye vicinity multiplied to 4.28 million ha, whilst manufacturing reached 12.12 million tons. This growth trend is indicative of rye's growing importance as a crop. The common grain yield of rye at some stage in this era was 3.01 heaps in line with hectare, which reflects enhancements in agricultural practices and probably favorable climatic situations. Rye is valued for its hardiness and capacity to thrive in bad soils where other crops may fail, making it an important staple in many elements of the sector. Its resilience and adaptability will be factors contributing to the located growth in yield and manufacturing. Moreover, rye is utilized in loads of products, from bread and whiskey to animal feed, and its straw may be applied as bedding fabric for cattle. The increase in rye cultivation will also be driven through its environmental advantages, inclusive of improving soil fitness and offering cowl crop advantages. As the sector keeps to are trying to find sustainable agricultural practices, vegetation like rye which might be flexible and environmentally friendly will probably emerge as even more large. The information on rye's average vicinity and manufacturing not best spotlight its modern popularity but additionally hint at its capacity for destiny growth inside the agricultural zone.

The international cultivation of rye has proven an advantageous fashion over a latest five-year duration, with the common location devoted to rye farming achieving 4.28 million ha. This expansion in rye cultivation has caused a great increase in manufacturing, hitting 12.12 million tons. Moreover, the average grain yield has been pretty incredible, standing at 3.01 t consistent with hectare, indicating now not best a boom in the amount of land used for rye farming however additionally an enhancement in farming strategies and crop yield performance (Popović et al., 2022).

Rye has been a vast crop in Russia, historically recognized for its adaptability to terrible soil first-class and bloodless temperatures. In latest years, Russia has maintained its role as one of the pinnacle rye manufacturers globally. In 2012, Russia changed into the 0.33-largest manufacturer with over 2 million metric heaps. However, production has visible fluctuations, with a first-rate decrease from the 13.9 million metric heaps produced in 1992. Despite those modifications, rye stays crucial to Russian agriculture, frequently grown in regions like Central, Volga, North Caucasus and Siberia. It serves various functions, together with bread making, animal feed and alcohol manufacturing. The state-of-the-art records suggests that within the marketing 12 months 2021/2022, Russia produced approximately 1.7 million metric heaps of rye, with expectancies to reach round 2 million metric tons within the following year.

Germany has vast records of rye production, which has been a staple within the USA agriculture for hundreds of years. As of 2022, Germany's rye manufacturing modestly extended to 3.4 million tons, marking a 2.8% upward push compared to the previous year. This increase is part of a steady upward trend found during the last five years, with a mean annual growth of 4.6%. Rye is valued in Germany now not only for its use in traditional breads and beers but also as animal fodder and for its adaptability to negative soil conditions and colder climates. The United States of America's understanding in rye cultivation has made it a key participant within the international market, contributing notably to the worldwide deliver of this flexible grain.

Ukraine has skilled a fluctuation in rye manufacturing over current years. In 2022, the United States of America saw a giant decline in rye manufacturing, with a lower of 11.5% to 525,000 t. Despite this, the general trend in rye production has been pretty stable, with a super increase in 2020. The harvested location for rye also noticed a reduction to 155,000 ha in 2022, which became a 9.8% decrease from the previous year. However, the average yield of rye in Ukraine has shown a constant growth of approximately 2.7% annually from 2017 to 2022, even though there has been a slight decrease of 1.8% within the yield in 2022. Ukraine stays one of the top rye producers globally, and the crop is particularly grown in regions like Kyiv, Chernihiv and Zhytomyr. Despite the challenges, Ukraine's grain exports have improved; with a 60.5% improve in grain exports considering the fact that the start of the 2024/2025 marketing year.

Poland is a big manufacturer of rye, known for its giant history and expertise in cultivating this cereal grain. In 2022, Poland's rye production considerably extended to 2.7 million tons, marking a 9.2% upward thrust from the preceding year. This growth is a part of a particularly stable fashion in rye production, with the most sizable boom recorded in 2020 at 21%. Despite fluctuations within the harvested location over the years, the yield in 2022 stepped forward modestly to 3.4 lots consistent with hectare, a 3.2% increase in comparison to 2021. The cost of rye production also noticed a mild expansion to approximately $503 million, envisioned at export prices. These figures underscore Poland's role as a pivotal player within the international rye market, contributing drastically to each domestic intake and global exports (Arseniuk & Oleksiak, 2002).

Rye production in China has skilled a duration of relative stability in latest years. In 2022, the United States noticed a moderate boom of 0.7% in rye manufacturing, achieving a quantity of 516,000 lots. This uptick followed two years of decline, but typical, the manufacturing stages have maintained a fantastically flat fashion. The maximum giant growth took place in 2019 with a 1.7% increase. Historically, the height of rye manufacturing turned into in 2017, with 527,000 heaps. However, considering then, manufacturing figures have no longer regained that momentum. The harvested place for rye has additionally remained solid, with 167,000 ha mentioned in 2022, reflecting a consistent trend through the years. The yield of rye in step with hectare has been pretty flat as well, with the common yield in 2022 being 3.1 lots in keeping with hectare, mirroring the yield of the preceding year. In phrases of monetary cost, rye manufacturing noticed a lower to $109 million in 2022, estimated at export costs. Despite these fluctuations, the production fee has normally accompanied a flat trend, with an excellent increase of 6.4% in 2021. The top manufacturing cost was $119 million in 2019, however there was a moderate decrease seeing that that year. These traits in rye manufacturing in China are indicative of a market that has reached a stable nation, with minor variations in manufacturing and fee over the past few years. This balance is essential for both home intake and the worldwide market, where China contributes as a sizeable participant.

Rye production in Canada has shown a trend of boom over the last few years, with a few fluctuations. In 2022, the manufacturing of rye modestly expanded to 480,000 heaps, marking a 1.6% rise from the preceding year. This growth is a part of a broader trend determined from 2017 to 2022, wherein rye production accelerated at a median annual rate of 7.1%. Despite this overall boom, there were intervals of decline, including a 1.5% lower in production compared to 2020. The harvested region for rye also noticed a modest increase to a 100 and 50,000 ha in 2022, developing with the aid of 2.1% from the previous 12 months. This enlargement in harvested location contributed to the overall boom in production, in spite of a few setbacks in yield figures. In phrases of monetary fee, rye manufacturing in Canada reached $122 million in 2022, reflecting the vast function of this crop in the agricultural zone. The production value had peaked at $130 million in 2020, and although there was a moderate decline after that, the sturdy marketplace for rye has endured to help its manufacturing. Canada's function as a primary producer and exporter of rye, especially to the USA, underscores the importance of this crop inside the countrywide agricultural landscape. Rye is more often than not grown within the Prairie provinces Alberta, Saskatchewan and Manitoba—in addition to in Ontario and Quebec. It serves numerous purposes, along with bread making, animal feed and alcohol manufacturing, showcasing its versatility and value. The resilience of rye, specifically in dry environments, has been stated by using agricultural professionals, which may additionally account for its sturdy overall performance even in the course of challenging climatic situations. The crop's capacity to tolerate drought and its efficient use of water and nitrogen make it a strategic desire for farmers, especially in regions at risk of such situations. The call for rye as a feed grain has also been on the upward thrust, further bolstering its production. With the rural quarter continuously adapting to converting market demands and environmental situations, the production of rye in Canada stays a dynamic and important part of the industry (Darambazar et al., 2022).

Denmark has visible a fluctuating fashion in rye production through the years, with remarkable increases and reduces influenced by using different factors. In 2022, the United States experienced a rise in rye production, reaching 686k heaps, marking the end of a two-year decline. These increases become enormous as compared to the mentioned increase in 2019 whilst manufacturing surged via 80.5%, hitting a height of 884k lots. However, the general fashion has been barely downward, with the harvested area and yield figures experiencing moderate contractions. Despite those fluctuations, the value of rye production saw a rapid expansion to $297 million in 2022, estimated at export costs. The average annual increase price from 2017 to 2022 become + 2.6%, despite the fact that the production price peaked at $313 million in 2020 and did no longer hold that momentum inside the following years. The harvested area of rye in Denmark additionally picked up by means of 3.2% in 2022, totaling 112k ha, yet this turned into part of a distinctly flat fashion pattern discovered over the years. The most substantial boom inside the harvested location took place in 2019 with a 58% increase from the previous 12 months. This information replicates the dynamic nature of agricultural production and the various factors that could impact crop yields and market values year on year.

Sweden has visible a fluctuating trend in rye production over time. In 2022, the United States of America skilled an upward push in rye production, attaining 167,000 heaps, marking the stop of a two-year decline and indicating a moderate enlargement in this quarter. This increase was in particular sizeable in 2019, with a 151% increase in production extent. However, the harvested region for rye confirmed some variability, with a fantastic growth in 2019 accompanied through a decrease in subsequent years. Despite those fluctuations, rye remains a critical crop in Sweden, representing approximately 4% of the overall grain harvest primarily based on a five-year common. The manufacturing fee of rye additionally reflects this fashion, with a peak in 2019 at $52 million, followed by a contraction to $42 million in 2022. These figures underscore the dynamic nature of agricultural production and the various factors which could impact crop yields and market values year over year.

4.12 CHARACTERIZATION TECHNIQUES OF RYE

Whole grain cereals' effect on health is becoming clear, as a number of epidemiological studies revealed protections against obesity, diabetes, inflammation, cardiovascular disease and multiple kinds of cancer (Kamal-Eldin et al., 2009). Cereal grains are gaining majority of interest from food manufacturing industries. In the food companies, rye grain mainly serves as a raw material for the production of various types of flour from which bread (rye, mixed) is made (Sułek et al., 2024). Rye (*S. cereale*) possesses a substantial amount of macromolecules, including starch, fiber and proteins, which stimulates researchers and industries to use it in a variety of applications such as bakery items, drinks and edible film production (Kaur et al., 2021). Consumers' interest toward creating foods high in fiber and phenolic compounds led researchers to compile thorough information on rye (Kaur et al., 2021). A number of techniques are implemented to characterize the rye, as listed below.

- Preliminary characterization
- Quantitative characterization
- Confirmatory characterization

4.12.1 Preliminary Characterization

During qualitative analysis, preliminary tests are used to identify the presence of specific features in an unknown material. It is an essential component of analytical chemistry. It is performed to estimate the physical properties before further analysis. It provides the details about the nutritional properties, chemical properties and physical properties like color, size and weight.

4.12.1.1 Physical Properties

Rye is widely renowned for its strong resistance to abiotic stressors and good adaptability to barren soil (Li et al., 2021). Rye starch granules exhibit identical shape and size to wheat starch granules (Dziki, 2022). Rye is tougher than wheat or barley. Rye is used to make crisp bread, flakes and cereal treats. Rye starch granules are lenticular and spherical in form and size, with an A-type starch fraction of 20–35 μm and a B-type starch fraction of 5–10 μm, comparable to wheat starch (Sluková et al., 2021).

4.12.2 Growth Parameters

Rye is a vital crop that grows well on sandy or peaty soils. Additionally, this cereal possesses climate-resilient abilities and hence can be grown in low-temperature and high-altitude zones. Rye's chemical profile is mostly monitored by genetic parameters, soil quality, climate and production circumstances (Dziki, 2022). Rye crop is thought to have originated in southwestern Asia and is widely grown in Europe, North America and Asia (Kaur et al., 2021).

4.12.2.1 Rye Milling

Rye requires a more stringent milling process, with fewer passes than wheat. One example of a specific milling method is the successful testing of creating finely granulated whole meal rye flour. This method is carried out in the Czech Republic utilizing a specialized impact mill with a vertically rotating shaft. The mill achieves disintegration by using revolving steel segments known as 'hammers' placed in numerous rows above each other and properly regulating the mill's inner housing. A screen with adjustable mesh size is situated on the upper side of the mill drum, and an air stream transports the material through the disintegration space until it passes through the specified screen. The material is carried by the air stream, which rises in the mill chamber. Compared to the standard milling procedure, it allows for significantly finer granulation of the outer layers while inflicting minimum harm to the endosperm's starch granules (Ikram et al., 2023).

4.12.3 Quantitative Characterization

Quantitative analysis refers to any process for determining the amount of a chemical in a sample. The quantity is always expressed as a number with the correct units. The chemical composition of cereal grains impacts their functional and technical properties (Cardoso et al., 2019). Rye grain consists mainly of carbohydrate (57.1%–65.6%), dietary fiber (14.7%–20.9%), protein (9.0%–15.4%) and ash (1.8%–2.2%) (Sułek et al., 2024).

4.12.3.1 Phenolic Content

The nutritional benefits of phenolics, such as phenolic acids, lignans and alkylresorcinols in the human diet have lately been explored. Rye's phenolic acids help prevent low-density lipoprotein degradation (Ikram et al., 2023).

4.12.3.2 Anti-Oxidant Activity

Bioactive substances with antioxidant qualities are the important factor for maintaining a healthy lifestyle since they are involved in mechanisms that slow down the detrimental effects created by free radicals and oxidative stress. Rye grains include a variety of bioactive chemicals (alkyl resorcinols, ferulic acid, catechol, sinapic acid, vanillin and vanillic acid) with antioxidant properties that may help boost immunity and combat age-related problems (Kaur et al., 2021). Rye shows high antioxidative activity as compare to wheat (Dziki, 2022). Fructans (including fructo oligosaccharides) are present in various levels in rye, ranging from 4.6% to 6.6%. Rye's DF concentration has increased even further as fructose are now recognized in several countries as a component of DF (Ikram et al., 2023).

4.12.3.3 Starch and Carbohydrates

Rye, like other cereal grains, contains a significant amount of starch (55%–70%). Starch is a major source of energy in the human diet. It contributes to the functional characteristics of food products (Kaur et al., 2021). Rye contains high levels of arabinoxylans (AX), β-glucans and resistant starches (Dziki, 2022). High performance liquid chromatography (HPLC), Gas chromatography-mass spectrometry (GC-MS), Total Carbohydrate Assay Kits and Enzymatic Assays are used to quantify and separate carbohydrates such as starch, sugars and fibers of Rye.

4.12.3.4 Protein

Rye is the only crop, after wheat that contains a high concentration of gluten proteins making it widely used in the food industry around the world to make bread, biscuits, rusk, flakes, beer and other items (Kaur et al., 2021). Rye proteins are full of lysine, though they cannot form a continuous gluten network as wheat proteins can (Dziki, 2022). Bradford Assay, HPLC, Dumas Method and Kjeldahl Method are used to identify proteins in Rye.

4.12.3.5 Fibers

Dietary fibers are important because they help with weight management and digestion. Rye grain contains primarily arabinoxylan, followed by fructan, β-glucan and lignins (Kaur et al., 2021). Rye grain has the greatest fiber content among popular grains (Sluková et al., 2021). HPLC, Near-Infrared Reflectance Spectroscopy (NIRS) and Enzymatic Assay are used to quantify fibers in Rye.

4.12.3.6 Enzymatic Activity

Enzymes, particularly amylases and endo-xylanases, are important rye ingredients that contribute to bread quality (Németh & Tömösközi, 2021). UV-Vis Spectroscopy, Colorimetric Assays, HPLC, GC-MS, Amylase activity, Protease activity, Lipase Activity and Amperometry are used depending upon the quantification on specific Enzyme.

4.12.4 Confirmatory Characterization

Understanding the characteristics of a newly discovered material is critical in determining its applications. Some traits are inherent to the compound and regarded part of its fingerprint, while others can be created by material design and engineering using various treatment processes. Whether inherent or engineered, the understanding of the chemical and structural properties of the material plays a vital role in its

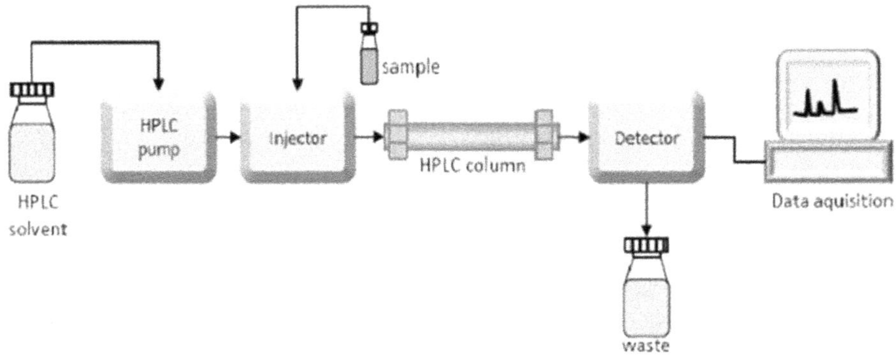

FIGURE 4.10 High performance liquid chromatography (HPLC) instrumentation.

utilization in various areas (Ortega et al., 2022). Following different characterization techniques are used to confirm the structure of Rye.

4.12.4.1 High Performance Liquid Chromatography

Rye's unique phenolic compounds are observed by using HPLC. HPLC consists of a liquid phase that is mobile and moves through a stationary phase. HPLC Instrumentation is shown in Figure 4.10. Different HPLC procedures can be classified according to their retention, separation or operating mechanism. HPLC is widely used in several fields to separate and quantify mixture components (Ortega et al., 2022).

4.12.4.2 Gas Chromatography-Mass Spectrometry

Rye's volatile compounds and distinct aroma profile is studied by using GC-MS. The GC-MS has proved to be a useful instrument for studying complicated materials. GC-MS instrumentation is shown in Figure 4.11. The GC and MS is a physicochemical process that separates a homogeneous mixture using a gas flow distributed

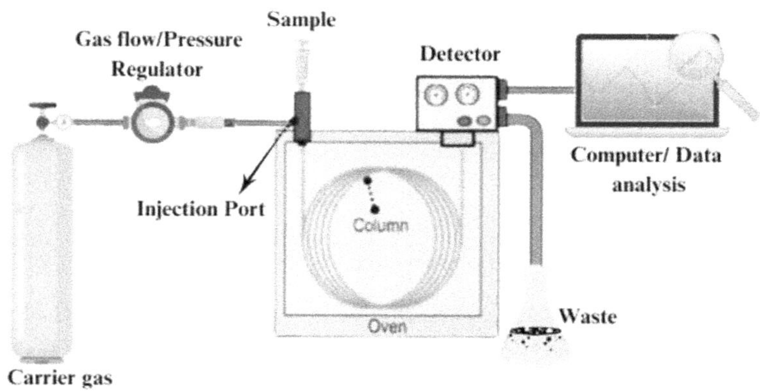

FIGURE 4.11 Gas chromatography-mass spectrometry (GC-MS) instrumentation.

between two phases. The stationary phase will remove molecules that are inclined to stay at a lower velocity, while the mobile phase will prefer molecules that have the inclination to go ahead (Ortega et al., 2022).

4.12.4.3 Atomic Mass Spectroscopy (AFM)

Atomic Mass Spectroscopy (AFM) is a method that provides high resolution. The AFM technique provides a topological mapping of the material's surface through the contact between a tip and a specimen (Ortega et al., 2022). Rye surface morphology and structure is studied by using this technique.

4.12.4.4 X ray Diffraction (XRD)

Mineral composition and crystalline structure of Rye is determined by using this technique. If the wavelength and the crystal's periodicity are of comparable magnitudes, the interaction of wavelengths with periodic structures produces diffraction effects. Though electrons or neutrons with the right energy can also be utilized for diffraction investigations on crystals, X-rays can be produced with ease at wavelengths that correspond to the unit cell dimensions of crystals. Because it can investigate both pure substances and mixtures in a powder-like, solid and liquid state, it produces qualitative and quantitative information about the phase composition, structure and texture of almost all types of samples (Ortega et al., 2022).

4.12.4.5 Scanning Electron Microscope (SEM)

Scanning electron microscopy (SEM) generates images by colliding a beam of electrons with the sample. The sample is placed on the specimen stage, which can be shifted to the side, tilted and rotated to obtain a better image. SEM is shown in the form of flow chart in Figure 4.12. It describes the size, shape texture, smoothness or roughness of Rye.

4.12.4.6 Transmission Electron Microscope (TEM)

The operation mechanism of the transmission electron microscopy (TEM) is similar to that of an optical microscope. TEM is elaborated in Figure 4.13. The most significant distinctions are the transition of the illumination source from light to electrons, the creation of electromagnetic/electrostatic lenses rather than glass lenses and the functioning of TEM in a high vacuum. The final point is critical when using the TEM because electron scattering collides with a remaining gas molecule, affecting image resolution (Ortega et al., 2022). It gives information about morphological information of Rye.

4.12.4.7 Nuclear Magnetic Resonance (NMR)

When a molecule is exposed to a strong magnetic field, its atoms' nuclei act similarly to magnets. A broad radio frequency spectrum applied to the molecule causes the atom's nucleus to resonant at a specific and distinctive frequency. NMR seeks to precisely determine the frequency of a nuclear resonance in this sense. It provides chemical features of the Rye sample. Structure elucidation can be done by using this technique (Ortega et al., 2022).

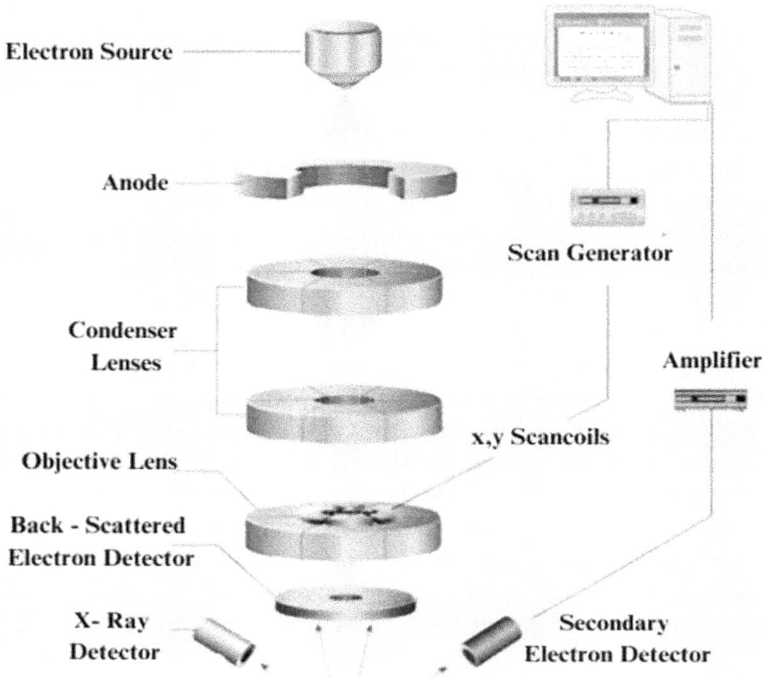

FIGURE 4.12 Scanning electron microscopy (SEM).

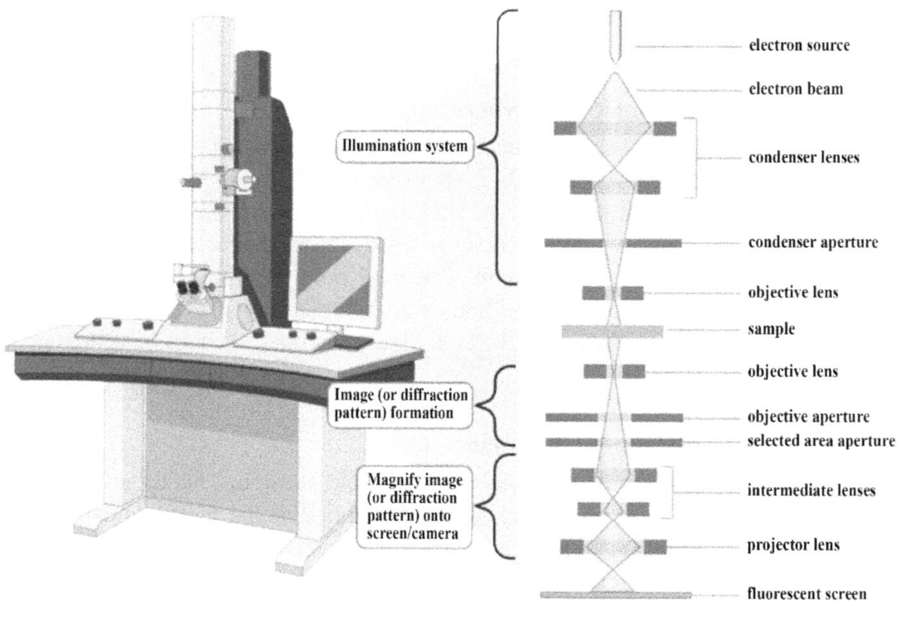

FIGURE 4.13 Transmission electron microscopy (TEM).

REFERENCES

Arseniuk, E., & Oleksiak, T. (2002). Production and breeding of cereals in Poland. In E. Arseniuk (Ed.), *Proceedings of the 5th International Triticale Symposium: Vol. I. Oral presentations* (pp. 11–18). Plant Breeding and Acclimatization Institute (IHAR).

Bahrani, H., Båga, M., Larsen, J., Graf, R. J., Laroche, A., & Chibbar, R. N. (2021). The relationships between plant developmental traits and winter field survival in rye (*Secale cereale* L.). *Plants*, *10*(11), 2455.

Barac, S., Biberdzic, M., Đikic, A., & Milenkovic, B. (2012). Effects of the threshing harvester device and its influence on the breakage and damage of buckwheat and rye depending of predefined parameters. *Poljoprivreda i Sumarstvo*, *58*(2), 7.

Biesiekierski, J. R. (2017). What is gluten? *Journal of Gastroenterology and Hepatology*, *32*, 78–81.

Bondia-Pons, I., Aura, A.-M., Vuorela, S., Kolehmainen, M., Mykkänen, H., & Poutanen, K. (2009). Rye phenolics in nutrition and health. *Journal of Cereal Science*, *49*(3), 323–336.

Briggle, L. W. (1959). *Growing rye*. US Government Printing Office.

Bushuk, W. (2001). Rye production and uses worldwide. *Cereal Foods World*, *46*(2), 70–73.

Cardoso, R. V., Fernandes, Â., Heleno, S. A., Rodrigues, P., Gonzaléz-Paramás, A. M., Barros, L., & Ferreira, I. C. (2019). Physicochemical characterization and microbiology of wheat and rye flours. *Food Chemistry*, *280*, 123–129.

Chmielewski, F.-M., & Köhn, W. (2000). Impact of weather on yield components of winter rye over 30 years. *Agricultural and Forest Meteorology*, *102*(4), 253–261.

Daly, E. J., Hernandez-Ramirez, G., Puurveen, D., Ducholke, C., Kim, K., & Oatway, L. (2022). Perennial rye as a grain crop in Alberta, Canada: Prospects and challenges. *Agronomy Journal*, *114*(1), 471–489.

Darambazar, E., Larson, K., Damiran, D., & Lardner, H. A. (2022). Evaluation of new fall rye cultivar 'Bono' in single and double cropping systems. *Agronomy*, *12*(6), 1382.

das Graças Costa, E., & de Souza, P. M. (2023). *Introduction to cereals cereal-based food products* (pp. 1–24). Springer.

Deleu, L. J., Lemmens, E., Redant, L., & Delcour, J. A. (2020). The major constituents of rye (*Secale cereale* L.) flour and their role in the production of rye bread, a food product to which a multitude of health aspects are ascribed. *Cereal Chemistry*, *97*(4), 739–754.

Dykes, L., & Rooney, L. (2007). Phenolic compounds in cereal grains and their health benefits. *Cereal Foods World*, *52*(3), 105–111.

Dziki, D. (2022). Rye flour and rye bran: New perspectives for use. *Processes*, *10*(2), 293.

Fattakhova, Z. F., Shakirov, S. K., Krupin, E. O., Bikchantaev, I. T., Shayakhmetova, L. N., & Askarova, A. A. (2020). The chemical composition, nutrition and fractional composition of winter rye grain proteins after various methods of exposure. *Carpathian Journal of Food Science & Technology*, *12*(1).

Francois, L., Donovan, T., Lorenz, K., & Maas, E. (1989). Salinity effects on rye grain yield, quality, vegetative growth, and emergence. *Agronomy Journal*, *81*(5), 707–712.

Ghafoor, A. Z., Karim, H., Studnicki, M., Raza, A., Javed, H. H., & Asghar, M. A. (2024). Climate change and rye (*Secale cereale* L.) production: Challenges, opportunities and adaptations. *Journal of Agronomy and Crop Science*, *210*(4), e12725.

Griffith, M., Ala, P., Yang, D. S., Hon, W.-C., & Moffatt, B. A. (1992). Antifreeze protein produced endogenously in winter rye leaves. *Plant Physiology*, *100*(2), 593–596.

Griffith, M., Antikainen, M., Hon, W. C., Pihakaski-Maunsbach, K., Yu, X. M., Chun, J. U., & Yang, D. S. (1997). Antifreeze proteins in winter rye. *Physiologia Plantarum*, *100*(2), 327–332.

Gruner, P., & Miedaner, T. (2021). Perennial rye: Genetics of perenniality and limited fertility. *Plants*, *10*(6), 1210.

Hackauf, B., Siekmann, D., & Fromme, F. J. (2022). Improving yield and yield stability in winter rye by hybrid breeding. *Plants, 11*(19), 2666.
He, H., Du, H., Liu, R., Liu, T., Yang, L., Gong, S., …, & Han, R. (2021). Characterization of a new gene for resistance to wheat powdery mildew on chromosome 1RL of wild rye *Secale sylvestre. Theoretical and Applied Genetics, 134*, 887–896.
Herbstritt, S., Richard, T. L., Lence, S. H., Wu, H., O'Brien, P. L., Emmett, B. D., …, & Malone, R. W. (2022). Rye as an energy cover crop: management, forage quality, and revenue opportunities for feed and bioenergy. *Agriculture, 12*(10), 1691.
Ikram, A., Saeed, F., Noor, R. A., Imran, A., Afzaal, M., Rasheed, A., …, & Naz, S. (2023). A comprehensive review on biochemical and technological properties of rye (*Secale cereale* L.). *International Journal of Food Properties, 26*(1), 2212–2228.
Isaksson, H., Sundberg, B., Åman, P., Fredriksson, H., & Olsson, J. (2008). Whole grain rye porridge breakfast improves satiety compared to refined wheat bread breakfast. *Food & Nutrition Research, 52*(1), 1809.
Johansson, A., & Hedberg, M. E. (2024). *Fermented rye bran as biocide replacement in paper production.* Paper presented at the 4th International Conference Strategies toward Green Deal Implementation Water, Raw Materials & Energy, Online conference, 14–15 December, 2023.
Jonsson, K., Andersson, R., Knudsen, K. E. B., Hallmans, G., Hanhineva, K., Katina, K., …, & Nordlund, E. (2018). Rye and health-where do we stand and where do we go? *Trends in Food Science & Technology, 79*, 78–87.
Kamal-Eldin, A., Lærke, H. N., Knudsen, K.-E. B., Lampi, A.-M., Piironen, V., Adlercreutz, H., …, & Åman, P. (2009). Physical, microscopic and chemical characterisation of industrial rye and wheat brans from the Nordic countries. *Food & Nutrition Research, 53*(1), 1912.
Kaur, P., Sandhu, K. S., Purewal, S. S., Kaur, M., & Singh, S. K. (2021). Rye: A wonder crop with industrially important macromolecules and health benefits. *Food Research International, 150*, 110769.
Klikocka, H., Podleśna, A., Narolski, B., & Podleśny, J. (2024). Spring rye as a source of biomass and carbon in the soil. *International Agrophysics, 38*(3), 243–255.
Korzun, V., Malyshev, S., Voylokov, A., & Börner, A. (2001). A genetic map of rye (*Secale cereale* L.) combining RFLP, isozyme, protein, microsatellite and gene loci. *Theoretical and Applied Genetics, 102*, 709–717.
Korzun, V., Ponomareva, M. L., & Sorrells, M. E. (2021). Economic and academic importance of rye. In *The rye genome* (pp. 1–12). Springer.
Krueger, E. S., Ochsner, T. E., Porter, P. M., & Baker, J. M. (2011). Winter rye cover crop management influences on soil water, soil nitrate, and corn development. *Agronomy Journal, 103*(2), 316–323.
Kulshreshtha, A., Malik, A., Khan, M., & Arya, V. (2020). Status of cereal processing and preservation in India. *International Journal of Emerging Technologies and Innovative Research (JETIR), 7*(6), 96–104. http://www.jetir.org/papers/JETIR2006015.pdf
Li, G., Wang, L., Yang, J., He, H., Jin, H., Li, X., …, & Han, X. (2021). A high-quality genome assembly highlights rye genomic characteristics and agronomically important genes. *Nature Genetics, 53*(4), 574–584.
Li, J.-J., Li, Z., Lü, B.-Y., Yu, F., Zhang, S.-F., Liu, S.-H., …, & Chen, X.-H. (2023). Development and characterization of a novel common wheat–Mexico Rye T1DL 1RS translocation line with stripe rust and powdery mildew resistance. *Journal of Integrative Agriculture, 22*(5), 1291–1307.
Linina, A., Kunkulberga, D., Kronberga, A., & Locmele, I. (2019). Winter rye grain quality of hybrid and population cultivars. *Agronomy Research, 17*(S2), 1380–1389.

Lorenz, K. (2000). *Rye handbook of cereal science and technology, revised and expanded* (pp. 223–256). CRC Press.

Németh, R., & Tömösközi, S. (2021). Rye: Current state and future trends in research and applications. *Acta Alimentaria*, *50*(4), 620–640.

Ortega, E. O., Hosseinian, H., Meza, I. B. A., López, M. J. R., Vera, A. R., & Hosseini, S. (2022). *Material characterization techniques and applications*. Springer.

Popović, D., Rajičić, V., Popović, V., Burić, M., Filipović, V., Gantner, V., …, & Božović, D. (2022). Economically significant production of *Secale cereale* L. as functional foods. *Agriculture and Forestry*, *68*(3), 133–145.

Quinn, D. J., Poffenbarger, H. J., Miguez, F. E., & Lee, C. D. (2023). Corn optimum nitrogen fertilizer rate and application timing when following a rye cover crop. *Field Crops Research*, *291*, 108794.

Rabanus-Wallace, M. T., Hackauf, B., Mascher, M., Lux, T., Wicker, T., Gundlach, H., …, & Guo, L. (2021). Chromosome-scale genome assembly provides insights into rye biology, evolution and agronomic potential. *Nature Genetics*, *53*(4), 564–573.

Rakha, A., Åman, P., & Andersson, R. (2010). Characterisation of dietary fibre components in rye products. *Food Chemistry*, *119*(3), 859–867.

Rosentrater, K., & Evers, A. (2018). Flour treatments, applications, quality, storage and transport. In *Kent's technology of cereals* (5th ed., pp. 515–564). Woodhead Publishing.

Sadeghpour, A., Adeyemi, O., Reed, B., Fry, J., & Keshavarz Afshar, R. (2022). Profitability of dual-purpose rye cover crop as influenced by harvesting date. *Agronomy Journal*, *114*(1), 627–640.

Sathya, G., Jayas, D., & White, N. (2008). Safe storage guidelines for rye. *Canadian Biosystems Engineering*, *50*(3), 1–8.

Schlegel, R. H. (2013). *Rye: Genetics, breeding, and cultivation*. CRC Press.

Schmidt, M., & Sciurba, E. (2021). Determination of FODMAP contents of common wheat and rye breads and the effects of processing on the final contents. *European Food Research and Technology*, *247*, 395–410.

Schreiber, M., Himmelbach, A., Börner, A., & Mascher, M. (2019). Genetic diversity and relationship between domesticated rye and its wild relatives as revealed through genotyping-by-sequencing. *Evolutionary Applications*, *12*(1), 66–77.

Schreiber, M., Özkan, H., Komatsuda, T., & Mascher, M. (2021). Evolution and domestication of rye. In M. T. Rabanus-Wallace & N. Stein (Eds.), *The rye genome* (pp. 85–100). Springer. https://doi.org/10.1007/978-3-030-83383-1_6

Sencer, H., & Hawkes, J. (1980). On the origin of cultivated rye. *Biological Journal of the Linnean Society*, *13*(4), 299–313.

Shamsaie, P., Hosseini, S. E., Asadi, G., & Sharifan, A. (2023). Production and characterization of a novel symbiotic plant-based beverage rich in antioxidant phenolic: Mung bean and rye sprouts. *Plant Foods for Human Nutrition*, *78*(3), 584–589.

Sharma, P., Singh, A., Kahlon, C. S., Brar, A. S., Grover, K. K., Dia, M., & Steiner, R. L. (2018). The role of cover crops towards sustainable soil health and agriculture—A review paper. *American Journal of Plant Sciences*, *9*(9), 1935–1951.

Silventoinen, P., Kortekangas, A., Ercili-Cura, D., & Nordlund, E. (2021). Impact of ultra-fine milling and air classification on biochemical and techno-functional characteristics of wheat and rye bran. *Food Research International*, *139*, 109971.

Sluková, M., Jurkaninová, L., Švec, I., & Skřivan, P. (2021). Rye–the nutritional and technological evaluation in Czech cereal technology–A review: Grain and flours. *Czech Journal of Food Sciences*, *39*(1).

Smeds, A. I., Jauhiainen, L., Tuomola, E., & Peltonen-Sainio, P. (2009). Characterization of variation in the lignan content and composition of winter rye, spring wheat, and spring oat. *Journal of Agricultural and Food Chemistry*, *57*(13), 5837–5842.

Strebeyko, P., & Rosiek, N. (1964). Dynamics of growth and development of spring rye. *Acta Agrobotanica*, *15*, 13–20.

Sułek, A., Cacak-Pietrzak, G., Studnicki, M., Grabiński, J., Nieróbca, A., Wyzińska, M., & Różewicz, M. (2024). Influence of nitrogen fertilisation level and weather conditions on yield and quantitative profile of anti-nutritional compounds in grain of selected rye cultivars. *Agriculture*, *14*(3), 418.

Tenhola-Roininen, T., Tanhuanpää, P., & Immonen, S. (2005). The effect of cold and heat treatments on the anther culture response of diverse rye genotypes. *Euphytica*, *145*, 1–9.

Wilson, M. L., Baker, J. M., & Allan, D. L. (2013). Factors affecting successful establishment of aerially seeded winter rye. *Agronomy Journal*, *105*(6), 1868–1877.

Wrigley, C. W., Corke, H., & Walker, C. E. (2004). *Encyclopedia of grain science*. Academic Press.

5 Harvesting and Postharvest Processing

Aalia Farooq, Shazia Akhter, Mehvish Mushtaq Banday, Aaruba Maqbool, Jahangir Ahmad Rather and Ruqaya Tariq

5.1 INTRODUCTION

Rye (*Secale cereale* L.) has been grown for over 2000 years and was the primary grain used for bread making until, in the 19th century, wheat became a more widely used cereal grain. It probably first grew in Southwestern Asia (Schreiber et al., 2021). This tall cereal is mainly found in cooler temperate regions. It can survive cold winters, sandy, poor-quality soil and arid environments (Poutanen et al., 2014). Rye also does well in acidic soils because it can handle aluminum toxicity. Also, it needs around 30% less water compared to wheat (Ryan, 2018). For less productive lands, rye is often the best crop to grow. There are two main kinds of rye: winter and spring, with winter rye being more common in the United States and globally. Rye can be cultivated for grain, forage, green manure or as a cover crop to combat erosion and compaction (Langdale et al., 1991). It thrives in low temperatures, poor soil fertility and dry conditions, making it the most productive cereal grain in these environments (Kottmann et al., 2016). Unlike wheat, rye requires less fertilizer and pesticides, making it ideal for organic farming. New hybrid rye cultivars yield grain comparable to wheat on fertile soils and superior on impoverished soils where wheat fails (Hansen & Schieberle, 2004). Around 90% of rye production globally occurs in European nations, making it the second most significant crop for bread production. Major producers include Russia, Poland, Germany, Finland, Ukraine and Denmark (Dziki, 2022).

Rye is rich in starch, fiber, protein, antioxidants and micronutrients. Its starch content ranges from 55% to 70%. Dietary fiber, particularly arabinoxylan, plays a vital role in weight management and digestive health (Kaur et al., 2021). The gluten content of rye makes it important in food production, with applications in bread, biscuits and beer. Scientific findings show that rye has a well-balanced amino acid profile compared to wheat (Arendt & Zannini, 2013). Rye grains contain several bioactive compounds (alkylresorcinols, ferulic acid, catechol, sinapic acid, vanillin, vanillic acid) with antioxidant properties that can enhance immune function and address age-related health issues (Jonsson et al., 2018). A balanced, nutritious diet may lower the risk of illness because micronutrients, which are just as vital as macronutrients, support several

DOI: 10.1201/9781003585817-5

biochemical functions within the body. Rye is considered to be naturally abundant in vitamins and minerals, including riboflavin, thiamin, magnesium and iron (Kaur et al., 2021). So, effective harvest and postharvest processing of rye are crucial for maintaining grain quality and ensuring safe storage. Timely harvesting at optimal moisture levels prevents spoilage and minimizes insect damage.

5.2 GROWTH AND MATURATION OF RYE

Rye consists of 12 stages in its development cycle (Abdimominova, 2022) (shown in Table 5.1). In the first stage, the growing point is not yet differentiated. By the second stage, stems, nodes and internode primordia start to form at the growing point. Rye is typically planted in autumn and enters winter at this stage. In the third stage, the growing point further differentiates into spikelet primordia. Adequate nitrogen during this period promotes the development of more spikelets, resulting in longer spikes with more flowers and grains (Schlegel, 2013). Stages 3 and 4see additional differentiation where flower primordia appear, particularly in early spring. As spikelet primordia form towards the top of the spike, flowers develop in the middle section (Bahrani, 2022). The plant transitions to stage 5 of organogenesis, which occurs fairly quickly under long daylight and limited nitrogen. In stages 6 and 7, meiosis occurs in pollen mother cells, forming tetrads, the embryo sac and the egg. Stage 7 is considered for significant shoot internode elongation. In stage 8, the plant forms ears and flowers, followed by fertilization and caryopsis maturation in the final stages (Mahamatovich, 2024).

The maturity of rye cereal is affected by various factors, including environmental conditions, genetics and farming practices (Ghafoor et al., 2024). Temperature is crucial, with warmer conditions generally speeding up growth, while cooler temperatures slow it down. The length of daylight is also important, as rye responds to photoperiod, with longer days usually leading to quicker development (Bhattacharya, 2022). Soil health and moisture are vital since adequate nutrients and water are

TABLE 5.1
Developmental Stages of Rye (*Secale cereale*)

Stage	Description	Key Features
1	Undifferentiated growing point	Seedling emerges with no differentiation
2	Formation of stems, nodes and internodes; winter dormancy	Early growth, enters the winter stage
3	Spikelet primordia formation; nitrogen supply affects spike length	Initial spike formation begins in spring
4	Flower primordia formation	Differentiation of floral buds in spring
5	Organogenesis and early elongation	Stem elongation under long days and low nitrogen
6	Meiosis in pollen mother cells	Cellular reproductive processes
7	Extensive elongation of shoot internodes	Rapid stem growth and internode elongation
8	Ear formation and flowering	Spikes fully develop and flowering occurs
9–12	Fertilization, caryopsis maturation and seed development	The mature plant produces seeds ready for harvest

required for full maturity. The rye variety chosen can influence maturity timing because some types are bred for early or late harvest. Additionally, effective management of planting schedules and pest control helps ensure rye matures as expected (Kantar & Porter, 2014).

5.3 HARVESTING OF RYE

Rye is typically harvested in late July or early August (Darby and Emick, 2022). It is common practice to harvest rye after barley and concurrently with wheat, despite its faster growth in the spring (Bushuk, 2001). Similar to other cereal grains, rye can be harvested when the moisture content is below 18–20%, but it is essential to reduce it to under 13% for effective storage (Gunasekaran, 2006). The optimal harvest period aligns closely with that of wheat. Allowing ripe crops to stand may lead to grain shedding. Maturity levels can vary, necessitating a thorough inspection of the entire paddock before harvesting (O'Keefe, 2021). A practical method to assess rye readiness for harvest is to apply pressure with a fingernail to the kernel; only a light indent should be made, indicating the soft-dough stage (Celestina et al., 2023). Rye tends to be tall and yields a lot of straw, producers may need to modify the cutting height while harvesting with a combination to avoid machine blockage. Taller grass stays in the field as a result of this modification (Fiennes, 2022). There exists a possibility of harvest losses due to lodging and shedding. Windrowing may be more advantageous for crops with high yields than for crops with low yields. It is not recommended for crops that are typically expected to generate less than 2 t/ha. Due to the requirement of so many components, harvesting windrowed grains takes a lot longer than direct heading. Windrowing should be avoided if the crop is scant or if the foliage is too small to sustain the windrow above the ground. Grain heads that come into contact with the ground are prone to sprout and harvesting these low-lying heads could contaminate the soil (Passaris et al., 2021). Windrowing can commence once the grain moisture level is below 35%, typically when the grain reaches the medium dough stage, characterized by hardness yet still dentable with a thumbnail. Early windrowing is advisable to mitigate losses from shedding and lodging; however, it should be avoided when the ground is wet from rain (Paulet, 2011). Caution must be taken not to windrow too early when the grain is underdeveloped, as this will result in small, pinched kernels. Although late windrowing might present convenience, the windrows of a fully ripe crop may fail to interlock adequately to withstand strong wind disturbances (Tautges et al., 2023).

Crops with larger row spacing, cutting should be done at a 45° angle or parallel to the planting direction to guarantee the windrow stays in place on the stubble. Windrowing is not recommended for paddocks with crop row spacing exceeding 25 cm. It is prudent to vary the locations of windrows from year to year to prevent nutrient concentration in a single area (Pratley & Stanton, 2003). The width of the windrow or cut should correspond with the header capacity. With a double-up attachment on the windrower or two windrows next to each other, a bigger capacity header will be needed and the residue will be concentrated in a smaller area of the paddock. The cutting height should be carefully managed to preserve adequate straw length, which is essential for holding the windrow together (minimum 30 cm), as well as to maintain sufficient stubble height that can support the windrow (Tresemer, 2021). The initial cutting height should start at 10–20 cm above ground level (approximately one-third of the crop height) and should be adjusted as needed to produce a uniform windrow comprising well-interwoven straws that sit above the soil. This method proves to be efficient for air flow and fast drying during rainfall (Vanbeveren et al., 2018).

5.3.1 Challenges in Harvesting of Rye

Rye cereal harvesting is considered to be a challenging task, especially when it includes double cropping systems. The main harvesting challenges are:

a. Getting the timing right for planting is a big hurdle for farmers who grow winter cereals alongside corn silage. Late-harvested corn silage and early frost present some challenges when planting winter cereals (small grains) in the fall (especially in the Northeast). This timing issue truly hinders the operability of the double-cropping system (Ketterings et al., 2015).
b. The extreme weather can greatly affect the yields of cereal rye. For example, the unusual weather in 2012 and 2013 negatively impacted the corn silage and hay production for many dairies in the northeastern United States. This has led to a greater interest in double-cropping winter cereals. Unpredictable weather patterns make it hard to consistently harvest good-quality rye forage (Ketterings et al., 2015).
c. Finding the right balance between the amount of cereal rye collected and its quality is another challenge faced during the harvest. While cereal rye typically reaches the swollen boot stage earlier and yields abundantly, it generally offers lower relative forage quality compared to barley (Liebert et al., 2023).

Although rye cereals present potential advantages as a double crop, farmers encounter challenges related to timing, climate volatility and the need to balance yield with quality. Nevertheless, a significant majority of farmers (83% in one survey) express their intention to continue cultivating double crops, suggesting that the potential benefits outweigh the encountered difficulties (Ketterings et al., 2015).

5.3.2 Equipment Utilized in Harvesting

Cereal rye is distinguished by its height and the enormous volume of straw causes the harvester to process a greater amount of material, which slows down the harvesting process. It is essential to clean all machinery and equipment both before and after harvest to prevent cross-contamination between cereal grains (Rudoy et al., 2021). A standard wheat header is appropriate for harvesting cereal rye; however, adjustments to the harvester settings are necessary to minimize grain loss and damage, as the grain is lighter and longer than that of wheat (Barać et al., 2011). Tall crops may lean or lodge, necessitating the use of crop lifters.

In 2014, experiments in the southern high-rainfall zone (HRZ) showed that for every 10-cm increase in harvest length, header output increased by 10%. Researchers compared three harvest heights of about 15 cm, 30 cm and 50 cm for both wheat and barley (noting that cereal rye grows taller than wheat). Additionally, the experiments showed that harvesting at an elevation of 15 cm considerably slows down the process; time costs increase and fuel consumption also rises. The speed of harvesting increased by about 25% when the harvest height was increased from 30 cm to 50 cm. A relevant rule of thumb suggests a 10% efficiency gain for every 10 cm elevation in harvest height. Thus, harvesting a 100 ha crop at 15 cm is expected to take about 20% longer than if harvested at 30 cm and 38% longer than at 50 cm (Spot et al., 2018).

To guarantee a small chop length (20–50 mm), forage harvesting tools intended for whole-crop silage (WCS) ought to prefer a precise cutting forage harvester. Minimizing the quantity of trapped air maximizes material compactness in a pit or stack and lowers dry matter (DM) and nutritional value losses. Throughout the early stages of fermentation, losses are caused by ongoing plant and microbial respiration. Effective compaction is extremely difficult because the majority of alternative forage-harvesting equipment, including self-loading wagons, shreds the material to different lengths, frequently surpassing 200 mm (Carey, 2023; Spot et al., 2018).

5.3.3 Advancements in Harvesting Technology

Each year, substantial grain losses occur during the harvesting and storage of grain. These quantitative losses, resulting from crushing and damage, significantly impact seed sowing quality and can also adversely influence the nutritional value when rye is utilized as animal feed. Enhancing the quality of harvesting rye and other grain crops necessitates the creation of new, highly efficient technological solutions. When selecting a harvesting technology for rye, it is imperative to consider factors such as crop ripeness, climatic conditions and field characteristics, as each technology presents its distinct advantages and disadvantages.

5.3.3.1 Direct Combining Technology

Direct combining technology has emerged as a vital approach for harvesting cereal grains, providing efficiency and cost-effectiveness for farmers. This system uses a combine harvester for concurrent cutting, threshing and cleaning of grain all at once while moving through the field (Los et al., 2018; Srivastava & John, 2002). The process initiates with the header of the combine harvester, which cuts the standing crop and feeds it into the machine. Inside, the threshing section works to separate the grain from the straw and chaff. After that, the clean grain is stored in the harvester tank, while any leftover material is either spread back onto the field or saved for other purposes (Srivastava & John, 2002). Direct combining is advantageous because it reduces the moisture in the grain at harvest. When the grain enters the combine, the airflow also provides some drying, reducing the amount of drying needed after. This is especially useful in places where drying facilities are scarce or where energy costs for drying are high. However, several things can impact how well direct combining works, like how mature the crop is, the moisture level and the weather conditions. Therefore, there is a need for farmers to cautiously plan their harvest time to get the best grain quality and avoid losses. Occasionally, farmers might still have a preference to cut and lay out their crops to dry before gathering them, particularly when the crops are not all ready at the same time or in places with unpredictable weather. New technology for direct combining has brought in smart farming tools like yield monitoring and Global Positioning System (GPS) systems. These improvements help farmers make the most of their harvests, save on fuel and reduce the amount of crops that go to waste.

5.3.3.2 Standing Stripping Technology

A new technology for harvesting and collecting crops has been created, based on designs for harvesting machines and research on how crops behave physically and mechanically. This technology features a header made up of a stripping drum, a reflector beater, an auger and a fixed reflective cover. As the machine moves forward, the suction airflow from the stripping drum and reflector beater causes the plant stems to bend into the casing. They are then directed to the shredding area, where they pass through working parts that look like combs with rectangular fingers on the drum. The combination of the motion of these parts and the airflow moves the shredded material to the auger, which then sends it into the feeder chamber of the combine. By using a stripper header, the productivity of the combine can be increased by 30–100% on average. The stripping mechanism increases production and reduces losses by delivering just grain, not straw, to the harvester. To separate grain from unthreshed spikes, the stripper can be attached to a normal harvester or installed on a specialist harvester. In other situations, it can be used to move a small grain–straw heap from the stripper to a bunker, discharge it onto a cart and move it to a fixed place. Harvesting grain through stripping from standing crops conserves 20–35% of fuel (Rudoy et al., 2021) and enables agricultural enterprises to organize grain harvesting at varying times, with some crops harvested at full ripeness for seed and food grain and others during the milk and wax ripeness phases through stripping without separating the grain heap to produce feed raw materials. This organization of harvesting allows for the preparation of raw materials for feed production enriched with higher protein and fat content, further increasing its utilization while preserving the non-cereal components of the ear (Zubrilina et al., 2019).

5.4 POSTHARVEST OPERATIONS

5.4.1 Cleaning and Sorting

As with other small grains, it is crucial to thoroughly clean rye. Various equipment is available for this purpose, including rotary cleaners, gravity tables and fanning mills (Lawson, 2022). Color sorting, which utilizes optical sorting technology to categorize kernels by color, has emerged as a novel cleaning method. Implementing grain cleaning at the farm level effectively enhances the value of the crop, prepares it for market and optimizes storage space. The presence of debris such as sticks, stones, husks, stems and weed seeds can adversely impact the quality and marketability of the crop (Rosentrater, 2022).

5.4.2 Drying Techniques

5.4.2.1 Sun Drying

Sun drying is among the most ancient and widely adopted methods for drying rye and other cereal grains. This approach entails spreading harvested grains in thin layers exposed to direct sunlight, allowing natural solar energy to evaporate moisture. Sun drying is cost-effective, requiring no specialized equipment, thus making it accessible for smallholder farmers (Chel & Kaushik, 2011). However, this method demands significant labor input and is heavily influenced by weather conditions, which can lead to inconsistent drying outcomes. Additionally, there is a chance of getting dirt, bugs or birds on the grains, and if not watched carefully, they could end up with different moisture levels (Alp & Bulantekin, 2021).

5.4.2.2 Pavement Drying

Pavement drying is spreading cereal grains on unique paved surfaces, which can help dry them quickly and prevent contamination to the crop. This approach allows for drying more grain than the usual sun drying method. The paved areas boost airflow, which helps the grains dry evenly. While this method is great for processing large amounts of grain, it does need good weather and careful management to avoid problems like uneven moisture or overheating. It is important to turn the grains regularly to get even drying results (Jimoh et al., 2023).

5.4.2.3 Field Drying and Stacking

Field drying and stacking help get harvested rye plants ready for threshing. In this method, the plants are cut and arranged in small piles right in the field. This helps reduce moisture naturally while the grains are still attached to their stalks. It is a budget-friendly choice for farmers, but it carries some risks. For instance, strong winds or animals can cause the grains to shatter. If the grains are left out too long, they can dry out too much, which can damage their quality when it is time to thresh. Keeping an eye on the timing and conditions is important to limit losses and keep the grains viable during this step (Sathya et al., 2008).

5.4.2.4 Other Drying Methods

One method of drying rye is in-bin drying, which happens right in the storage bins. Here, cool air is blown through the grains to take away moisture. In-bin drying is similar to batch drying, where working with a specific amount of grain at once is done. It is useful in conserving energy and limiting excessive handling of grains, therefore preventing their quality decay (Homayoonfal & Malekjani, 2023). In batch drying, a fixed quantity of rye is dried at one time. The grains are loaded into the dryer, dried to the appropriate moisture content, and then unloaded. This approach ensures that all grains can be dried within the required conditions, but it must be considered and even altered depending on the variety of grains and their moisture content from the beginning. For effective moisture control, different types of dryers are used for batch drying, such as the fluidized bed dryers (Zahed et al., 1995). Conversely, continuous flow drying enables rye grain to pass through the drier in

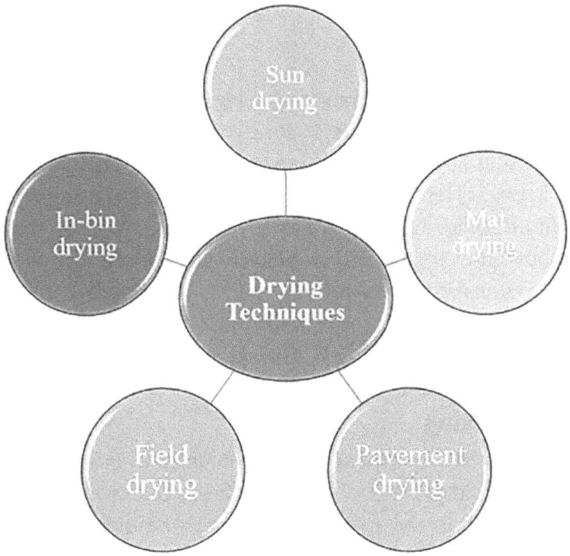

FIGURE 5.1 Different drying methods used in rye cereals.

a constant flow. Agitating tanks are a great option for larger operations since they tend to be faster and use less energy. These dryers can operate in different types like counter-flow, parallel-flow and cross-flow, which have the advantages in drying efficiency and keeping the quality of the grain (Myhan & Markowski, 2022). The mathematical models can be applied to adjust batch and continuous flow drying in a manner to enhance the quality of the grain and minimize energy consumption. These models also examine air temperature, airflow and dryer capacity to correlate energy use, drying time and end product quality (Myhan & Markowski, 2022). Either method can be used with intermittent drying, which alternates between drying and resting. So far, it appears to protect grain from damage at harvest time and improve its storability. Corn research shows that periodic drying can maintain its composition and limit damage to starch granules (Franco et al., 2020; Mabasso et al., 2023). Different drying methods are shown in Figure 5.1.

5.5 STORAGE CONDITIONS AND TYPES OF STORAGE FACILITIES

Storing rye cereals in proper conditions is essential to maintain them fresh, and if they are to be stored for a long time, to control moisture, temperature and air movement to minimize spoilage and extend their shelf life. The optimum moisture content for rye is 12% or lower, so storing it at this moisture greatly minimizes the risk of mold and spoilage (Bala, 2016). Better still is lowering the moisture to 10% or less. Elevated temperatures can hasten spoilage and cause mold problems. Temperatures of around 20–30°C are particularly ideal for mold to grow, and anything above 25°C can lead to the grains warming themselves through their own processes (Cui et al., 2022). Air movement is critical for temperature and humidity control in storage.

5.5.1 Types of Storage Facilities

5.5.1.1 Silos and Bins

Silos are vertical storage structures used to store large volumes of grain. They are tall, rounded structures that are used to hold large quantities of grains. They can come in many variants, reinforced concrete, metal or in developing countries, even from clay obtained from termite mounts (Omobowale et al., 2015). Metal silos are prevalent, and adding insulation helps protect the grain from outside temperature changes (Kyrpa, 2023). Even some of these silos are equipped with temperature and humidity sensors, which makes it easier to follow the climatic conditions in the storage (Basciftci et al., 2021). Then a setup was designed that provides the best environment for cereal storage, thus preventing waste and preserving the grain. Insulation is commonly added to these large metal silos when used for storage to protect the grains inside from temperature extremes (Kyrpa, 2023). These can be made using clay, muds or even jute or polypropylene bags. Some containers are composed of natural plant materials or clay pots (Trematerra et al., 2003). In some places, wooden platforms are temporarily used (Trematerra et al., 2003), and smaller items are sometimes stored in gourds or calabashes. Silos and bins are differentiated mostly by size; silos are also typically taller and larger (Kumar & Reddy, 2023). Grain bags are the next item. Think of them like miniature silos that help manage the storage of grain. They work well in conjunction with a larger grain management system or for temporary storage. It is essential to monitor the temperature and moisture content of these bags (Boumans, 2012).

A second option is flat storage, which refers to piling up grain in warehouses or on concrete slabs. This method might be more challenging to control than using silos or bins, so it requires close monitoring of moisture and air circulation (Friedmann & Maier, 2020). This process reduces the risk of insect damage by stabilizing the crop's natural breathing processes through lowering the stored grain temperature from approximately 25°C to approximately 15°C. In fact, insects experience almost no growth when temperatures dip below 17°C. This method has proven successful in reducing losses and maintaining quality (Demito et al., 2019). Before the air circulates through the stored goods, it is dried and cooled using specialized equipment. It can take a while for everything to cool down, particularly in taller silos, because the process is gradual and produces a cooling effect that permeates the grains (Ziegler et al., 2021).

5.5.1.2 Hermetic Storage

Hermetic storage relies on the naturally occurring atmosphere created by living organisms. When aerobic organisms in the stored goods breathe, they create an environment with less oxygen and more carbon dioxide (Navarro & Navarro, 2016). This method allows insects and other aerobic life in the grain to lower oxygen levels while raising carbon dioxide levels through their breathing. The result is an atmosphere that contains about 1–2% oxygen and around 20% carbon dioxide. The effectiveness of hermetic storage in controlling insects is similar to that of traditional fumigants, achieving over 99.9% effectiveness. Moreover, it keeps losses from insect activity very low. A low oxygen and high carbon dioxide setting kills pests and stops the growth of aerobic fungi (Said & Pradhan, 2014).

5.5.2 Storage Challenges

Keeping cereal grains, such as rye, can be tough if not managed properly; it can lead to huge losses. The biggest dangers to stored grains are insects, rodents and birds, which can cause damage and contaminate the crops. Some farmers faced a loss of around 70–80% of their harvests within an eight– to nine-month span due to these pests. The major loss in stored rye grains is considered to be caused by insects such as *Latheticus oryzae*, *Trogoderma granarium* and *Tribolium confusum* Jacquelin du Val (Athanassiou et al., 2005). In order to prevent losses brought out mostly by insects, like weevils, beetles and moths, appropriate grain harvesting and storage are essential (Mishra et al., 2012). After harvest, the main element that affects grain degradation is moisture level, which enhances the growth of fungal species such as *Aspergillus* and *Penicillium* (Gunasekaran, 2006). In addition to moisture content, other variables that may affect the quality of rye grain include temperature, relative humidity and storage time. So, drying is the main stage in controlling these variables that leads to the disintegration of grains (Ziegler et al., 2021). Qualitative losses, which include the existence of foreign material, contaminants, biochemical and enzymatic processes, if not managed appropriately, may lead to nutritional losses and economic value as well as potential for the development of hazardous toxins in the grains that are stored (Demito et al., 2019).To assist in maintaining the qualities of rye grains for an extended duration, adequate ventilation, frequent examination using concise techniques and disinfection must be carried out during storage of rye grains, resulting in obtaining high-quality grains.

5.6 PROCESSING OF RYE

5.6.1 Milling

For the production of white rye flour, roller mills are generally used (Drakos et al., 2017). The milling of rye grains to produce refined flour is less complicated in comparison to the milling of wheat flour, and hence, fewer categories of rye flour are produced. However, the different flour fractions can also be obtained from various milling streams. Rye is mainly processed into white or wholemeal flour using roller mills. In preparation for milling, rye grains are initially tempered. In this step, the pericarp and germ layers are strengthened, which makes the isolation of the endosperm easier. However, since the outer parts of the rye kernels adhere closely to the endosperm, the extent of the maximum milling is severe, but unlike wheat flour milling, it is achieved within a few processes (Kołodziejczyk et al., 2018; Sluková et al., 2021). The chemical structure of flour obtained from various streams in the milling stages is all different because the other structures within the kernel are not evenly mixed. Through milling, the rye kernels are broken down into various fractions comprising the pericarp/testa, aleurone and endosperm (Glitsø & Knudsen, 1999). They have different chemical compositions in terms of the amount of starch, which varies from 3.0 g/100 g of DM in the pericarp/testa to 78.2 g in the endosperm. It is also observed that the content of the dietary fiber ranged widely from 6.5 g/100 g of DM in the endosperm to 73.3 g in the pericarp/testa (Glitsø & Knudsen, 1999). However,

Miller can impact the distribution of bioactive compounds within rye. For instance, a significant portion of key phytochemicals, such as alkylresorcinols (ARs), which are present in rye, are located in the bran (Ross et al., 2004). The total AR in the milling fractions corresponds closely to the ash content and can be used as a marker for the level of bran in flour (Ross et al., 2003). It is rather surprising that the intensity of rye milling can affect the physical and chemical properties of the products derived from the rye (Dziki, 2022). There is an increasing trend of seeking to produce fiber-rich wholemeal flours through the use of specialized milling machines that yield fine wholemeal rye and dark rye flour (Ikram et al., 2023). For example, stearylferulates are not spread out evenly throughout rye grains, which causes differences in the end products of milling. Their amounts can range from very small amounts in low-ash flours to 20 mg/100 g in rye bran (Hakala et al., 2002). Also, it has been classified that the plant sterols present in rye depend on the ash content of the milled rye products (Piironen et al., 2002).

5.6.2 Fermentation

The sourdough process is one of the oldest biotechnological processes used for leavening in cereal food production. Rye sourdough is a mixture of water and rye flour fermented by lactic acid bacteria (LAB) and yeasts (De Vuyst et al., 2009). Sourdough fermentation has proved to be crucial in the production of rye bread by ensuring the dough is acidified, inhibiting the activity of amylase and increasing the ability of starch to bind water (Weckx et al., 2010). Furthermore, in rye bread, it improves its nutritional value by increasing the levels of bioactive compounds (Weckx et al., 2010) and aroma-active compounds (Kirchhoff & Schieberle, 2002). It has been observed that fermentation and baking of rye bread enable the conversion or breakdown of carbohydrates without lowering the total dietary fiber content (Koj & Pejcz, 2023). It has been proven that sourdough fermentation triggers the conversions of lipids, macromolecules and several phytochemicals, including folates, phenolic acids and bioactive compounds (Koistinen et al., 2018). During sourdough fermentation, LAB produces a number of metabolites, e.g., organic acids, exopolysaccharides and/or enzymes, which have a positive effect on the texture and staling of rye bread (Arendt et al., 2006). LAB increases proteolysis and hydrolyzes flour biopolymers such as gluten and starch, thus enhancing the texture and increasing the shelf-life of bakery products. Yeast improves the loaf volume and aroma of the bread by producing CO_2 and ethanol. In addition, microorganisms present in sourdough synthesize bioactive compounds and a few vitamins, as well as increasing the bioavailability of minerals, which improves the nutritional value as compared to other bread products (Fekri et al., 2024).

5.6.3 Extrusion

Extrusion is a high-temperature and short processing time ongoing procedure that results in highly productive and nutrient-rich products (Makaravicius et al., 2012). Rye, a widely grown cereal grain, possesses a tarty flavor produced by some phenolic substances and short peptides, which makes its application for consumption confined (Heiniö et al., 2012). Therefore, extrusion is identified as one of the useful processing methods for reducing the strong flavor of rye. Extruding rye wholegrain at elevated temperature with minimal moisture content also enhances the antioxidant capacity and results in high phenolics and improves stability by lowering the build-up in oxidation of lipids when rye grain is being stored (Nordlund et al., 2013). Rye grain, despite being primarily utilized as feed for livestock, is considered to be abundant in fiber content, which makes it useful for the health benefits of humans. Therefore, the extrusion process has the potential to improve fiber content as well as sensory attributes of extrudates of rye grain (Andersson et al., 2017). Rye flour undergoes several modifications throughout extrusion that impact its ability to retain water, the general organization of dough, and its consistency which causes changes in the output of the product, GI (glycemic index) and its overall quality (Ikram et al., 2023). In addition, the process breaks down toxic substances, weakens unwanted enzymes, deactivates harmful nutrients and sterilizes the finished products. To stop the degradation of rye residue in storage, the process of extrusion is very crucial because it may decrease the microbial activity (Castells et al., 2005). Thus, the use of novel ingredients in the process of extrusion creates opportunities for the advancement of innovative nutritional food products with particular scientific qualities as well as for expanding the variety of products produced by extrusion.

The extruded rye products become more palatable and nutrient-dense because of denatured proteins, gelatinization of starch and alteration in nutritional fibers caused by the elevated pressure and temperature in the extruder (Gulati et al., 2020). Regulating the process of extrusion may improve the dietary content of products made from rye, e.g., employing short-type barrel extruders might result in a relatively increased Vitamin B complex preservation of 44–62% (Athar et al., 2006). Extrusion is regarded as the process that makes proteins solubilize and digest easily; however, it is crucial to preserve the accessible lysine (Gulati et al., 2020). Precisely modifying the parameters of extrusion can result in products with proper physical properties while maintaining their dietary value (Omwamba & Mahungu, 2014; Wójtowicz et al., 2015). Using rye and various cereals via the extrusion process develops great accessibility to produce novel and nourishing goods that satisfy customers' shifting tastes as they seek out alternative snacks (Espinoza-Moreno et al., 2016).

5.7 QUALITY ASSURANCE AND LOSS PREVENTION

Over the years, consumers have become used to demanding grains with certain specific qualities. While assessing grain quality, test weight is an indicator of milling quality and it is based on grain size and the lack of shriveled grains. Another determining factor is the variety or varietal mix. These also 'build in' quality attributes established by the breeder and are of prime importance when assessing harvested grain quality. Traditionally, methods for identifying variety have entailed an expert's visual inspection (Jarman, 1995). For definite identification, however, laboratory techniques are required, such as capillary electrophoresis or gel electrophoresis, to determine the grain-protein composition (Uthayakumaran et al., 2006). Common requirements for grain to be admitted to a particular grade or class are its moisture content and grain protein concentration. The standard method for figuring out both of these parameters is to use near infrared (NIR) spectroscopy of whole or powdered grain. Since rye is susceptible to pre-harvest sprouting, grain soundness (absence of sprouting) is particularly crucial. Visual examination will be effective enough in advanced stages of sprouting, where some roots and shoots are present or splitting of the germ cover has started. The absence of moldy, frosted or broken grains is another indicator of grain soundness that may be seen with the naked eye. Deliveries of rye are especially vulnerable to ergot bodies. Unless they have been broken up during harvesting and transportation, the ergots are visible. Given that postharvest grain deterioration is a global problem that frequently happens during storage and transit (Garcia-Mosqueda et al., 2019; Xue et al., 2017). One of the main biotic factors that causes grains to deteriorate while being stored is insect infestation (Banga et al., 2020; Hiruy & Getu, 2018). Proper pre- and postharvest management employing cost-effective and safe methods can help control it (Hiruy & Getu, 2018). Proper sanitation and cleaning are crucial preventative steps. Before being accepted, infestation methods must be in line with market requirements and safety regulations (Paul et al., 2020). On-farm drying of grains is essential for ensuring the quality of the produced grain. Compared to natural drying, power-driven drying methods have several advantages. Solar energy-assisted dryers are another alternative, with low-cost and simple design for small- to medium-scale applications. In industrialized countries, NIR-based dryers, microwave-assisted dryers and convective hot air dryers are both quality and market value assured despite the high-cost installation (Abdoli et al., 2018; Nosrati et al., 2022). To increase the storability of grains, various plant elements, including oils, ashes and synthetic chemicals, are also utilized (Manandhar et al., 2018). Phosphine, sulfuryl fluoride, ethylene formate, methyl bromide and ozone are the most common fumigants used to control insects in grains during storage; on the other hand, the most common synthetic pesticides used are organophosphates, organochlorines, carbamates and pyrethroids (Paul et al., 2020). Since synthetic chemicals have numerous side effects (Ayalew, 2020), some processing techniques, such as soaking, steaming, fermentation and distillation, minimize the amount of pesticide residues stored in grain products (Han et al., 2016). Recently developed technologies that are being advocated in developing nations include multilayer bagging systems and hermetic metallic silos (Manandhar et al., 2018). Postharvest losses are undoubtedly decreased by better storage facilities with real-time temperature and R.H. alarms (Shisiali, 2018). Insect-desiccant dust (powders added to grains during storage) and hermetic packing are both much safer and offer products with superior storage properties (Kalsa et al., 2019; Schmidt et al., 2018). The use of chemical insecticides, extreme temperature treatments, ozone, irradiation and dielectric heating, metabolic stress disinfestation, non-thermal plasma, high hydrostatic pressure processing, ultra-sonication, microwave treatments, ultraviolet light and modified atmosphere packaging are common novel approaches to protecting grains from pest and microbial contaminations (Paul et al., 2020; Schmidt et al., 2018).

5.8 ECONOMIC IMPORTANCE OF RYE

Rye cereal plays an important role in the economy of the country. It is used for the manufacturing of biofuels, used as a food, as well as animal fodder. This cereal can flourish in colder climates and on less fertile soils, which makes it vital for those countries that do not have very productive land. As more people become aware of the high fiber content of rye breads, their popularity is growing. This fiber can help control blood sugar levels and is excellent for digestion. Rye is used in many different foods, including pasta, porridge, ginger cake, pastries and breakfast cereals, in addition to bread. Additionally, rye can be fermented to make alcoholic drinks. Rye is a valuable alternative raw material because of its special quality. Rye and its derivatives find industrial applications in the production of drilling materials, adhesives and coatings (Seibel & Weipert, 2001). Wholemeal rye flour-based foods are high in vitamins, minerals, phytoestrogens and dietary fibers, all of which have health benefits (Poutanen, 1997). Eating more rye is linked to benefits for digestion and a lower risk of heart disease, high cholesterol, obesity and type 2 diabetes (Figure 5.2) (Hallmans et al., 2003; Smith et al., 2003). It may also offer protection against certain hormone-related cancers (Hallmans et al., 2003; Smith et al., 2003). Research by Liukkonen et al. (2003) highlights that many beneficial compounds in whole-meal rye remain stable during processing, and processing methods like making sourdough bread can even increase their levels.

Besides benefiting humans, rye is also an important animal feed ingredient and has promise for biofuel production. As a green plant, rye can provide pasture for livestock and serve as green manure in crop rotations. There is growing interest in using rye straw for biofuels. The chemical properties of straw allow industries to convert its sugars into useful chemicals (Munck, 2016). Rye straw is well-liked for use in thatching and building. In drought areas, rye provides dependable feed for livestock because it grows well and produces a lot. Most of the studies also indicate that winter rye can be a good choice for energy crops and may be used to create biogas, making it important for renewable energy.

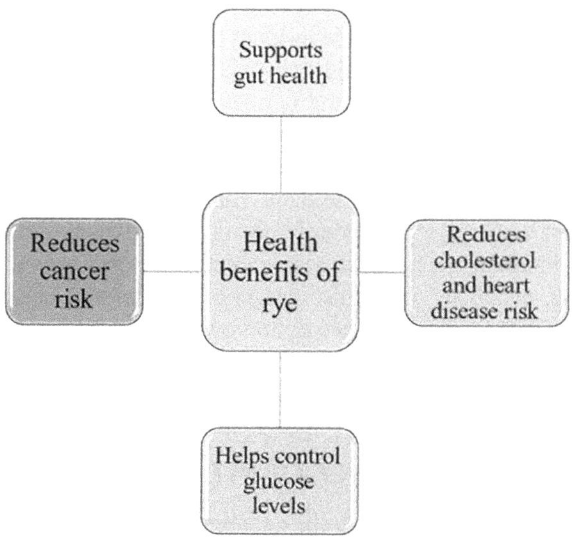

FIGURE 5.2 Health benefits of rye cereal.

5.9 CONCLUSION

Storing and processing rye cereals properly is important for keeping them safe, nutritious and high-quality. Using methods like silos and bins, along with air-tight storage, can help reduce waste and protect the structure of grains. It is also key to control moisture levels and handle pests properly to prevent spoilage and contamination. New ways of processing, like fermentation and extrusion, open up possibilities for creating tasty and healthy rye products. Improving how we store and process rye is vital to meet the growing demand worldwide and to support food security and sustainable farming.

REFERENCES

Abdimominova, M. A (2022). Phases of development and growth of cereals. *Economics and Society*, *2–2*(93), 3–9.

Abdoli, B., Zare, D., Jafari, A., & Chen, G. (2018). Evaluation of the air-borne ultrasound on fluidized bed drying of shelled corn: Effectiveness, grain quality, and energy consumption. *Drying Technology*, *36*(14), 1749–1766.

Alp, D., &Bulantekin, Ö. (2021). The microbiological quality of various foods dried by applying different drying methods: A review. *European Food Research and Technology*, *247*(6), 1333–1343.

Andersson, A. A., Andersson, R., Jonsäll, A., Andersson, J., & Fredriksson, H. (2017). Effect of different extrusion parameters on dietary fiber in wheat bran and rye bran. *Journal of Food Science*, *82*(6), 1344–1350.

Arendt, E. K., Ryan, L. A., & Dal Bello, F. (2006). Impact of sourdough on the texture of bread. *Food Microbiology*, *24*(2), 165–174.

Arendt, E. K., & Zannini, E. (Eds.). (2013). *Cereal grains for the food and beverage industries*. Woodhead Publishing.

Athanassiou, C. G., Kontodimas, D. C., Kavallieratos, N. G., & Veroniki, M. A. (2005). Insecticidal effect of NeemAzal against three stored-product beetle species on rye and oats. *Journal of Economic Entomology*, *98*(5), 1733–1738.

Athar, N., Hardacre, A., Taylor, G., Clark, S., Harding, R., & McLaughlin, J. (2006). Vitamin retention in extruded food products. *Journal of Food Composition and Analysis*, *19*(4), 379–383.

Ayalew, A. A. (2020). Insecticidal activity of Lantana camara extract oil on controlling maize grain weevils. *Toxicology Research and Application*, *4*, 1–10.

Bahrani, H. (2022). *Genetic analysis of developmental traits associated with enhanced winter survival in autumn-seeded rye (Secale cereale L.)* [Doctoral dissertation, University of Saskatchewan]. USask Research Works.

Bala, B. K. (2016). *Drying and storage of cereal grains*. John Wiley & Sons.

Banga, K.S., Kumar, S., Kotwaliwale, N., & Mohapatra, D. (2020). Major insects of stored food grains. *International Journal of Chemical Studies*, *8*(1), 2380–2384.

Barać, S., Vuković, A., Milenković, B., Biberdžić, M., & Dimitrijević, A. (2011). Qualitative analysis of wheat combine harvester work effects in rye harvest. *Research Journal of Agricultural Science*, *43*(1), 163–169.

Basciftci, F., Unlu, T., & Dasdemir, A. (2021). Monitoring grain silos instantly with IoT based control system. *The Eurasia Proceedings of Science Technology Engineering and Mathematics*, *14*, 15–23.

Bhattacharya, A. (2022). *Physiological processes in plants under low temperature stress* (p. 734). Springer.

Boumans, G. (2012). *Grain handling and storage*. Elsevier.
Bushuk, W. (2001). Rye production and uses worldwide. *Cereal Foods World*, *46*(2), 70–73.
Carey, R. E. (2023). *Short-season high-moisture corn products in western Canadian beef cattle diets* [Doctoral dissertation, University of Saskatchewan].
Castells, M., Marin, S., Sanchis, V., & Ramos, A. J. (2005). Fate of mycotoxins in cereals during extrusion cooking: A review. *Food Additives and Contaminants*, *22*(2), 150–157.
Celestina, C., Hunt, J., Brown, H., Huth, N., Andreucci, M., Hochman, Z., …, & Kohout, M. (2023). Scales of development for wheat and barley specific to either single culms or a population of culms. *European Journal of Agronomy*, *147*, 126824.
Chel, A., & Kaushik, G. (2011). Renewable energy for sustainable agriculture. *Agronomy for Sustainable Development*, *31*, 91–118.
Cui, H., Wang, S., Yang, X., Zhang, W., Chen, M., Wu, Y., …, & Wang, S. (2022). Predictive models for assessing the risk of *Fusarium pseudograminearum* mycotoxin contamination in post-harvest wheat with multi-parameter integrated sensors. *Food Chemistry: X*, *16*, 100472.
Darby, H., & Emick, H. (2022). *Rye harvest date*. University of Vermont Extension.
Demito, A., Ziegler, V., Goebel, J. T. S., Konopatzki, E. A., Coelho, S. R. M., & Elias, M. C. (2019). Effects of refrigeration on biochemical, digestibility, and technological parameters of carioca beans during storage. *Journal of Food Biochemistry*, *43*(7), e12900.
De Vuyst, L., Vrancken, G., Ravyts, F., Rimaux, T., & Weckx, S. (2009). Biodiversity, ecological determinants, and metabolic exploitation of sourdough microbiota. *Food Microbiology*, *26*(7), 666–675.
Drakos, A., Kyriakakis, G., Evageliou, V., Protonotariou, S., Mandala, I., & Ritzoulis, C. (2017). Influence of jet milling and particle size on the composition, physicochemical and mechanical properties of barley and rye flours. *Food Chemistry*, *215*, 326–332.
Dziki, D. (2022). Rye flour and rye bran: New perspectives for use. *Processes*, *10*(2), 293.
Espinoza-Moreno, R. J., Reyes-Moreno, C., Milán-Carrillo, J., López-Valenzuela, J. A., Paredes-López, O., & Gutiérrez-Dorado, R. (2016). Healthy ready-to-eat expanded snack with high nutritional and antioxidant value produced from whole amarantin transgenic maize and black common bean. *Plant Foods for Human Nutrition*, *71*, 218–224.
Fekri, A., Abedinzadeh, S., Torbati, M., Azadmard-Damirchi, S., & Savage, G. P. (2024). Considering sourdough from a biochemical, organoleptic, and nutritional perspective. *Journal of Food Composition and Analysis*, 125, 105853.
Fiennes, J. (2022). *Land healer: How farming can save Britain's countryside*. Random House.
Franco, C. M., de Lima, A. G., Farias, V. S., & da Silva, W. P. (2020). Modeling and experimentation of continuous and intermittent drying of rough rice grains. *Heat and Mass Transfer*, *56*, 1003–1014.
Friedmann, M. A., & Maier, D. E. (2020). Advances in bulk storage of cereals and grains. In *Advances in postharvest management of cereals and grains* (pp. 125–170). Burleigh Dodds Science Publishing.
Garcia-Mosqueda, C., Salas-Araiza, M. D., Ceron-García, A., Estrada-García, H. J., Rojas-Laguna, R., & Sosa-Morales, M. E. (2019). Microwave heating as a post-harvest treatment for white corn (*Zea mays*) against *Sitotroga cerealella*. *Journal of Microwave Power and Electromagnetic Energy*, *53*(3), 145–154.
Ghafoor, A. Z., Karim, H., Studnicki, M., Raza, A., Javed, H. H., & Asghar, M. A. (2024). Climate change and rye (*Secale cereale* L.) Production: Challenges, opportunities and adaptations. *Journal of Agronomy and Crop Science*, *210*(4), e12725.
Glitsø, L. V., & Knudsen, K. B. (1999). Milling of whole grain rye to obtain fractions with different dietary fibre characteristics. *Journal of Cereal Science*, *29*(1), 89–97.
Gulati, P., Brahma, S., & Rose, D. J. (2020). Impacts of extrusion processing on nutritional components in cereals and legumes: Carbohydrates, proteins, lipids, vitamins, and minerals. In *Extrusion cooking* (pp. 415–443). Woodhead Publishing.

Gunasekaran, S. (2006). *Safe storage guidelines for rye and canola* [Master's thesis, University of Manitoba]. MSpace. http://hdl.handle.net/1993/3789

Hakala, P., Lampi, A. M., Ollilainen, V., Werner, U., Murkovic, M., Wähälä, K., …, & Piironen, V. (2002). Steryl phenolic acid esters in cereals and their milling fractions. *Journal of Agricultural and Food Chemistry*, *50*(19), 5300–5307.

Hallmans, G., Zhang, J. X., Lundin, E., Stattin, P., Johansson, A., Johansson, I., & Adlercreutz, H. (2003). Rye, lignans and human health. *Proceedings of the Nutrition Society*, *62*(1), 193–199.

Hansen, A., & Schieberle, P. (2004). Generation of aroma compounds during sourdough fermentation: Applied and fundamental aspects. *Trends in Food Science & Technology*, *16*(1–3), 85–94.

Heiniö, R. L., Nordlund, E., Poutanen, K., & Buchert, J. (2012). Use of enzymes to elucidate the factors contributing to bitterness in rye flavour. *Food Research International*, *45*(1), 31–38.

Hiruy, B., & Getu, E. (2018). Insect pests associated to stored maize and their bio rational management options in sub-Sahara Africa. *International Journal of Academic Research and Development*, *3*(1), 741–748.

Homayoonfal, M., & Malekjani, N. (2023). Drying of cereal grains and beans. In *Drying technology in food processing* (pp. 459–489). Woodhead Publishing.

Ikram, A., Saeed, F., Noor, R. A., Imran, A., Afzaal, M., Rasheed, A., …, & Kinki, A. B. (2023). A comprehensive review on biochemical and technological properties of rye (*Secale cereale* L.). *International Journal of Food Properties*, *26*(1), 2212–2228.

Jarman, R. J. (1995). Visual identification by grain characteristics. In C. W. Wrigley (Ed.), *Identification of food-grain varieties* (pp. 35–54). AACC International.

Jimoh, K. A., Hashim, N., Shamsudin, R., Man, H. C., Jahari, M., & Onwude, D. I. (2023). Recent advances in the drying process of grains. *Food Engineering Reviews*, *15*(3), 548–576.

Jonsson, K., Andersson, R., Knudsen, K. E. B., Hallmans, G., Hanhineva, K., Katina, K., Kolehmainen, M., Kyrø, C., Langton, M., Nordlund, E., Lærke, H. N., Olsen, A., Poutanen, K., Tjønneland, A., & Landberg, R. (2018). Rye and health - Where do we stand and where do we go? *Trends in Food Science & Technology*, *79*, 78–87.

Kalsa, K. K., Subramanyam, B., Demissie, G., Mahroof, R., Worku, A., & Gabbiye, N. (2019). Evaluation of postharvest preservation strategies for stored wheat seed in Ethiopia. *Journal of Stored Products Research*, *81*, 53–61.

Kantar, M., & Porter, P. (2014). Relationship between planting date, growing degree days and the winter rye (*Secale cereale* L.) variety "Rymin" in Minnesota. *Crop Management*, *13*(1), 1–9.

Kaur, H., Singh, B., & Singh, A. (2021). Comparison of dietary fibers obtained from seven Indian cereal grains. *Journal of Cereal Science*, *102*, 103331.

Ketterings, Q. M., Ort, S., Swink, S. N., Godwin, G., Kilcer, T., Miller, J., & Verbeten, W. (2015). Winter cereals as double crops in corn rotations on New York dairy farms. *Journal of Agricultural Science*, *7*(2), 18–25.

Kirchhoff, E., & Schieberle, P. (2002). Quantitation of odor-active compounds in rye flour and rye sourdough using stable isotope dilution assays. *Journal of Agricultural and Food Chemistry*, *50*(19), 5378–5385.

Koistinen, V. M., Mattila, O., Katina, K., Poutanen, K., Aura, A. M., & Hanhineva, K. (2018). Metabolic profiling of sourdough fermented wheat and rye bread. *Scientific Reports*, *8*(1), 5684.

Koj, K., & Pejcz, E. (2023). Rye dietary fiber components upon the influence of fermentation inoculated with probiotic microorganisms. *Molecules*, *28*(4), 1910.

Kołodziejczyk, P., Makowska, A., Pospieszna, B., Michniewicz, J., & Paschke, H. (2018). Chemical and nutritional characteristics of high-fibre rye milling fractions. *Acta Scientiarum Polonorum Technologia Alimentaria*, *17*(2), 149–157.

Kottmann, L., Wilde, P., & Schittenhelm, S. (2016). How do timing, duration, and intensity of drought stress affect the agronomic performance of winter rye? *European Journal of Agronomy*, *75*, 25–32.

Kumar, D. U., & Reddy, V. M. (2023). Analysis and design of reinforced concrete silo by conventional method. *E3S Web of Conferences*, *391*, 01200. EDP Sciences.

Kyrpa, M. Y. (2023). State and development of grain storage engineering and technologies in Ukraine. *The Scientific Journal Grain Crops*, *6*(2), 69–76. https://doi.org/10.31867/2523-4544/0234

Langdale, G. W., Blevins, R. L., Karlen, D. L., McCool, D. K., Nearing, M. A., Skidmore, E. L., Thomas, A. W., Tyler, D. D., & Williams, J. R. (1991). Cover crop effects on soil erosion by wind and water. In W. L. Hargrove (Ed.), *Cover crops for clean water* (pp. 15–22). Soil and Water Conservation Society.

Lawson, T. (2022). *Historic mills of West Virginia*. 35th Star Publishing.

Liebert, J., Cherney, J. H., Ketterings, Q. M., Mirsky, S. B., Pelzer, C. J., & Ryan, M. R. (2023). Winter cereal species, cultivar, and harvest timing affect trade-offs between forage quality and yield. *Frontiers in Sustainable Food Systems*, *7*, 1067506.

Liukkonen, K. H., Katina, K., Wilhelmsson, A., Myllymaki, O., Lampi, A. M., Kariluoto, S., …, & Poutanen, K. (2003). Process-induced changes on bioactive compounds in whole grain rye. *Proceedings of the Nutrition Society*, *62*(1), 117–122.

Los, A., Ziuzina, D., & Bourke, P. (2018). Current and future technologies for microbiological decontamination of cereal grains. *Journal of Food Science*, *83*(6), 1484–1493.

Mabasso, G. A., Siqueira, V. C., Resende, O., Quequeto, W. D., Schoeninger, V., Simeone, M. L. F., …, & Crippa, D. S. (2023). The effect of intermittent drying with variable resting times on quality parameters of corn obtained after storage. *LWT*, *182*, 114855.

Mahamatovich, T. A. (2024). The Influence of the Timing and Norms of Sowing Winter Rye on the Cycles of Growth and Development. *Web of Agriculture: Journal of Agriculture and Biological Sciences*, *2*(3), 68–71.

Makaravicius, T., Basinskiene, L., Juodeikiene, G., van Gool, M. P., & Schols, H. A. (2012). Production of oligosaccharides from extruded wheat and rye biomass using enzymatic treatment. *Catalysis Today*, *196*(1), 16–25.

Manandhar, A., Milindi, P., & Shah, A. (2018). An overview of the postharvest grain storage practices of smallholder farmers in developing countries. *Agriculture (Switzerland)*, *8*(4), 1–21.

Mishra, A., Prabuthas, P., & Mishra, H. N. (2012). Grain storage: Methods and measurements. *Quality Assurance and Safety of Crops and Foods*, *4*(3), 136–158.

Munck, L. (2016). Utilization of the whole cereal plant to maximize its value. In *Reference Module in Food Science*. Elsevier. https://doi.org/10.1016/B978-0-08-100596-5.03418-7

Myhan, R., & Markowski, M. (2022). Generalized mathematical model of the grain drying process. *Processes*, *10*(12), 2749.

Navarro, H., & Navarro, S. (2016). Organic treatments of stored cereals and grains. In *Proceedings of the 15th International Cereal and Bread Congress* (p. 6).

Nordlund, E., Heiniö, R. L., Viljanen, K., Pihlava, J. M., Lehtinen, P., & Poutanen, K. (2013). Flavour and stability of rye grain fractions in relation to their chemical composition. *Food Research International*, *54*(1), 48–56.

Nosrati, M., Zare, D., Nassiri, S. M., Chen, G., & Jafari, A. (2022). Experimental and numerical study of intermittent drying of rough rice in a combined FIR-dryer. *Drying Technology*, *40*(10), 1967–1979.

O'Keefe, C. L. (2021). *Validating the stage of maturity at harvest for barley, oat and triticale for swath grazing* [Doctoral dissertation, University of Saskatchewan].

Omobowale, M., Mijinyawa, Y., Armstrong, P., Igbeka, J., & Maghirang, E. (2015). Performance evaluation of termite-mound clay, concrete and steel silos for the storage of maize grains in the humid tropics. *Journal of Stored Products and Postharvest Research*, *6*(4), 56–65. https://doi.org/10.5897/JSPPR2015.0187

Omwamba, M., & Mahungu, S. M. (2014). Development of a protein-rich ready-to-eat extruded snack from a composite blend of rice, sorghum and soybean flour. *Food and Nutrition Sciences*, *5*(14), 1309–1317.

Passaris, N., Flower, K. C., Ward, P. R., & Cordingley, N. (2021). Effect of crop rotation diversity and windrow burning of residue on soil chemical composition under long-term no-tillage. *Soil and Tillage Research*, *213*, 105153.

Paul, A., Radhakrishnan, M., Anandakumar, S., Shanmugasundaram, S., & Anandharamakrishnan, C. (2020). Disinfestation techniques for major cereals: A status report. *Comprehensive Reviews in Food Science and Food Safety*, *19*(3), 1125–1155.

Paulet, R. (2011). *The integration of livestock and cropping in high rainfall zones: Jack of all trades, master of none* (Project No. 1004). Nuffield Australia. https://www.nuffieldscholar.org/sites/default/files/reports/2010_AU_Rowan-Paulet_The-Integration-Of-Livestock-And-Cropping-In-High-Rainfall-Zones-Jack-Of-All-Trades-Master-Of-None.pdf

Piironen, V., Toivo, J., & Lampi, A. M. (2002). Plant sterols in cereals and cereal products. *Cereal Chemistry*, *79*(1), 148–154.

Poutanen, K., Katina, K., & Heiniö, R. L. (2014). Rye. In W. Zhou (Ed.), *Bakery products science and technology* (2nd ed., pp. 75–87). Wiley-Blackwell.

Poutanen, K. (1997). Rye bread: Added value in the world's bread basket. *Cereal Foods World*, *42*(8), 682–683.

Pratley, J., & Stanton, R. (2003). Crop establishment and management. In *Principles of field crop production* (pp. 321–356). Oxford University Press.

Rosentrater, K. A. (2022). Overview of storage systems for cereal grains and grain products. In *Storage of cereal grains and their products* (pp. 41–80). Woodhead Publishing.

Ross, A. B., Kamal-Eldin, A., & Aman, P. (2004). Dietary alkylresorcinols: Absorption, bioactivities, and possible use as biomarkers of whole-grain wheat–and rye–rich foods. *Nutrition Reviews*, *62*(3), 81–95.

Ross, A. B., Shepherd, M. J., Schupphaus, M., Sinclair, V., Alfaro, B., Kamal-Eldin, A., & Aman, P. (2003). Alkylresorcinols in cereals and cereal products. *Journal of Agricultural and Food Chemistry*, *51*(14), 4111–4118.

Rudoy, D., Egyan, M., Kulikova, N., & Chigvintsev, V. (2021, December). Review and analysis of technologies for harvesting perennial grain crops. In *IOP conference series: Earth and environmental science* (Vol. 937, No. 2, p. 022112). IOP Publishing.

Ryan, P. R. (2018). Assessing the role of genetics for improving the yield of Australia's major grain crops on acid soils. *Crop & Pasture Science*, *69*(3), 242–264.

Said, P. P., & Pradhan, C. R. (2014). Food grain storage practices: A review. *Journal of Grain Processing and Storage*, *1*(1), 1–5.

Sathya, G., Jayas, D. S., & White, N. D. G. (2008). Safe storage guidelines for rye. *Canadian Biosystems Engineering*, *50*(3), 1–8.

Schlegel, R. H. (2013). *Rye: Genetics, breeding, and cultivation*. CRC Press.

Schmidt, M., Zannini, E., & Arendt, E. K. (2018). Recent advances in physical postharvest treatments for shelf-life extension of cereal crops. *Foods*, *7*, 1–22.

Schreiber, M., Özkan, H., Komatsuda, T., & Mascher, M. (2021). Evolution and domestication of rye. In M. T. Rabanus-Wallace & N. Stein (Eds.), *The rye genome* (pp. 85–100). Springer.

Seibel, W., & Weipert, D. (2001). Animal feed and industrial uses. *Rye Production, Chemistry, and Technology, American Association of Cereal Chemist, St. Paul, Minnesota, 221.*

Shisiali, A. S. (2018). Creativity and innovation for sustainable development. *Proceedings of Kibabii University 3rd Interdisciplinary International Scientific Conference*, 1.

Sluková, M., Jurkaninová, L., Švec, I., & Skřivan, P. (2021). Rye–the nutritional and technological evaluation in Czech cereal technology–A review: Grain and flours. *Czech Journal of Food Sciences*, *39*(1).

Smith, A. T., Kuznesof, S., Richardson, D. P., & Seal, C. J. (2003). Behavioural, attitudinal and dietary responses to the consumption of wholegrain foods. *Proceedings of the Nutrition Society*, *62*(2), 455–467.

Srivastava, V. K., & John, J. (2002). Deep bed grain drying modeling. *Energy Conversion and Management*, *43*(13), 1689–1708.

Tautges, N., Detjens, A., & Jungers, J. M. (2023). *Kernza grower guide*. University of Minnesota Digital Conservancy. https://hdl.handle.net/11299/255778

Trematerra, P., Gentile, P., & Djikoloum, T. (2003). Traditional cereal storage and insect pests in some villages of southern Chad. In P. F. Credland, D. M. Armitage, C. H. Bell, P. M. Cogan, & E. Highley (Eds.), *Advances in stored product protection: Proceedings of the 8th International Working Conference on Stored Product Protection* (pp. 281–287). CABI Publishing.

Tresemer, D. (2021). *The scythe book: Mowing hay, cutting weeds, and harvesting small grains with hand tools*. Rowman & Littlefield.

Uthayakumaran, S., Batey, I. L., & Wrigley, C. W. (2005). On-the-spot identification of grain variety and wheat-quality type by lab-on-a-chip capillary electrophoresis. *Journal of Cereal Science*, *41*(3), 371–374.

Vanbeveren, S. P., De Francesco, F., Ceulemans, R., & Spinelli, R. (2018). Productivity of mechanized whip harvesting with the Stemster MkIII in a short-rotation coppice established on farmland. *Biomass and Bioenergy*, *108*, 323–329.

Weckx, S., Van der Meulen, R., Maes, D., Scheirlinck, I., Huys, G., Vandamme, P., & De Vuyst, L. (2010). Lactic acid bacteria community dynamics and metabolite production of rye sourdough fermentations share characteristics of wheat and spelt sourdough fermentations. *Food Microbiology*, *27*(8), 1000–1008.

Wójtowicz, A., Mitrus, M., Oniszczuk, T., Mościcki, L., Kręcisz, M., &Oniszczuk, A. (2015). Selected physical properties, texture and sensory characteristics of extruded breakfast cereals based on wholegrain wheat flour. *Agriculture and Agricultural Science Procedia*, *7*, 301–308.

Xue, L., Liu, G., Parfitt, J., Liu, X., Van Herpen, E., Stenmarck, A., O'Connor, C., Ostergren, K., & Cheng, S. (2017). Missing food, missing data? A critical review of global food losses and food waste data. *Environmental Science and Technology*, *51*(12), 6618–6633.

Zahed, A. H., Zhu, J. X., & Grace, J. R. (1995). Modelling and simulation of batch and continuous fluidized bed dryers. *Drying Technology*, *13*(1–2), 1–28.

Ziegler, V., Paraginski, R. T., & Ferreira, C. D. (2021). Grain storage systems and effects of moisture, temperature and time on grain quality-a review. *Journal of Stored Products Research*, *91*, 101770.

Zubrilina, E., Markvo, I., Novikov, V., Beskopylny, A., Vysochkina, L., Rudoy, D., & Butovchenko, A. (2019, December). Precise seeding planter with automated monitoring and control system. In *IOP conference series: Earth and environmental science* (Vol. 403, No. 1, p. 012063). IOP Publishing.

6 Industrial Applications of Rye Cereals

Mujahid Hassan Khan

6.1 INTRODUCTION

Human health and well-being depend greatly on cereal, which serves as an important source of macro and micronutrients essential for human nutrition. Today, they are staple foods throughout the world, especially in developing regions where they contribute to food security and dietary diversity (Badora et al., 2014). In cereals, rye cereals are nutritiously significant and adaptable to various conditions. As a major crop for both human and animal consumption with high content of essential amino acids, dietary fiber and bioactive compounds, they are a significant crop. Advances in rye breeding aimed at reducing the concentration of low–pentosan rye have increased its value in food production and as components for livestock diets (Kobylyansky et al., 2021). Due to the superior nutritional profile of rye compared to other cereals, high in lysine, methionine and valine (essential amino acids), the yields of rye make increasing its use in different food products (bread, pasta, breakfast cereals) (Allai et al., 2022). Rye utilization in many food products such as bread, pasta and breakfast cereals is expanding because of its unique flavors and health benefits (Ikram et al., 2023). Innovations in processing, such as sprouting and extrusion, have improved the nutritional quality of rye-based products (Lima et al., 2023).

Special processing techniques for the rye cereals may enhance their utility as a food material by improving the nutritional and technological properties of rye grains. Rye products, such as flour, flakes and malt, are very influenced by processing methods in terms of composition and quality. Milling in the processing of rye cereal produces different flour types, affecting components such as starch and fiber. Sourdough fermentation techniques increase nutritional properties, while enzyme actions alter dough properties, water absorption and glycemic index of final products (Sluková et al., 2021). In addition, the rye cereals have diverse industrial uses in food processing and manufacturing because of their high-fiber content and nutritional properties. More significantly, they are used as a key ingredient in many foods, such as bakery and breakfast products. Rye is the second most important raw material for bread after wheat, and rye-based bread is known for its high fiber content and dense texture (Németh & Tömösközi, 2021). Rye is added to breakfast cereals or snack foods to capitalize on consumer demand for high-fiber products. Key applications of rye cereals in other industries, such as animal feed, beverage, paper, fruit juice and alcoholic beverage production, are covered in this chapter.

DOI: 10.1201/9781003585817-6

6.2 PROCESSING OF RYE FOR INDUSTRIAL APPLICATIONS

Techniques of rye harvesting and storage are important to keep the seed viable and good. Managing seed moisture content and temperature during storage and spoilage is an effective practice. To prevent spoilage and have the best seed quality, rye should be harvested when the moisture content is about 14%–20% (Gunasekaran, 2006). Mechanical harvesting is common, but care must be taken not to damage the seed, which affects germination rates. Ideally, seeds should be stored at 5%–7% moisture content. Higher moisture can result in mold growth and decreased levels of germination (Zadorozhna et al., 2020). Long-term storage at low temperatures (e.g., –15°C) has minimal loss over the time of germination, and short periods of storage at 20°C are also recommended (Sathya et al., 2008). Longevity is affected by the atmosphere in which seeds are stored, and germination loss is reduced under nitrogen environments compared to air (Specht & Börner, 1998).

Several milling and refining techniques play a prominent role in changing the rye grain's nutritional and functional properties during the processes. Producing high-quality rye flour fit for a range of culinary applications necessarily involves these processes. The dominant process for making rye flour is roller milling, incorporating a gradual reduction passing mechanism, with the result of producing multiple flour streams. It also impacts the moisture, starch and protein content of the final flour (Gómez et al., 2009). Multistage grinding, however, is the method that combines the hammer and the roller mill to optimize energy consumption and particle size distribution. This approach refines the flour and decreases the energy requirements (Hameed Hassoon & Dziki, 2017). Milling can yield fractions that are rich in dietary fiber, with 50.0 and 36.0 g dietary fiber per 100-g dry matter, respectively. These are health beneficial, which offer low available saccharide fractions (Kołodziejczyk et al., 2018). Rye protein, lipids and dietary fiber contents are very different for various rye varieties and milling fractions, which affects their functional properties. Two important bread quality parameters are water absorption capacity and viscosity of dough, and these two are both significantly affected by the milling process (Gómez et al., 2009). In addition, increased lipid oxidation decreased β-glucan content, while higher β-glucan content led to better dough properties. Despite the milling and refining improving the nutritional profile of the rye, excessive refining is raised as something concerning because it is usually associated with the loss of some micronutrients. The challenge of rye processing is to reconcile the benefits of fiber enrichment with the retention of essential nutrients.

Rye-based ingredients, including flours and bran, are known for their nutritional benefits and some degree of applicability in the food industry. The milling processes of rye, especially rye grain, which are particularly valuable because of their high fiber content and the content of bioactive compounds, significantly affect the quality and characteristics of end products. Due to their functional properties, rye flour and bran are used in many food products, such as bread, pasta and breakfast cereal (Sibakov et al., 2013).

6.3 APPLICATIONS IN THE FOOD INDUSTRY

6.3.1 Rye Bread

Use of rye as an important ingredient in the food sector, mainly due to its features and health benefits in the making of rye bread, has seen considerable developments in recent years. Rye bread is becoming more and more popular due to its fiber content and bioactive compounds. Like other breads, rye bread is also popular in Europe and North America. It is very appealing due to the high fiber content and nutritional goodness and processing techniques further improve its quality and versatility. Dietary fiber from rye bread promotes a feeling of fullness and helps digestion (Ikram et al., 2023). Different rheological properties, such as water absorption of the dough and flour singularity, are important for the bread quality (Ikram et al., 2023). Rye is becoming a popular functional food in culinary products apart from traditional bread. Although rye bread has become known for its health benefits and its distinct qualities, many consumers still choose products that are wheat-based because they like the taste, or because they find the taste familiar.

6.3.2 Rye in Crackers and Biscuits

Rye is becoming more popular for its health benefits and applications in the preparation of crackers and biscuits. However, these products, having rye incorporated, are not only enhanced regarding their dietary properties but also in the total quality, and consumers find them appealing. Biscuits made with rye contain less carbohydrate, and more polyunsaturated fatty acids, which are healthy for the heart (Садыгова et al., 2022). The rye seaweed biscuit is just one of several rye-based snacks that are fortified with extra vitamins B and beta-carotene, making these snacks even more healthful. Other innovative recipes include rye-and-oat biscuits and rye soda crackers, which include various kinds of ingredients to give a more palatable taste and increase satiety, to reach health-concerned consumers (Sluková & Skřivan, 2020). Rye flour is used in combination with other ingredients, mainly honey and fruit juices products to enhance their nutrition and taste (Nguyen et al., 2023). Rye biscuits' sensory qualities such as taste and aroma have been positively evaluated, showing a growing acceptance of rye products by consumers (Садыгова et al., 2022).

6.3.3 Rye in Beer and Whiskey Production

Because of its own unique properties and potential benefits, rye is being used in beer and whiskey production. Both beverages have distinct flavors and aromas of rye, making the difference in the sensory profile. Though it possesses high hemicellulose and pentosan contents, they can complicate fermentation processes and thus need special treatment to boost alcohol yields. It has been determined through research that the alcohol yields from rye can be greatly improved by pre-treatment with β-glucanase. Using β-glucanase at optimal conditions improved yields by 187.1–406 LA/ton dry weight (lower than yields for barley and maize) (Morris et al., 2024). Irish Whiskey Association emphasizes that the use of locally sourced grains, such as rye, in the distilling industry is vital to sustainability (Morris et al., 2024). Specific mashing and malting techniques give rye beer its flavor, which has a similarity to rye bread. This special taste contributes to the unique drinking experience (Almaguer et al., 2023). Innovations include rye beer powder, enabling simpler preparation of rye beer by bringing together several ingredients to form a fermented drink with alcohol and carbonation (Bamforth, 2016).

6.3.4 Rye-Based Fermented Drinks

Fermented rye-based drinks such as kvass have become more popular recently as people have found them healthy with a unique tangy flavor. The process of production is fermentation of rye malt, and other fermentation processes can be improved by adding some additives or adjusting fermentation conditions. This synthesis focuses on all important aspects of rye fermented drinks, such as their modes of production, health benefits and sensory properties. The primary ingredient is usually rye malt, but we are adding variations of malt types (barley, wheat, oat) to make the flavor unique and improve the nutritional value. Drinks made from rye are rich in polyphenols and amino acids, making them antioxidants (Rokhsartalab-Azar et al., 2020). For example, a study showed that kvass with black chokeberry juice had high antioxidant activity (Kaszuba et al., 2024). Sensory evaluations report that rye drinks attain a pleasant flavor, which is enhanced when sweetened from natural sources, for example, apple concentrate (Rokhsartalab-Azar et al., 2020). While fermented rye drinks have many benefits, there are still associated production challenges in terms of standardized methods and the reproducibility of results.

6.4 NON-FOOD APPLICATIONS

6.4.1 Rye as a Raw Material for Biofuels

Rye serves as a promising raw material for biofuels due to its favorable biomass production characteristics and low environmental impact. Its properties, such as high carbon content and energy yield, make it an attractive option for sustainable energy solutions. Rye exhibits high biomass production, with certain genotypes yielding over 25 tons per hectare (Milunović et al., 2022). The carbon content in rye straw averages around 47.93% with a high heating value of 19.03 GJ Mg^{-1} (Stolarski et al., 2024). Winter rye can produce significant biomass while also providing high-value food and feed from summer crops (Shao et al., 2015). Rye cultivation as an energy cover crop enhances ecosystem services without necessitating land-use change. The carbon footprint of rye cultivation can be carbon negative, particularly when integrated with other crops. While rye presents numerous advantages as a biofuel feedstock, challenges such as the need for optimized cultivation practices and market fluctuations may impact its overall viability in the bioenergy sector.

6.4.2 Extracting Biopolymers from Rye

There are methods of rye biopolymer extraction, focused on the extraction of hemicelluloses, glutenins or proteins. Rye-based products have significant applications as food and non-food products, utilizing these biopolymers in various applications.

Extraction of hemicelluloses from rye bran can be carried out by means of high temperature treatments jointly with an enzymatic process, which allows recovery with high-molecular-weight polymers (Sárossy et al., 2011). Water as well as alkali extraction yield products that have different arabinoxylan and beta-glucan ratios, where water-extracted materials are more highly substituted than the alkali. Transgenic rye expression of wheat high-molecular-weight glutenin subunits considerably increases the polymeric glutelin fraction, and thereby enhances bread-making quality. Data from the growth experiment revealed that the transgenic rye had three times as much polymerized glutelins as wild-type rye in response to genetic engineering aimed at improving baking properties (Altpeter et al., 2004). Protein extracts from winter rye can form stable gels through freeze-thaw cycles, with the major component identified as Rubisco. The gelation process is influenced by pH and specific treatments, suggesting potential applications in food products (Lim, 2011).

6.4.3 Rye as a Component in Livestock Diets

Rye is increasingly recognized as a valuable component in livestock diets, offering various nutritional benefits and potential improvements in animal health. Its unique composition, particularly the high levels of non-starch polysaccharides (NSP), promotes gut health and enhances nutrient digestibility across different livestock species. Rye has shown comparable digestibility to traditional grains like wheat in pigs, with significant post-ileal digestibility improvements, particularly for nitrogen-free extracts (Hartung et al., 2024). Rye and its bran contribute to gut health by fermenting NSP into short-chain fatty acids, which can enhance the intestinal barrier and potentially reduce *Salmonella* prevalence in pigs (Homann et al., 2023). Incorporating rye bran in sow diets did not negatively impact performance metrics, and it may improve colostrum intake in piglets, particularly those with lower birth weights (Homann et al., 2024). Recent studies indicate that hybrid rye can replace corn in cattle diets without adversely affecting growth performance, suggesting its viability as a sustainable feed option (Tobin et al., 2023). Moreover, the applications of rye in food and non-food industries have been summarized in Table 6.1.

TABLE 6.1
Applications of Rye in Food and Non-Food Industries

Category	Applications
Food industry	• Bread and bakery products (e.g., sourdough, crisp bread, loaf bread). • Breakfast cereals, porridges, pasta and snack products. • Functional foods with health benefits, like improved gut health and metabolic risk reduction. • Brewing industry for malting and beer production.
Non-food industry	• Production of starch for use as gelling and thickening agents. • Raw material in biodegradable plastics and insulation materials. • Animal feed is especially important in areas with unfavorable soil and climate conditions. • Use in bioactive compound extraction for pharmaceuticals and nutraceuticals.

6.5 INNOVATIVE APPLICATIONS AND EMERGING TRENDS

Due to the rising demand from people with celiac disease and gluten intolerance, gluten-free rye products have come into the limelight as a development of gluten-free products. It is established that incorporation of rye flour in gluten-free formulations can yield good results in respect to both nutritional and sensory qualities of baked goods. However, being gluten-free is no longer enough for manufacturers to challenge continuing consumer expectations for taste and texture. As a result of its properties and potential to preserve food, rye is becoming a more recurrent ingredient in sustainable food packaging. On the other side, the films of rye starch, especially when they are mixed with natural extracts, for example with rosehip, exhibit a great antioxidant activity and may be used for increasing the shelf life of food products and decreasing the negative influence on the environment. The base material of biodegradable films is rye starch with promising potential to replace traditional plastic packaging. Inclusion of natural extracts, such as rosehip, to enhance these films further improves their flexibility and mechanical properties while also providing the antioxidant benefits (Go & Song, 2019). Shift to use renewable resources like rye is a matter of concern because it helps to reduce the ecological crisis created by plastic waste. Rye-based edible packaging not only increases waste management but also adds to the concept of the circular economy by valorizing agricultural byproducts. Rye-based packaging offers many benefits and with this solution comes the cost and scalability challenges. The emerging application is shown in Figure 6.1: the use of rye in different fields.

Nutritional profile and health benefits of rye make it important in alternative proteins and plant-based diets. Rye is a high fiber grain that can make an important contribution to dietary diversity and supplies bioactive compounds with health-enhancing potential. Rye is around 51% in protein, and its hydrolyzates have antioxidant, antihypertensive and anti-diabetic potential (Islas-Martínez et al., 2023). It is a good source of dietary fiber, specifically NSP: arabinoxylans and fructans, which benefit

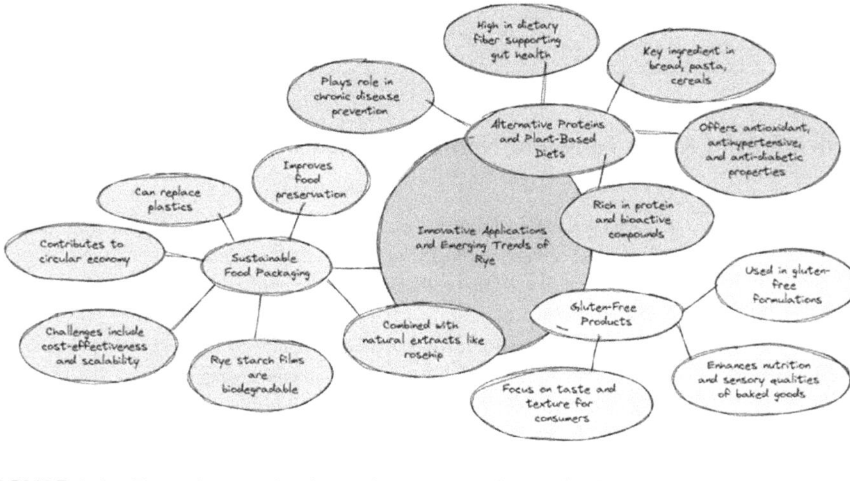

FIGURE 6.1 Emerging applications of rye across diverse fields.

gut health and satiety. Rye is an essential ingredient that has been used to make bread, pasta and breakfast cereals and is often found in plant-based diets (Ikram et al., 2023). Including it in diets is also part of plant-based nutrition and can help cut down the risk of developing chronic diseases.

6.6 CONCLUSION

Rye cereals have much potential in a range of industrial applications from food processing to applications outside of the food area. It is attractive in nutrition and in sensory attributes and is used in the food industry in bakery products such as bread, crackers, biscuits and beverages, for example, beer, whiskey and fermented drinks. Rye is explored in novel uses such as sustainable food packaging, and its use in plant-based diets showcases the rye application to combat market health-conscious trends and promote sustainability. Due to its uses, rye proves to be an integral part as different industries aim to be more eco-friendly toward the environment and more inventive with solutions. Multifaceted applications of rye have affirmed its importance in food security, advanced renewable energy materials and environmentally friendly materials.

REFERENCES

Allai, F. M., Azad, Z., Gul, K., & Dar, B. (2022). Wholegrains: A review on the amino acid profile, mineral content, physicochemical, bioactive composition and health benefits. *International Journal of Food Science & Technology*, *57*(4), 1849–1865.

Almaguer, C., Kollmannsberger, H., Gastl, M., & Becker, T. (2023). Comparative study of the impact of malting on the aroma profiles of barley (*Hordeum vulgare* L.) and rye (*Secale cereale* L.). *Food Chemistry*, *427*, 136694.

Altpeter, F., Juan, C. P., & Wieser, H. (2004). Stable expression of 1Dx5 and 1Dy10 high-molecular-weight glutenin subunit genes in transgenic rye drastically increases the polymeric glutelin fraction in rye flour. *Plant Molecular Biology*, *54*, 783–792.

Badora, A., Kozłowska-Strawskaa, J., Domańska, J., & Filipek, T. (2014). Cereals–health or disease. *Problemy Ekorozwoju*, *9*(2), 87–98.

Bamforth, C. W. (2016). *Brewing materials and processes: A practical approach to beer excellence*. Academic Press.

Садыгова, М., Крупнова, О., Ермолаева, Т., & Левчук, О. (2022). РАЗРАБОТКА РЕЦЕПТУРЫ БИСКВИТНОГО ПОЛУФАБРИКАТА НА ОСНОВЕ МУКИ ИЗ СВЕТЛОЗЕРНОЙ РЖИ С ПОНИЖЕННЫМ СОДЕРЖАНИЕМ УГЛЕВОДОВ. *Сурский вестник* (3), 50–55.

Go, E. J., & Song, K. B. (2019). Antioxidant properties of rye starch films containing rosehip extract and their application in packaging of chicken breast. *Starch-Stärke*, *71*(11–12), 1900116.

Gómez, M., Pardo, J., Oliete, B., & Caballero, P. A. (2009). Effect of the milling process on quality characteristics of rye flour. *Journal of the Science of Food and Agriculture*, *89*(3), 470–476.

Gunasekaran, S. (2006). *Safe storage guidelines for rye and canola*.

Hameed Hassoon, W., & Dziki, D. (2017). *The study of multistage grinding of rye*.

Hartung, C. B., Visscher, C., Grone, R., & Kamphues, J. (2024). The ileal, total tract and postileal digestibility of compound feeds either rich in wheat or rye in an ileo-caecal fistulated minipig model. *Archives of Animal Nutrition*, *78*(3), 285–297.

Homann, C., Eckey, I., Chuppava, B., Teich, K., Buch, J., Zimmermann, A., Kaltschmitt, M., Grone, R., Wilke, V., & Visscher, C. (2023). Rye and rye bran as components of diets in piglet production—Effects on salmonella prevalence. *Animals*, *13*(14), 2262.

Homann, C., Wilke, V., Eckey, I., Chuppava, B., Kaltschmitt, M., Zimmermann, A., & Visscher, C. (2024). Rye bran as a component in the diets of lactating sows—Effects on sow and piglet performance. *Animals*, *14*(3), 380.

Ikram, A., Saeed, F., Noor, R. A., Imran, A., Afzaal, M., Rasheed, A., Islam, F., Iqbal, A., Zahoor, T., & Naz, S. (2023). A comprehensive review on biochemical and technological properties of rye (*Secale cereale* L.). *International Journal of Food Properties*, *26*(1), 2212–2228.

Islas-Martínez, D., Ávila-Vargas, Y. N., Rodríguez-Serrano, G. M., González-Olivares, L. G., Pérez-Flores, J. G., Contreras-López, E., Olloqui, E. J., & Pérez-Escalante, E. (2023). *Multi-bioactive potential of a rye protein isolate hydrolysate by enzymatic processes.* Biology and Life Sciences Forum.

Kaszuba, J., Jańczak-Pieniążek, M., Migut, D., Kapusta, I., & Buczek, J. (2024). Comparison of the antioxidant and sensorial properties of kvass produced from mountain rye bread with the addition of selected plant raw materials. *Foods*, *13*(3), 357.

Kobylyansky, V., Solodukhina, O., & Nikonorova, I. (2021). Morphological features of rye grain with low pentosan content. *Proceedings on Applied Botany, Genetics and Breeding*, *182*, 123–130.

Kołodziejczyk, P., Makowska, A., Pospieszna, B., Michniewicz, J., & Paschke, H. (2018). Chemical and nutritional characteristics of high-fibre rye milling fractions. *Acta Scientiarum Polonorum Technologia Alimentaria*, *17*(2), 149–157.

Lim, Z. L. (2011). Examination of a novel proteinaceous extract from winter rye (*Secale cereale* L. cv Musketeer). University of Saskatchewan.

Lima, C. T., Santos, T. M. d., Neves, N. d. A., Lavado-Cruz, A., Paucar-Menacho, L. M., Clerici, M. T. P. S., Meza, S. L. R., & Schmiele, M. (2023). New breakfast cereal developed with sprouted whole ryegrass flour: Evaluation of technological and nutritional parameters. *Foods*, *12*(21), 3902.

Milunović, I., Popović, V., Rakaščan, N., Ikanović, J., Trkulja, V., Radojević, V., & Dražić, G. (2022). Genotype × year interaction on rye productivity parameters cultivated on sandy chernozem soil. *Genetika-Belgrade*, *54*(2), 887–905.

Morris, S., Byrne, J., Whelan, S., Carroll, J., & Ryan, D. (2024). The impacts of a β-glucanase step pre-cooking on alcohol yields during Irish rye whiskey production. *Journal of the American Society of Brewing Chemists*, *82*(3), 258–270.

Németh, R., & Tömösközi, S. (2021). Rye: Current state and future trends in research and applications. *Acta Alimentaria*, *50*(4), 620–640.

Nguyen, S. N., Drawbridge, P., & Beta, T. (2023). Resistant starch in wheat-, barley-, rye-, and oat-based foods: A review. *Starch-Stärke*, *75*(9–10), 2100251.

Rokhsartalab-Azar, S., Jafari, P., Tukmechi, A., & Malekinejad, H. (2020). A new rye-based beverage with high polyphenols and amino acids: Production and study of physicochemical and sensory properties. *Journal of Food Processing and Preservation*, *44*(3), e14349.

Sárossy, Z., Egsgaard, H., & Plackett, D. (2011). Isolation and chemical characterization of hemicelluloses from rye bran. 19th European Biomass Conference and Exhibition.

Sathya, G., Jayas, D., & White, N. (2008). Safe storage guidelines for rye. *Canadian Biosystems Engineering*, *50*(3), 1–8.

Shao, X., DiMarco, K., Richard, T. L., & Lynd, L. R. (2015). Winter rye as a bioenergy feedstock: Impact of crop maturity on composition, biological solubilization and potential revenue. *Biotechnology for Biofuels*, *8*, 1–10.

Sibakov, J., Lehtinen, P., & Poutanen, K. (2013). *Cereal brans as dietary fibre ingredients.*

Sluková, M., Jurkaninová, L., Švec, I., & Skřivan, P. (2021). Rye—The nutritional and technological evaluation in Czech cereal technology—A review: Grain and flours. Czech Journal of Food Sciences, *39*(1).

Sluková, M., & Skřivan, P. (2020). Food reformulation and bakery products. *Eat & Live Correctly*, 17.

Specht, C.-E., & Börner, A. (1998). An interim report on a long term storage experiment on rye (*Secale cereale* L.) under a range of temperatures and atmospheres. *Genetic Resources and Crop Evolution*, *45*, 483–488.

Stolarski, M. J., Welenc, M., Krzyżaniak, M., Olba-Zięty, E., Stolarski, J., & Wierzbicki, S. (2024). Characteristics and changes in the properties of cereal and rapeseed straw used as energy feedstock. *Energies*, *17*(5), 1243.

Tobin, C. T., Carlson, Z. E., Swanson, K. C., Hoppe, K., Rusche, W. C., & Smith, Z. K. (2023). 355 Effects of processing method of hybrid rye for cattle backgrounding diets. *Journal of Animal Science*, *101*(Supplement_3), 255–256.

Zadorozhna, O., Yehorov, D., & Zhmurko, V. (2020). Effects of different storage regimens on the rye seed germination and abscisic acid content. *Селекція і насінництво* (117), 68–79.

7 Rye Cereals and Their By-Products

Samreen Ahsan, Atif Liaqat, Tariq Mehmood, Muhammad Adnan Nasir, Adnan Khaliq, Nimra Sameed, Assam Bin Tahir, Muhammad Adil Farooq, Ayesha Siddiqa and Muhammad Farhan Jahangir Chughtai

7.1 INTRODUCTION

The name 'cereals' refers to nine species of the Gramineae family: wheat (*Triticum aestivum*), barley (*Hordeum vulgare*), rye (*Secale cereale*), oat (*Avena sativa*), millet, rice (*Oryza sativa*), corn (*Zea mays*), sorghum (*Sorghum*) and triticale, a wheat/rye hybrid (Papageorgiou & Skendi, 2018). A significant amount of dietary energy and nutrients comes from cereals and cereal-derived products, which are staple foods for the majority of people's diet in developed and developing nations. They contribute more than 50% of the world's energy supply and are made up of approximately 75% carbs, mostly starches, and 6%–15% protein. The fact that cereal production, which amounts to approximately 2600 million tons annually, is the primary factor influencing global food security further supports the significance of cereals and cereal products (Laskowski et al., 2019).

The food and pharmaceutical industries are becoming more interested in cereal grains because of their potential to lower the prevalence of several chronic illnesses. Since they have several health benefits, most people concentrate on eating a diet high in particular nutrients and bioactive compounds. Scientists worldwide are investigating the untapped potential of various cereal grains (Kaur et al., 2021a).

Rye (*S. cereale* L.) is a cereal crop, unlike other cereals such as wheat and rice, that often grows under harsh environmental circumstances. Rye is mostly produced in Europe, Germany, Poland, Russia, Ukraine and Belarus leading the way (El-Mahis et al., 2023). It is widely used in Europe for making breads such as pumpernickel and crispbread. It may be grown in places that are not suitable for other cereals because of its minimal requirements for soil and fertilizer as well as its strong overwintering ability. The northern, central and eastern regions of Europe produce more than 90% of the world's rye. Due to its high dietary fiber and lysine (Lys) content, wholemeal rye flour, in particular, is of significant importance for utilization (Deleu et al., 2020).

Although it is more difficult to produce than wheat, rye, which is closely related to wheat, has been grown in Europe since before the Common Era (more than 1000 BCE). The complicated chemical composition of rye provides a variety of

 DOI: 10.1201/9781003585817-7

nutritional and technical benefits. Rye kernels are physiologically similar to wheat grains; however, they do not belong to the same species. The pericarp and testa, which wrap the germ and endosperm, including the starchy endosperm and aleurone, are the outermost layers of the rye kernel. Rye-flour bread, black and sour bread, hot bread, morning cereals and porridge flakes are among the most popular rye-based dishes. Rye is becoming a more and more popular component for munching and snacking. Owing to the nutritional and medicinal advantages of whole grains, such as their health-promoting qualities and gut-friendly metabolism ascribed to bioactive components, consumers are increasingly choosing them (Ikram et al., 2023b).

It is generally recognized that the outer layers of the kernel contain a considerable amount of nutritionally beneficial components that are not or only partially absorbed by lighter (whiter) flours (Jaksics et al., 2022). The primary components of rye grain are protein (9.0%–15.4%), ash (1.8%–2.2%), dietary fiber (14.7%–20.9%) and starch (57.1%–65.6%). Rye starch granules resemble wheat starch granules in size and form (Dziki, 2022).

Rye contains fibers (arabinoxylan, lignin, β-glucan, fructan and cellulose) that promote a healthy weight and lower the risk of diabetes, heart issues and cancer. Rye has the potential to be used to produce nutritious, fiber-rich dishes. Rye is traditionally used for baking and as an excellent source of energy and nutrition due to its high content of carbs, protein, dietary fibers, vitamins and minerals, particularly magnesium and zinc. The increasing interest in cereal grains is related to their bioactive components and possible health and nutritional advantages (El-Mahis et al., 2023). The most important rye fiber components are fructans (3%–7%), ß-glucans (1%–3%) and arabinoxylans (7%–12%), all of which may have a number of positive health impacts (Jaksics et al., 2022).

In addition to dietary fiber, rye grains are a rich source of phenolic acids, lignans, alkylresorcinols and other phytochemicals (Pihlava et al., 2018). Rye's protein composition can vary greatly (8%–15%) based mostly on growing circumstances. Their primary proteins, albumins, provide 29%–40% of their overall protein concentration. Globulins provide 8%–11%, but prolamins and glutelins, the rye storage proteins (also known as secalins), account for 17%–19% and 9%–15% of the total protein composition, respectively. According to Osborne fractionation, between 21% and 30% of proteins cannot be extracted. Rye proteins have a somewhat better amino acid composition than wheat because they contain more lysine, but they are deficient in tryptophan and isoleucine (Németh & Tömösközi, 2021). Approximately 42% of rye is used to feed animals, 31% to make processed foods such as rye bread for human use and 27% for other reasons (Rani et al., 2021).

Numerous approaches and technologies have been put forth and used in research to create functional cereal-based meals with higher levels of dietary fiber and bioactive substances. The need for innovative food production techniques that will provide easy, ready-to-eat, safe and always-available food items has arisen due to the growing global population and the needs of modern consumers.

7.2 COMPOSITION OF RYE CEREAL

7.2.1 Macronutrient Profile: Carbohydrates, Proteins and Fats

The value of rye is determined by the significant content of proteins (12.8%) and carbohydrates (69.1%) in the grain (Kyrylchuk et al., 2024). Rye is chiefly composed of starch, which constitutes the majority of its macronutrient content (Kataria et al., 2024). Rye seeds are composed of approximately 80%–85% starch in the endosperm, 2%–3% starch in the embryo and 10%–15% starch in the outer layers. Rye starch granules are reported to exist in two different forms: large lenticular A type granules and smaller polygonal B-type granules (El-Mahis et al., 2023). The starch contents in rye seeds differ across different tissues, with the testa containing approximately 3 g/100 g dry matter, and the endosperm containing an appreciably higher concentration of 78.2 g/100 g dry matter (Wrigley & Bushuk, 2017). The amylose content of rye starch has been reported to be 20%–30%, which is notably lower than that of amylopectin. This amylose-to-amylopectin ratio results in a lower gelatinization temperature for rye starch, which is attributed to the higher proportion of short amylopectin chains and relatively lower levels of longer chains. In addition to starch, non-starch polysaccharides, including β-glucan, lignin, cellulose and arabinoxylan, contribute to the nutritional value and potential health benefits of rye (El-Mahis et al., 2023).

Depending on factors such as growing conditions and genotype, the protein content of rye varies between 6.5% and 14.5%. Notably, the rye variety 'Virgiai' demonstrates the highest crude protein content, while 'Palazzo' exhibits the lowest (Alijošius et al., 2016). Rye, in processed food products, protein acts as binders to enhance flavor (Banu et al., 2024). The quality of rye protein is determined by the profile of essential amino acids (Surówka & Rzepka, 2024). Key essential amino acids found in rye proteins include lysine, threonine, valine, leucine, phenylalanine, tyrosine, proline and alanine (Ikram et al., 2023a). The amino acid profile of rye is considered to be more favorable than that of wheat and other cereals because of its higher levels of lysine, proline and glutamine (Rani et al., 2021). Compared to wheat flour, rye flour contains significantly higher levels of non-storage proteins, such as albumins and globulins, and lower levels of polymerized storage proteins, such as glutelins (Wojciechowski, 2022). Rye endosperms contain gluten protein at levels higher than those found in wheat endosperms (Xhaferaj et al., 2023). This limits its use in individuals with celiac disease due to gluten sensitivity (Yalcin, 2024). Rye also contains various prolamins known as secalins, which are proteins with different molecular weights (Pastrana-Pastrana et al., 2025). Rye flour consists of albumins, globulins and secalins, comprising 31.7%, 11.4% and 57.0%, respectively, of the total protein content (Redant et al., 2017).

Rye oil yields are approximately 1.4%, consisting predominantly of unsaturated fatty acids, including linolenic acid (6.73%), palmitoleic acid (0.3%), oleic acid (19.42%) and linoleic acid (55.01%). It also contains saturated fatty acids such as myristic acid (0.11%), palmitic acid (15.92%), stearic acid (0.87%) and arachidic acid (0.18%) (Kan, 2015). Additionally, the concentrations of oleic, linoleic and linolenic acids in various annual ryegrass cultivars have been found to range between 0.89–1.34, 3.97–4.01 and 15.22–20.56 g/kg, respectively (Ozdemir et al., 2024).

7.2.2 Micronutrients: Vitamins and Minerals

Rye is recognized as a valuable source of B-complex vitamins, including thiamine (B1), riboflavin (B2), nicotinic acid (B3), nicotinamide, pantothenic acid (B5), pyridoxine (B6) and pyridoxal (Ghafoor, Wijata et al., 2024). Additionally, rye grains have a relatively high-folate content, ranging from 630 to 780 µg/kg, with six different forms of folate identified. Among all vitamins, formyl folates and 5-CH_3H_4-folate are the most abundant in rye (El-Mahis et al., 2023).

Macrominerals, such as magnesium (Mg), phosphorus (P) and potassium (K), as well as several microminerals, particularly zinc (Zn), iron (Fe) and manganese (Mn), are present in rye. Rye consists of 5567 ppm potassium, 3512 ppm phosphorus, 820 ppm calcium, 1908 ppm magnesium, 30 ppm iron, 28 ppm zinc and 2 ppm copper (Bağcı et al., 2019).

7.2.3 Unique Bioactive Compounds in Rye

7.2.3.1 Phenolic Acids

The primary phenolic compounds in rye are phenolic acids, which are generally concentrated in the aleurone layer and the bran. These phenolic acids belong to two main groups: p-hydroxybenzoic acid and its derivatives and p-coumaric acid and its derivatives (Dynkowska, 2019). Although free phenolic acids account for less than 3% of the total phenolic content in rye, they also play a significant role in the antioxidant activity of rye flour (Pihlava et al., 2015). The concentration of phenolic acids in the rye grains varied from 65 to 300 mg/100 g. Rye is particularly rich in gallic, p-hydroxybenzoic and caffeic acids compared to other cereal grains. The most abundant phenolic acids in rye were ferulic acid (85%–90%), sinapic acid (9%–10%) and p-coumaric acid (3%–5%), with smaller amounts of caffeic, gallic, protocatechuic, vanillic and syringic acids also present in rye grains (Dziki, 2022).

7.2.3.2 Flavonoids

Various studies have reported the flavonoid content in refined rye flour, wholemeal flour and rye bran (Pihlava et al., 2018). These studies revealed that the flavonoid content ranged from 46 mg/kg in wholemeal flour to 168 mg/kg in bran, with refined rye flour containing less than 5 mg/kg of flavonoids. The flavone content varied between 57 and 137 µg/g depending on the rye variety. The predominant flavones were found to be O-glycosides (50%–68%), which were primarily derivatives of apigenin, chrysoeriol and tricin (Ravisankar et al., 2020).

7.2.3.3 Rye Lignans

Rye contains the highest concentration of lignans among cereal grains, with their content in rye grains ranging from 11.4 to 67.0 mg/kg (Zanella et al., 2017). Similar to other bioactive compounds in rye, lignans are predominantly concentrated in bran (Luthria et al., 2015). Seven dietary lignans have been identified in rye grains. The most abundant lignin was secoisolariciresinol, which accounted for approximately 80% of the total lignan content. Other lignans, including medioresinol and pinoresinol, contribute roughly 10%, each with smaller amounts of secoisolariciresinol, matairesinol and 7-hydroxymatairesinol also present in rye grains (Dziki, 2022).

7.2.4 Comparison of Rye with Other Cereal Grains

The physicochemical properties of various cereal grains exhibit notable differences. Wheat showed a higher lactic acid-solvent retention capacity (95.86%–111.92%) than rye (53.78%–67.97%) and barley (50.24%–67.12%), indicating stronger gluten strength in wheat. In contrast, rye had a higher sucrose-solvent retention capacity and sodium carbonate-solvent retention capacity than wheat and barley flours. Barley and rye cultivars also contain a higher proportion of essential amino acids compared to wheat, leading to a higher biological value (BV) in these grains, largely because of their greater lysine content (Rani et al., 2021). Additionally, 48% and 62% of the rye's total root biomass were found in the top 15-cm soil depth, following corn and soybean, respectively (Patel et al., 2024). Nutritional composition and health benefits of rye have shown in Figure 7.1.

The composition of dietary fiber in cereal grains varies significantly among different types of cereals. Rye is at the top and contains total dietary fiber (15.0–25.0 g), followed by hulled oats (11.8–19.1 g), wheat (10.0–17.0 g), hulled

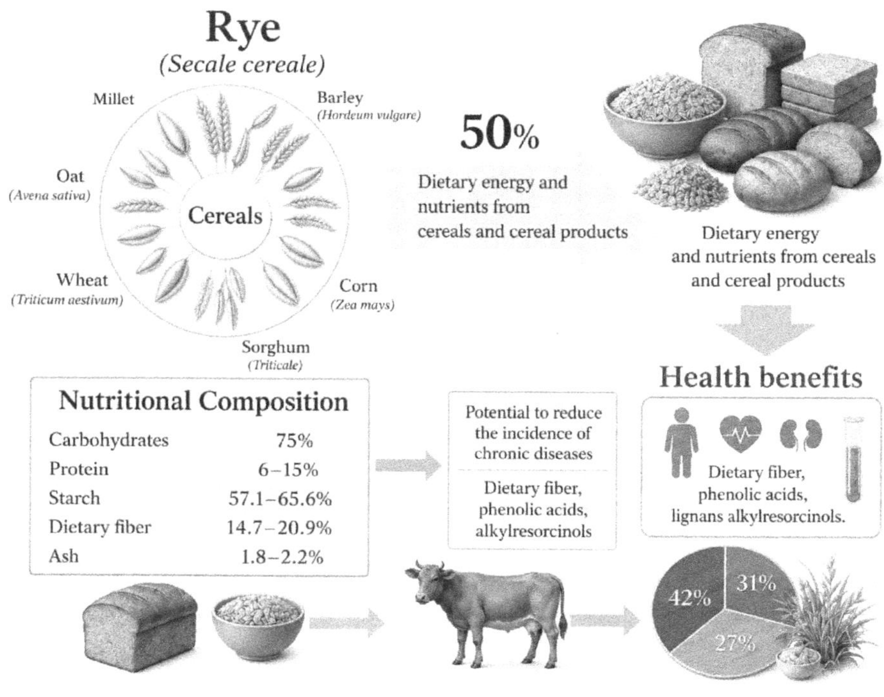

FIGURE 7.1 Nutritional composition and health benefits of rye.

TABLE 7.1
Comparison of Rye Mineral Content (mg/kg) with Other Cereals

Mineral	Rye	Wheat	Barley	Triticale	Oats
P	1806–4220	1500–5400	2700–3600	3210–3850	4510–5020
K	3480–6148	2900–6200	3190–4400	4660–4700	3890–4250
Ca	157–1447	370–1220	400–2700	350–400	500–590
Mg	920–1602	900–2900	660–1800	1100–1530	1220–1580
Fe	27–129	28–42	30–76	26–37	43–47
Zn	21–52	19–32	14–23	25–27	29–37
Mn	20–75	5–49	13–22	26	41–45
Cu	3–13	4–7	3–4	3	3–5
Zn	21–52	19–32	14–23	25–27	29–37

barley (8.0–16.2 g) and triticale (7.5–16.0 g). However, soluble fiber, oats, rank highest (3.5–12.3 g), greater than barley (4.1 g), wheat (3.6 g), triticale (2.6–3.2 g) and rye (2.5 g). In contrast, insoluble fiber was most abundant in rye (13.2 g), followed by barley (12.1 g), wheat (9.4–12.0 g), triticale (11.4–12.0 g) and oats (6.8–8.3 g). Rye also leads in arabinoxylan (6.0–11.5 g) and water-extractable arabinoxylan (1.05–4.0 g), with wheat, barley and triticale containing lower amounts. For cellulose, rye had the highest content (2.9 g), whereas oats contained less cellulose (1.3 g). Oats are notable for their high β-glucan levels (3.9–7.1 g), surpassing barley (3.2–4.6 g). Rye contains high levels of fructans (3.6–6.6 g) in comparoats, triticales and barley have lower quantities. Wheat has the highest arabinogalactan content (0.89–1.09 g), whereas Klason lignin is most abundant in oats (1.4–5.9 g) (Németh & Tömösközi, 2021).

The mineral and vitamin contents of various grains, such as rye, wheat, barley, triticale and oats, vary significantly. Minerals comparison of rye with other cereals is given in Table 7.1 (Biel et al., 2020).

In terms of vitamins the comparison of rye with other cereals is shown in Table 7.2 (Németh & Tömösközi, 2021).

TABLE 7.2
Comparison of Rye Vitamin Content (mg/kg) with Other Cereals

Vitamin	Rye	Wheat	Barley	Triticale	Oats
B1 (Thiamin)	4.0–4.6	5.0–12	2.0–2.6	3.8–9.8	5.0–8.0
B2 (Riboflavin)	1.8–1.9	1.0–3.1	0.9–1.0	1.3	1.0–1.4
B3 (Niacin)	12–15	41–64	45–50	29	9.6–16
B5 (Pantothenic Acid)	10	7.7–9.1	3.5	6.5–8.8	13.4
B6 (Pyridoxine)	3.0–3.4	3.0–4.7	3.0–3.2	4.0	2.0–2.4
B9 (Folate)	0.48–0.52	0.35–0.56	0.19–0.25	0.7	0.45–0.60
E (α-Tocopherol)	10–12	5–12	3.4–4.0	9.0	8.0–18

7.3 PROCESSING RYE CEREALS

Techniques for processing the grain can enhance its nutritional characteristics, such as modifying its fractional composition and protein solubility (Fattakhova et al., 2020). Numerous physicochemical events occur during the processing of rye and rye flour, which alter the accessibility, content and characteristics of components that are significant for both nutrition and technology. Modifications impact the glycemic index of rye bread, dough form, consistency, quality, production and the capacity of rye flour to retain moisture. Contemporary development in rye milling and processing is the development of flour with an increased fiber level of wholemeal flour. Finely granulated wholemeal and dark rye flours are often produced using unconventional customized milling machinery. Rye fiber and other bioactive substances can be more readily bioavailable owing to alteration, which reduces starch degradation, lowers the product's glycemic index and provides enough substrate for fermentation (Sluková et al., 2021).

7.3.1 Harvesting and Primary Processing Techniques

Scientists discovered that year of the harvest, not the genetic makeup of the crop, significantly impacted differences in grain-processing processes and baking qualities. Compared to wheat flour, rye flour has a higher percentage of fiber, but comparatively fewer proteins and carbohydrates (Dziki, 2022). A modified conventional baking test and numerous rye samples have been utilized to assess rye harvests over the past decade. The procedure is tedious, costly and not pertinent to the baking and milling industries (Oest et al., 2020). Harvesting and processing of rye and advancement in rye-processing technologies are shown in Figure 7.2.

The primary ingredient of flour and rye grains is starch. Rye starch granules are comparable to wheat starch granules in terms of both shape and size. The cores of the larger rye starch granules had cracks or voids. Compared to wheat starch, rye starch gelatinization occurs across a smaller temperature range. Rye flour and endosperm have higher fiber than wheat but comparatively less protein and carbohydrate. Food processing can alter the amount of fiber in food, employing extrusion or baking. The resistant starch content of extruded or deep-frozen rye products frequently gets elevated, and the soluble β-glucan content of extruded goods is often enhanced as well (Sluková et al., 2021).

7.3.2 Milling of Rye and Flour Types

The conventional milling technique of rye produces white and particularly dark bread flours, whereas rye is typically utilized in human nutrition as an initial component in the milling industry (Sluková et al., 2021).

There are two theoretical methods for producing regular wheat and rye wholemeal flour. A basic impact mill serves as the foundation for the disintegration process in the first, most typical scenario and a rolling mill for granulation treatment may come next. Reconstructing wholemeal flour from basic mill products is the second production technique. This approach is employed when the mill's primary

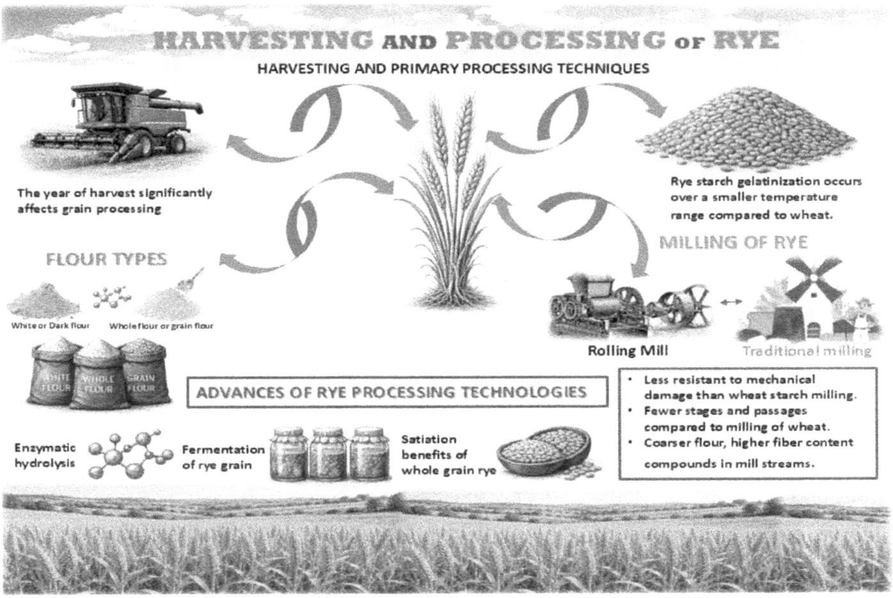

FIGURE 7.2 Harvesting and processing of rye and advancement in rye-processing technologies.

production program consists of typical compositional goods such as baking flour and whole-grain flour represents an insignificant item. Graham wheat flour serves as an illustration of this sort of good (Skřivan et al., 2021).

In order to transform grains into edible nutritious foods, milling is an essential step in the process (Kaushal & Kumar, 2022). To reinforce the outermost coat of bran and germ and to render it easier to separate the endosperm, rye grain has been tempered prior to milling. The rye grain grinding procedure is considerably more vigorous and covers fewer phases than the milling process for wheat flour because the outermost layers of the rye kernels are firmly linked to the endosperm. Due to the arbitrary arrangement of chemical components within the grain, flour derived from various streams exhibits varying chemical characteristics. To produce various flours with desired qualities, various varieties of ground rye flour were combined. Crucially, rye flour starch granules are less resistant to mechanical damage throughout the milling phase than wheat starch particles (Dziki, 2022).

The two foremost significant methods for regulating the amounts of bioactive compounds during milling involve selecting certain mill streams to combine to produce various flour types and regulating the milling extraction rate. Lower rates of extraction condense the majority of the bioactive substances into the brans, whereas higher extraction rates facilitate the retention of significant quantities of vitamins, minerals and fiber in flours. With the aleurone layer being particularly abundant in nutritional components, the accumulation of bioactive chemicals in the mill streams is correlated with the extent to which various structural components contribute to their development (Banu et al., 2021).

The idea behind the most popular rye-milling technique is similar to that of milling wheat. It is a sequence of recurring actions, or passages, that include separation and disintegration. The process of milling begins with the wetting and tempering of the grains. The goal of these procedures is to remove the endosperm and separate the germs and outer layers. In rye, it doesn't seem possible to separate the endosperm from the outermost layers, just like it is in wheat. As compared to wheat, rye grains are longer and thinner, and the outermost layers are more tightly affixed to the endosperm. Rye's milling method is less environmentally friendly, more comprehensive, and involves fewer passages than wheat. Steel sections, or 'hammers', rotate in many rows above one another, and the mills properly adjust the inner housing to disintegrate the substance as it flows higher in the mill chamber with the surrounding airflow. The material travels through the disintegration space in the stream of air until the particles enter the selected display, which is attached to the higher portion of the mill drum and has an adjustable mesh width. In comparison to the conventional milling technique, it permits a far finer breakdown or granulation of the outermost surfaces while avoiding additional degradation to the endosperm's starch particles. The bread offers superior sensory qualities because the end product is ultrafine whole-grain flours that are simpler to work with than regular whole-grain flours (Sluková et al., 2021).

A method has been developed for separating wholemeal flour by sifting and grinding fiber-rich rye flour in a ball mill before grinding it in a roller mill. For the purpose of facilitating protein separation, the jet-milling procedure for processed rye flour has been used. A distinct approach has also developed for grinding rye bran, in which ultra-fine grinding (UFG) classifies the bran according to its air content. To make the bran particles smaller, the scientists employed a fine impact mill and a pin disc mill. According to their research, the protein level in rye bran prior to grinding was twice as high as the protein-rich sections produced by the pin disc and air classification techniques (Dziki, 2022).

However, there are a few noteworthy variations between the two milling processes for rye and wheat that are comparable. For example, rye requires corrugated rollers for reduction rather than smooth rolls, and the tempering phase is faster. To prevent clumping, rye flour additionally demands a wider sifting surface. These are clarified by the fact that rye's endosperm is weaker than wheat's (Németh & Tömösközi, 2021). The development of increased fiber and whole wheat flour represents a new development in the grinding and processing of rye. Fine-granulated wholemeal and dark rye flours tend to be made with unconventional, specially designed mill machinery (Ikram et al., 2023a). In roller mills, rye is usually ground into white and wholemeal flour. Hansen et al. examined 19 rye cultivars for their characteristics. A number of 19 rye cultivars were examined for their characteristics by Hansen et al. (Dziki, 2022). The consistency, nutritional fiber and ash percentage of the three primary rye flour processing fractions vary—white flour, bran and whole-grain flour. The bran and other fractions with a substantial AX content possess a lower viscosity, and various portions are combined to provide distinct nutritive as well as sensory characteristics (Brzozowski et al., 2023).

Central European rye flours have ash contents that vary between 0.5% for white flour and 1.5% for dark-colored flour. Additionally, w/w in dry matter, wholemeal rye flours with an ash concentration of up to 2.0% are developed. In the Czech Republic, T 930 bread flour, which contains an ash concentration of up to 1.1%, is the most widely utilized rye flour. In addition to flour, other products include rye flakes, fermented grains and grains that have undergone various hydrothermal treatments. These commodities are primarily utilized in specific kinds of grains and breads (Sluková et al., 2021).

7.3.3 Advances in Rye-Processing Technologies

Due to its small quantity of fiber and elevated protein level in comparison with other cereal grains, novel rye cultivars represent a desirable option for commercial use. Barley is inferior to rye grain in terms of crude protein and carbohydrates. Rye grain has lower levels of ash, fats and dietary fiber than barley. The rye may be a nutritious food for people who are not gluten intolerant. During the processing of rye and rye flour, a number of physicochemical activities occur that influence the accessibility, content and properties of nutrients and key technical variables. The glycemic index, the viscosity and structure of the dough, the bread's grade, production and the rye flour's capacity of absorbing moisture are all impacted by multiple alterations. Using suitable methods for milling and categorizing milled goods makes it easier to change their chemical structure and functional attributes. Rye bran is given particular consideration because of its many commercial applications and its application as a culinary component. Furthermore, enzymatic hydrolysis and rye grain fermentation produce a variety of physiologically active compounds while increasing the quantity and bioavailability of phytochemicals (Ikram et al., 2023a).

It has been advised that rye and rye-based products ought to be included in a regular diet plan in order to achieve an active, disease-free way of life. Additionally, the satisfaction benefits of whole-grain rye as well as its derivatives have been the subject of several investigations. Since these promote satisfaction and enhance the efficiency of the digestive and metabolic systems, cereal grains loaded with protein and fiber might be used to satisfy appetite needs (Kaur et al., 2021a).

7.3.4 Quality Parameters in Rye Products

Rye dough behavior is greatly influenced by starch, the key ingredient in rye grain and flour, coupled with amylase activity. Compared with wheat starch, the starch grains of various rye flours exhibit more enzymatic and mechanical degradation while also having a greater capacity for swelling, resulting in a softer, more sticky dough (Dziki, 2022). In bread, using more 'high extraction' as well as wholemeal flours might be one way to enhance the incorporation of chemicals having a high-nutritional profile. Compared to whole-grain wheat flour and any other flour developed from non-bread cereals, rye flour is regarded as one of the greatest raw materials for manufacturing bread since it contains a substantially greater dietary fiber concentration. In addition, wholemeal rye soft bread is preferred by customers in Northern and Eastern Europe, especially Poland, due to its mouthwatering flavor as well as exquisite scent. In Finland, for instance, this sort of bread is regarded as the most reliable source of dietary fiber. As a result, novel varieties of rye bread need to be created that integrate appealing sensory qualities with enhanced nutritive content and health advantages (Kołodziejczyk et al., 2020).

To bioprocess rye bran, Koistinen et al. combined fermented yeast with hydrolyzing enzymes (xylanase and ferulic acid esterase). Wheat bread was subsequently enhanced with the bran. The dried and fermented bran fractions were used by the researchers to substitute 35% of the wheat flour and the amount of rye bran was altered to produce bread with a comparable amount of overall fiber. According to their findings, the bioprocessing of rye bran allows the bran granules' cell walls to partially degrade, which releases phenolic compounds and raises the amount of liberated PA in the bread. Ferulic acid esterase enzyme activity is more certainly accountable for that since this enzyme liberates ferulic acid inside the bran matrix and raises the bread's availability (Dziki, 2022).

Considering the growing trend in munching, conventional bread and grain items offer a lot of potential for usage in providing customers with nutritious snack alternatives. Rye-based goods, for instance, could be made using extrusion techniques. Extruded rye bran snacks are said to expand better when the particle size of the bran is reduced to 440 → 28 μm, which might enhance the texture and flavor. The Nordic market has experienced the development of rye flour chips and baked sweet and savory snacks along with extruded goods, likely due to positive customer feedback (Jonsson et al., 2018).

The percentage of WSP (water-soluble):WIP (water-insoluble) and the amount of pentosans both affect the characteristics of flour and bread, according to research by Cyran and Cygankiewicz. According to the aforementioned investigation, an elevated WSP:WIP ratio is indicative of high-quality rye flour. Rye flour has a greater proportion of water-soluble pentosans, especially one having a significant overall pentosan concentration, and is preferred for baking bread. The qualities of dough related to pentosan content and the functioning of the enzymes that break down cell walls are ascertained using the swelling curve test. Considering the following criteria, the quality of rye bread was evaluated: yield, volume, crumb hardness and moisture content (Stępniewska et al., 2019). According to Stępniewska et al. (2018), the swelling curve test can indicate the standard of rye milling products. The research revealed that crumb firmness was adversely associated with an alteration in the consistency in the swelling curve test, but this modification had a positive correlation with crumb moisture. The proportion of insoluble as well as soluble pentosans has a substantial effect on the variances in the swelling curve (Oest et al., 2020).

Lactic acid bacteria ferment the sourdough required to bake primitive rye bread. Due to a decline in phytates caused by sourdough fermentation, the absorption of minerals is increased significantly. Additionally, rye bread made from sourdough has been demonstrated to have superior antioxidant qualities compared to wheat bread (Warechowska et al., 2019).

7.4 RYE-BASED PRODUCTS

7.4.1 Traditional Rye Products: Bread, Crackers and Beverages

Rye is famous for its name 'the poverty grain', as it has the power to grow in soil that is nutritionally poor for the growth of other grains. The grain of rye is locally applicable in the production of flat and fermented breads. The result of the breadcrumb is dark brown in appearance along with a sour taste. Its flour is utilized in the production of crackers, cookies, breads that are sour to taste and cereals for breakfast (Barak, 2018). Rye flour contains non-starch polysaccharides such as β-glucan, pentosans, lignin and a material called cellulose as well as arabinoxylans, which improves its water-absorbing characteristics (Wrigley, 2019).

Rye is the second most important ingredient after wheat to produce bakery goods and breads, as well as in cooking items such as porridge, cereals for breakfast, pasta varieties and snacks. People frequently utilize rye grain for the production of crisp bread (Knaeckebrot) that is manufactured without yeast. Due to the manufacturing method, the rye flour employed for this particular purpose requires low production of alpha-amylase. Lighter rye breads as well as bread rolls are created using grated rye and wheat germ. Rye flour is the primary ingredient in American rye bread (60%–80%), alongside wheat flour serving as an additional substitute. Rye, rather than fermented bread, is used to make a variety of biscuits along with crackers. Cereal is an excellent way to begin the day. Rye is frequently used within cereals in the form of breakfast and muesli snacks, including rye chips, extruded rye, along with mashed corn being healthier options (Ikram et al., 2023b).

Rye beverages are produced from rye grain, of which the majority market is alcoholic beverages. In alcoholic category rye beer and whiskey are the main contenders. Rye beer originated in south Germany and rye malt was utilized in its production, which is approximately 61%. It results in the best fermented beer with a pleasing taste (Barak, 2018). The processing as well as extracting of rye to produce beverages such as rye whiskey is a distinctive application of grain (Ikram et al., 2023b). Rye whiskey is the second alcoholic beverage that stems from Canadian roots but is also available in American flavors. It is usually aged for two years or more with at least 50% of rye in it. Rye whiskey aroma is pungent with a spicy and grainy aftertaste (Barak, 2018).

7.4.2 Modern Innovations in Rye-Product Development

Recent innovations in rye product development have expanded its applications beyond the traditional use, with greater focus on functional and health benefits as well as its sustainability. The advancement of novel rye-based products is executed through distinctive elements and goals. As customers evolved into more health-conscious beings, the need for more nutritious products with greater dietary fiber along with bioactive compound contents grew. As a result, more studies aim to develop novel fiber-rich rye milling products and bakery products. Heat treatment may be able to contribute toward optimizing baking reliability. Another reliable option is the addition of quality-enhancing enzymes in the processing of the dough of rye bread (Németh & Tömösközi, 2021).

Roasted rye is also considered a coffee replacement. Rye is becoming increasingly popular as a snack food ingredient. Rye products like grain rye-based oatmeal, soft bread, along with crisp bread help to improve fullness and reduce appetite. Rye milling goods also include whole rye grain and the flour, sifted rye flour, rye bran, rye flakes, even toasted rye flakes. New rye products include white rice blends and already-prepared rye grains combining wheat and rye, quick rye flakes paired alongside dried berries for breakfast, and snack items (El-Mahis et al., 2023).

Development of innovative, rich-in-fiber products of rye milling is the main objective as they promote gut health via prebiotic effect. Whereas rye bran fermentation elevates the level of active biological compound in the bread. Development of high protein and fiber snacks, including milk concentrate and whole-grain rye flour using 3D printing techniques based on extrusion method is an interesting but currently difficult approach for creating new rye-based items (Németh & Tömösközi, 2021).

7.4.3 Gluten-Free and Functional Foods from Rye

Developing gluten-free meals from rye is fundamentally problematic considering rye's gluten concentration. Rye includes secalin, a kind of gluten protein, which renders it inappropriate for persons suffering from celiac disorder or gluten intolerance. Rye, on the other hand, is an essential component in the creation of functional foods because of its curative properties. Rye is considered a low glycemic index food. Rye-based foods have exhibited a tendency to induce lower postprandial insulin responses, which are proven to be beneficial for blood sugar management. Rye bread has been demonstrated to cause a decreased spike in insulin through the postprandial phase compared to other whole-grain foods, while having no effect on the glucose response. This type of phenomenon is identified by the term 'rye factor' (Iversen, Jonsson, & Landberg, 2022).

Rye's fiber content (arabinoxylan, β-glucan, lignin, cellulose, fructan) promotes good weight management and lowers risk of diabetes, heart issues and cancer. Phytonutrients extracted from rye include phenolic acids as well as phytosterols, along with benzoxazinoids, which mediate a variety of beneficial effects on health. *S. cereale*'s chemical composition positions it as one of the most significant grain throughout European countries (El-Mahis et al., 2023).

The whole-grain rye cereal products typically contain significant levels of alkylresorcinols. Because they are only found in the outer sections of these cereals, they have been proposed to be utilized as particular dietary biomarkers for whole-grain rye consumption. Alkylresorcinols are ingested throughout the small intestine, carried by lipoproteins to be dispersed in red blood cell surfaces and adipose tissue, possibly even in other body compartments. Alkylresorcinols have been speculated to disrupt the micelle solubility of cholesterol within the digestive system, which might lower cholesterol absorption, an aspect that may contribute to the proposed cholesterol-lowering action of rye.

7.4.4 Consumer Preferences and Market Trends

Regardless of how nutritious rye is, it must be available, selected and consumed in order to have a positive impact in reality. As a result, recent research in Sweden has assessed patterns of consumption, health-related and sensory aspects and consumer evaluations of commercial rye breads. Consumers' perceptions of healthy breads largely matched what was truly healthy, although the healthiest loaves were not always selected. Efforts must be made to offer an extensive variety of nutritious breads, which cater to different categories of customers, assisted by more research into what people want and prefer to choose (Jonsson et al., 2018).

The synthesis of additional fiber in whole-grain flour has recently become a trend in rye milling and processing. Whole-grain flour and darker rye flours employing fine granularity are frequently made with non-traditional, specialized milling equipment. This change results in a reduced incidence of starch breakdown, a reduced glucose index, better access to rye fiber plus other useful compounds, as well as sufficient medium for the process of fermentation. Adopting appropriate processes for crushing and classifying mill products makes it easier to change their biochemical structure and functional properties. Rye bran is well-known for its implementation as a food ingredient as well as its numerous industrial applications. In comparison to other grains, it is exceptionally rich in lignans, phenolic acids, and fiber (Ikram et al., 2023b).

To reduce the ecological impact of worldwide food consumption, everyone worldwide is being encouraged to eat a diet that includes more plants, which presents a prospect for rye to expand into new product categories. Several food firms are already working to address the obstacle of boosting the consumption of nutritious rye foods. It is vital to grasp consumer preferences and integrate them with new processing methods and crop breeding techniques. Furthermore, for new market segments, novel ideas that involve fresh-packed as well as bake-off goods alongside unique approaches to combining rye with other ingredients should be used in product development (Jonsson et al., 2018).

7.5 TYPES OF BY-PRODUCT IN RYE PROCESSING

7.5.1 Rye Bran

A by-product of rye is rye bran, which is obtained during the production of refined rye flour, and it is the rye flour's wholemeal ingredient. Over the last decade the production of rye bran worldwide is approximately 16–17 million tons (Demirel et al., 2022). Rye bran is a valuable source of bioactive compound, and it is composed of dietary fiber, protein, lipids and starch. The richest source of rye is ferulic acid that is commonly used in the pharmaceutical, foods and cosmetics industries. It includes anti-inflammatory, anti-cancer and antioxidant properties (Drăgan et al., 2018). Rye bran and its by-products have been shown in Figure 7.3.

Rye bran has various applications in many industries. The production of microbial enzymes, cellulose, amylase, pectinase and xylanase, has the potential use of the substrates of rye bran. The growth of selected strains of microorganisms are very useful substrates of rye bran and the production of various enzymes that are used for the hydrolysis of different types of waste material to obtain bioethanol (Ahmad et al., 2021). Rye bran is suitable for the production of xylanase in the presence of a high content of antioxidants in rye bran as compared to the other substrates. In addition, the approach to obtain the industrially important enzymes is found to be more feasible and economical as compared to the other commercial production of enzymes.

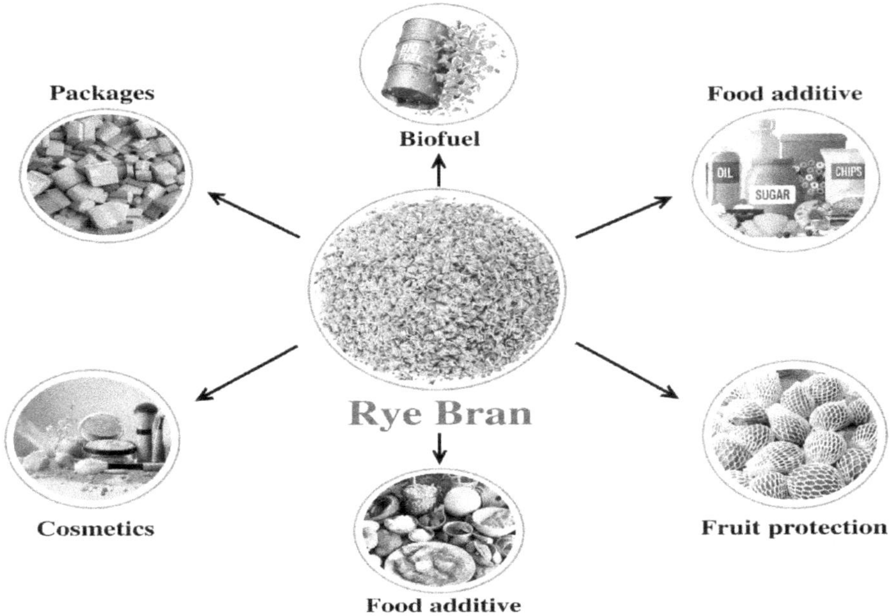

FIGURE 7.3 Rye bran and its by-products.

The hydrolysis of lignocellulosic waste material is a very slow process and it is a more economical process because no additional chemicals are needed (Mihajlovski et al., 2021).

7.5.2 Rye Straw

Rye straw and its components are shown in Figure 7.4.

The second generation of feedstocks used for bioethanol production are lignocelluloses containing raw material like straw from rye, and the cellulose in lignocelluloses is not accessible for enzymes (Talebnia et al., 2010). By the specific pretreatment, lignin and hemicelluloses are removed to make enzymatic degradation of cellulose possible. The sugar is fermented with the yeast to ethanol. The solution can be distilled and rectified and used as a blend with the gasoline to make E5, E10 and E85 (Baeyens et al., 2015).

There are three different types for the classification of pretreatment methods: thermophysical, acid-base and alkaline methods. Thermophysical methods such as pretreatment steam, steam hydrothermolysis open the structure and make most of cellulose accessible to the enzyme and solubilize hemicelluloses (Eisenhuber et al., 2013). Among all of the pretreatment methods, steam hydrothermolysis seems to be the state of technology of art for the commercial pilot and for the bioethanol demonstration plant for the second generation. In this unknown weather, the hydrothermolysis is needed in the pretreatment methods. For the necessity of pressure release, there are few analyses that compare and examine the pretreatment of

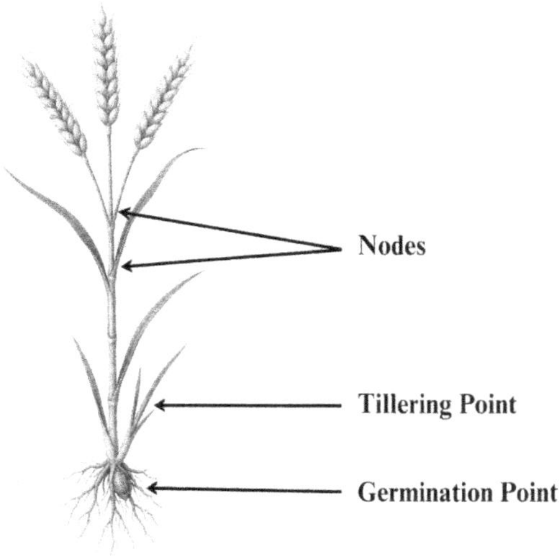

FIGURE 7.4 Rye straw and its components.

rye straw by means of steam treatment and the steam hydrothermolysis method (Lajvardi et al., 2010).

Rye straw is pretreated in the two different ways. In both ways, the rye straw is heated up to 160–200°C and then incubated for 10–20 minutes. Then the pressure is released suddenly and the temperature is decreased slowly. Different pretreatment rye straws are dry, and they are analyzed by scanning electron microscopy and fermented to determine the concentration of glucose and xylose, in addition to examine the lignocellulosic residuals for the production of bioethanol like straw from rye. The yield of bioethanol of plant is determined by the standard experimental design of steam explosion, enzyme hydrolysis and fermentation with yeast. Based on the quantity of feedstock available, we can calculate the total production of bioethanol capacity.

7.5.3 Rye Germ

The rye germ is a potential candidate of raw material for healthy and functional foods. The potential use is to add the variety of rye germ in foods that are available for human consumption and increase the level of dietary fiber and bioactive compounds (Kamal-Eldin et al., 2008). Rye germ is the major constituent that contains enzyme especially starch-degrading amylases that attack to play a key role in the baking quality of flour. The amylase activity is measured by the falling number (FN) method. The rye germ is sound and not damaged. The sprout has an FN in the range of 150 s and in the dry years it is FN 30 s and above (Boskov Hansen et al., 2002). The minimum requirement of FN for intervention of rye germ in the EU is 100 s. FN is too low in pasty and unacceptable bread. Low FN in rye germ problems can be reduced due to the breeding of new cultivars with better resistance to sprouting pre harvest (Boskov Hansen et al., 2002).

TABLE 7.3
Nutritional and Functional Properties of By-Products

Rye By-Products	Nutritional Properties	Functional Properties
Rye Bran	Rye bran is proven to be an excellent substrate of fungal solid-state fermentation by the synthesis of polyunsaturated fatty acids and the β-carotene (Čertík et al., 2013).	Phenolic compounds have the antioxidant activity in rye bran that neutralizes free radicals and protects against chronic disease and oxidative stress (Mishra et al., 2017).
Rye Straw	Polyphenols in rye straw are helpful in the managing of various physiological disorders like diabetes mellitus, hypertension and hypercholesterolemia; they prevent the oxidation of low-density lipoproteins (LDLs) and are also helpful for the improvement of gastrointestinal health (Luithui et al., 2019).	High cellulose content in rye straw is the renewable energy source that enables its conversion into biofuels like bioethanol, biogas and biomass. Lignocellulosic compositions are helpful for the fermentation processes and energy generation (Torres Rivera, 2022).
Rye Germ	Phytosterols in rye germ are useful in lowering the cholesterol level. Lignans in plant-derived compounds with antioxidant properties are potentially reducing the risk of cancer (Öncü Öner, 2023).	Anti-inflammatory properties in bioactive compounds include phenolic acids that are helpful in reducing inflammation associated with chronic diseases like arthritis and metabolic disorders (Dynkowska, 2019).

The main chemical composition of rye germ is starch, protein, mineral matter and dietary fiber. The starch content is limited due to the endosperm, and contents of dry matter between 57.1 and 65.6 g/100 g are reported in rye germ (Coda et al., 2015). The dietary fiber components are found in the cell wall of the kernel, and the total contents are reported in the rye germ are between 14.7 and 20.9 g/100 g of dry matter. Approximately, 25% of dietary fiber components are water extractable (Hansen et al., 2004).

Rye is a major food source in the human diet and generates a large quantity of by-products throughout the processing chain. Rye is the staple food and the major nutritional source throughout the world. By-products of rye obtained from milling contain multiple health benefits that are rich in components like minerals, dietary fibers, vitamins, phytosterols and polyphenols (Luithui et al., 2019). The important role of rye and particularly rye by-products of dietary fiber in human nutrition can enhance health and lifestyle that are related to non-communicable diseases and prevention in a large number of laboratory, epidemiological and clinical studies (Kołodziejczyk et al., 2020). Table 7.3 shows nutritional and functional properties of by-products.

7.6 FOOD INDUSTRY APPLICATIONS

Rye by-products have been used in crop rotation. When rye straw reaches maturity, it is gathered and useful for animal bedding, particularly for horses as well as hay, and feed for ruminants (Dziki, 2022). Research is being done on the utilization of straw to produce biofuel. Chemical industries have the opportunity to convert straw pentosane to xylose. Rye straw has been used for building materials. The rye bran can be used in a variety of ways. The greatest value addition is turning bran into flour for baking to produce a variety of foods (Valli, 2024).

7.6.1 Fiber-Enriched Baked Goods

For all the cereals including the rye bran, using whole-grain flour for baking is recommended due to the fact that rye bran is abundant in fiber, vitamins, minerals and bioactive phytochemical (Korzun et al., 2021). When compared to wheat rye, bran has a clear nutritional benefit due to its high fiber content. The pentosane concentration of rye bran is frequently double that of an identical wheat bran–based diet (Németh & Tömösközi, 2021).

The author from Dziki (2022) substituted dried and fermented bran fraction for 35% of the wheat flour and modified the amount of rye bran so that the bread had a comparable amount of fiber. Ikram et al. (2023a) discovered that the bioprocessing of rye bran can cause the bran particle cell wall become partially degraded, releasing phenolic chemicals and raising the bread's free PA content. This has most likely been caused by the ferulic acid esterase enzyme activity that boosts bread's bioaccessibility by releasing ferulic acid from the bran matrix. Because bread weakens the starch gluten matrix, it affects bread qualities negatively even though it is a rich source of fiber and numerous beneficial substances (Wrigley, 2019).

7.6.2 Natural Food Additives

In order to increase the textural qualities and stabilize the structure of rye bran wheat–based bread, a food additive based on iron oxide nanoparticles can be added; its composition contains significant nutrients that enable it to improve the nutritional profile of rye bran wheat bread (Tsykhanovska et al., 2024).

Amanov et al. (2021) added high fiber extruders to fermented and acidified rye bran. Rye bran structure and texture were enhanced by both acidification and fermentation by itself and had a less noticeable impact. The research found that when lactic acid bacteria are present during the fermentation of rye bran, pH is an important factor in the extrusion process. Better-quality extruders have been produced as a result, and the starch's ability to bind water has enhanced. While making corn snack, Auerbach et al. (2020) have substituted rye bran for corn grits by 21%–32%, respectively. When substitution limited to 21%, the resulting product displayed good sensory qualities and an enhanced level of dietary fiber (Tsykhanovska et al., 2024).

The usage of collagen and rye bran together has been shown to improve the taste of the meat. Recently, a water-extractable AX fraction that has separated from rye bran is used as an innovative, bioactive carrier material for microencapsulated honey. Because it improves the stability and bioavailability of phytochemical during digestion, microencapsulation has gained a lot of popularity recently, in comparison to the antioxidant activity of microcapsule (Amanov et al., 2021).

7.6.3 Animal Feed Applications

Rye straw is an important feed for animals in the world. In contrast to food consumption, it has produced in areas where other grain cannot produce high yield, which contribute to its purpose. Rye by-product have therefore suppressed the demand of the regional milling and baking sectors necessitating the utilization of the apparent substitute, namely feed applications (Dziki, 2022). Even in Germany a traditional producer of rye bran bread over two-thirds of the by-product is used for animal feed, usually in combination with other feed ingredients (Jansson et al., 2021).

A study conducted by Khiaosa-ard et al. (2024) concluded that a potential new supply of roughage for the equine industry and as well as potential partial substitute for open-field burning are provided by diets that include ryegrass straw. The storage, handling, transportation, bulkiness and expense of traditional baled hay and grain-feeding methods for horses may be addressed for urban horse owners by using densified physical forms of straw (Karasu et al., 2023). Ryegrass straw combined with extra protein and other vital components gives the frugal ration maker countless options (Maguire, 2024). Additionally, there are opportunities to combine entire corn plant pellets with straw and extra protein, vitamins and minerals. More investigation is required to determine how to include these subpar feed stuff into realistic light horse diet, which assists a severe solid waste disposal issue (Khiaosa-ard et al., 2024).

7.6.4 Non-Food Industrial Applications

7.6.4.1 Biofuel Production

A combination of rye bran and sawdust was utilized by some authors in Herbstritt et al. (2022) to create fuel pellets. The result demonstrated that, in comparison to sawdust-derived granules, rye bran functions as a superior binding medium and yields granules of superior quality. Additionally Drzymała et al. (2020) discovered that the rye bran–prepared granules helped to increase the calorific value of the pellets, which leads to a low amount of ash following burning. As the amount of rye bran increased, the binding capacity improved, and the granulator power consumption during the pelleting process dropped. Rye bran and powdered palm stones were combined by Magdziarz et al. (2024) to create pallet and that wood has been replaced with biofuel.

The study conducted by Edelmann et al. (2024) describes that advanced biofuel, which has produced from the biomass of inedible plant, offers an intriguing substitute for fossil fuel and first-generation biofuel while also dramatically lowering CO_2 emission (Mihajlovski et al., 2021). The ideal circumstances of rye straw pretreatment and enzymatic hydrolysis have been identified in this research. Rye straw–processed parameters produced the maximum output of reducing sugar. Rye straw processed with ionic liquid under ideal circumstances and then exposed to enzymatic hydrolysis produced a decreasing sugar (Herbstritt et al., 2022). Compared to the straw sample that have not pretreated with ionic liquid, the reduced yield of rye straw that was pretreated increased twisted. This study finding offers important information for upcoming investigation into the efficiency of ethanol production from lignocellulose biomass (Magdziarz et al., 2024).

7.6.4.2 Packaging Material

A study performed by Kärkönen et al. (2022) concluded that food shelf life has increased by adding active ingredients to starch films that have antibacterial or antifungal properties. One of the primary reasons why food quality deteriorates is lipid oxidation. Chicken breast is an example (Go & Song, 2019). Polyunsaturated fatty acid-containing lipids are also an example. Color, taste and texture all get affected when these lipids have oxidized (Fallah et al., 2023). Antioxidant films have been used to postpone lipid oxidation in food packaging. Since natural antioxidants are not harmful to health, they can be used in food packaging film rather than manufactured one. Rosehip, which is higher in antioxidants such as ascorbic acid and carotenoids, extends the shelf life of chicken breast. Creating a biodegradable film with antioxidant properties to replace synthetic plastic was the goal of the project by Qian et al. (2021). In order to prevent lipid oxidation and to extend the shelf life of chicken breast, in this research, an antioxidant rye-starch film containing rosehip extract has been created. The study additionally evaluated the new film worth and described the created film while chicken breast were being stored (Aslaner et al., 2021).

Rye is a cereal grain that comes from the seed of ryegrass. It was one of the first grains to be cultivated 13,000 years ago but was not widely used until the Middle Ages in Europe. Although rye grains look similar to wheat, they differ slightly. Rye is lower in calories, protein and fat than wheat and contains more fiber. When ground into flour, rye is darker in color and rich in flavor. Rye contains less gluten than wheat, so rye-baked goods are denser. Rye can taste slightly sour and is often mixed with wheat flour for a less dense, neutral-tasting product (Kaur et al., 2021b).

7.7 HEALTH BENEFITS OF RYE AND ITS BY-PRODUCTS

Rye-based foods have demonstrated superior benefits on insulin metabolism compared to wheat bread under isocaloric and standardized carbohydrate conditions, indicating potential for diabetes prevention. These effects are partially attributed to modulations in blood glucose flux. Additionally, rye consumption has been associated with enhanced satiety, suggesting a role in weight management. Emerging evidence also supports its positive impact on inflammation and blood lipid profiles. A key challenge remains: improving the palatability and consumer acceptance of rye products, underscoring the need for innovative product development and effective communication strategies to promote their consumption (Jonsson et al., 2018).

7.7.1 Impact on Digestive Health

Rye has long been recognized for its beneficial effects on digestive health, largely due to its high and diverse dietary fiber content (Ghazvini et al., 2025). Unlike many refined grains, rye contains a substantial amount of both soluble fibers, such as arabinoxylans, and insoluble fibers that work synergistically to promote healthy digestion (Feng, 2019). The soluble fibers in rye increase the viscosity of the intestinal contents, which slows down digestion and nutrient absorption, allowing for a more gradual release of energy and better blood sugar regulation (Pržulj et al., 2024). Insoluble fibers add bulk to stool and enhance intestinal motility, thereby preventing constipation and supporting regular bowel movements (Carter, 2025). Beyond mechanical effects, rye's fiber serves as an important food source for the gut microbiota (Kim et al., 2025). Through fermentation in the colon, these fibers are converted into short-chain fatty acids (SCFAs) like butyrate, acetate and propionate, which play critical roles in maintaining gut barrier integrity, reducing inflammation, and providing energy to colonocytes (cells lining the colon) (Faubel et al., 2025). Butyrate, in particular, has been extensively studied for its protective effects against colorectal cancer and its ability to modulate immune responses within the gut (Dutta et al., 2025). Additionally, a well-balanced gut microbiome fostered by rye consumption is linked to improved overall digestive function (Yu et al., 2025) and may contribute to systemic health benefits, including enhanced immune regulation and reduced risk of inflammatory bowel diseases (Almonajjed et al., 2025). Taken together, these factors highlight how incorporating rye into the diet can support not only regular digestive function but also long-term gastrointestinal health and disease prevention (Yılmaz et al., 2025).

7.7.2 Role in Blood Sugar Management and Satiety

Rye-based foods have a significant impact on postprandial glycemic regulation and appetite control, largely attributed to their distinct structural and nutritional properties (Åberg, 2024): the high concentration of viscous soluble fibers, particularly arabinoxylans, which increase the viscosity of intestinal contents and slow the enzymatic digestion and absorption of carbohydrates in the small intestine (Lu et al., 2024). This delayed glucose release results in a lower glycemic response and attenuated insulin secretion, contributing to improved glycemic control (Ying et al., 2024). Additionally, rye's ability to preserve the integrity of grain particles during processing further slows starch accessibility and enhances metabolic benefits (El-Mahis et al., 2023). Beyond its influence on glucose kinetics, rye has been shown to modulate the secretion of appetite-regulating gut hormones (Åberg, 2024) such as glucagon-like peptide and peptide, which are involved in promoting satiety by slowing gastric emptying and sending satiety signals to the brain (Chusak et al., 2021). Another contributing factor is the fermentation of rye-derived dietary fibers in the colon by gut microbiota, which produces SCFAs like acetate, propionate and butyrate. These SCFAs not only play a role in maintaining gut health but also influence glucose metabolism and appetite regulation through gut–brain and endocrine signaling pathways (Åberg, 2024).

7.7.3 Antioxidant Properties and Chronic Disease Potential

Rye is valued not only for its fiber and glycemic benefits but also for its impressive antioxidant properties (Kaur et al., 2021b). It contains a range of bioactive compounds, including phenolic acids like ferulic and caffeic acid, as well as lignans (Benito et al., 2025). These antioxidants help neutralize free radicals, thereby reducing oxidative stress (Din et al., 2023). The lignans in rye are also converted into enterolignans (Dziki, 2022), which have been linked to protective effects against hormone-related cancers and inflammation (Plaha et al., 2022). Taken together, the antioxidant-rich profile of rye supports its role as a functional grain that may contribute to long-term health and chronic disease prevention when consumed as part of a balanced diet (El-Mahis et al., 2023).

7.7.4 Potential Allergenic Concerns Risks

The most significant anti-nutrient is phytic acid, which is present in most grains and potent enough to compound multi-charged metal ions, particularly Zn, Ca and Fe, making them inaccessible for human usage (Pandey et al., 2023). Phytase enzymes break down phytate salt during germination, supplying phosphate for the growing seedling. Due to its potent capacity to combine highly charged metal ions (Ikram et al., 2023b), phytic acid has long been regarded as an anti-nutrient; as a result, eating a lot of food with a high phytic acid content may result in a deficiency in the absorption of several dietary minerals (Li et al., 2024). Though, by limiting the use of phytic acid-rich grain products, the harmful effects of phytic acid could be avoided (Li et al., 2024). In particular, the fermentation of germinated rye, which also led to a lower pH when compared to native rye flour, increased folates, free phenolic acids, lignans, total phenolic compounds and alkylresorcinols. The folate level increased up to seven times and that of free phenolic acids by up to ten times. The bioactive potential of wholemeal rye can thus be further increased through fermentation (Ikram et al., 2023a). Rye contains specific storage proteins, such as secalins and certain gliadin-like compounds, which can provoke immune responses in susceptible individuals (Wang et al., 2024). In rare cases, rye may cause true allergic reactions, including respiratory issues or, in severe instances, anaphylaxis (Goodman et al., 2025). Furthermore, prolonged exposure to rye flour dust, particularly in occupational settings like bakeries, can lead to allergic airway conditions known as baker's asthma (Cullinan et al., 2021). While rye offers valuable nutritional properties, its consumption should be carefully managed in individuals with gluten intolerance or grain allergies (Parrinello et al., 2025).

7.8 SUSTAINABILITY IN RYE PRODUCTION AND BY-PRODUCT UTILIZATION

7.8.1 Environmental Impacts of Rye Cultivation

Rye is a versatile cereal crop cultivated in temperate regions and is known for its resilience and adaptability to adverse growing conditions. However, as global temperatures and atmospheric CO_2 concentrations raise, the effects of climate change on rye growth, yield and grain quality become increasingly apparent (Ghafoor, Karim et al., 2024). Rye cultivation presents a relatively low environmental footprint compared to other cereals (Alijošius et al., 2016). Rye's drought tolerance and low water requirements make it ideal for regions with limited rainfall (Studnicki et al., 2021). Additionally, rye can grow on marginal lands, reducing the need for land conversion and helping preserve natural ecosystems (Purohit et al., 2025).

7.8.2 Circular Economy in Rye Processing

This approach emphasizes the proficient utilization of all biomass components, transforming by-products (Kumari et al., 2025), transformation of low value uses into high-value functional ingredients (El-Mahis et al., 2023). Residual materials from rye-based ethanol and beverage production, such as stillage, can be repurposed as nutrient-rich animal feed or as substrates for biogas generation, contributing to renewable energy development (Swamy et al., 2025). Rye straw, often considered agricultural waste, can be diverted into sustainable applications such as organic mulch, soil enhancers or processed into fiberboard and biodegradable composites (Mirski et al., 2021). Agronomically, rye plays a pivotal role in sustainable crop rotation systems by enhancing soil structure, suppressing weeds and reducing the need for chemical inputs (Ghafoor, Karim et al., 2024). The adoption of an integrated biorefinery model enables the fractionation of rye into food-grade, feed-grade and industrial-grade components, supporting the development of a closed-loop production system (Herbstritt et al., 2022). Overall, the circular economy framework in rye processing not only enhances environmental sustainability but also creates new economic opportunities across the agro-food and bio-industrial sectors (Korzun et al., 2021).

7.8.3 Waste Minimization Strategies

Waste minimization in rye cultivation and processing is a critical component of sustainable agri-food systems, and it can be effectively addressed through a combination of agronomic, technological and industrial strategies. At the production level, the adoption of precision agriculture techniques and optimized harvesting schedules reduces grain losses and ensures maximum yield efficiency (Chudzik et al., 2025). In milling and food processing, whole-grain utilization and minimal refining help retain the nutritional value of the rye kernel while reducing by-product generation (Chiş & Fărcaş, 2025). Residual materials such as bran, husk and screenings can be valorized into high-value applications, including dietary fiber supplements, animal feed and substrates for fermentation or mushroom cultivation (Danciu et al., 2023). In distillery operations, the nutrient-rich stillage can be transformed into biogas through anaerobic digestion (Segura et al., 2025). Rye straw, a typically underutilized biomass, can be converted into mulch, bio-char, paper pulp or biodegradable packaging materials, supporting zero-waste objectives (Nadiya et al., 2025).

7.9 CHALLENGES AND FUTURE PERSPECTIVES

7.9.1 Issues in Rye Cultivation and Supply Chain

Rye is relatively strong but still prone to intense weather conditions, such as heavy rainfall, prolonged droughts with changing climate patterns. Rye is hardy and grows well in poor soils, extremely fertile but long-time wet soils can lead to lodging, which reduces yield quality (Ghafoor, Karim et al., 2024).

Rye is susceptible to fungal diseases like ergot, which is caused by *Claviceps purpurea*; it can contaminate grain and cause serious health problems. Pests including aphids and nematodes can reduce yield by damaging crops. Rye yield is lower than wheat or barley, which makes it less economical and unattractive to farmers in some areas (Сафонова et al., 2019). Rye is harvested one time in a year; if harvesting is poor or if demands increase, then it leads to potential supply shortages. If rye is not properly dried and stored especially in humid conditions, then rye grains become prone to spoilage (Geiger & Miedaner, 2009).

Transportation of rye should be careful, ensuring the grain's quality and preventing the contamination. Fluctuations in price ranges and market demands of rye can discourage farmers from planting it, which impacts the long-term supply (Saitov et al., 2020). Wheat and corn are more profitable crops than rye; that's why farmers may prefer to grow beneficial crops, which causes a decrease in the production of rye and rye supply. Quality should be ensured while growing rye; hence, it becomes a challenge across the supply chain, especially when dealing with contamination and fungal infections (Geiger & Miedaner, 2009).

7.9.2 Technological Barriers in By-product Utilization

The by-products of rye include bran, husks, straw and residues with significant potential for high-value applications in food, feed and bioproducts. However, technological barriers often cause hinder in their efficient utilization (Ikram et al., 2023b).

7.9.3 Underdeveloped Processing Technologies

Technologies for processing rye by-products into high-value products are dietary-fibers, bioethanol or bioplastics are often under-developed or inefficient. Many by-products such as rye straw need pretreatment technique to break down lignocellulose structures for further processing. There should be an increase in research and development of enzymatic or chemical pretreatment technologies. Investment should be uncreased in research of biorefinery models that contain multiple processes, for example, the extraction of bioactive compounds and biofuel production (Ran et al., 2019).

7.9.4 Challenges in Extracting Bioactive Compounds

Bioactive compounds which are valuable are present in rye bran and husks such as arabinoxylans, phenolic acids and antioxidants. However, efficient extraction of these compounds at a commercial level is complex. Advanced extraction technologies such as supercritical fluid-extraction or ultrasound-assisted extraction should be used to enhance yield and purity (Ikram et al., 2023b).

7.9.5 Improper Infrastructure for Processing

Lack of awareness and local processing abilities for rye by-products often cause underutilization or wastage. High value and increase costs of setting up special equipment for processing rye by-products disturb investments. Limited partnership between industries (e.g., food, feed and bioenergy sectors) restricts the production of integrated utilization strategies. Rye by-products are used for food, and feed must follow strict quality standards. Contamination with toxins like ergot alkaloids cause decrease in their efficiency. The development of high detoxification methods can ensure safety such as thermal treatments or microbial degradation. There should be regulatory guidelines for innovative uses of rye by-products (Ran et al., 2019).

7.9.6 Opportunities for Innovation and Research

Rye genomes can be edited using CRISPR/Cas9 technology, to enhance their ability to resist biotic stresses, such as pests and pathogens (Ahmar et al., 2023). Also, combining the advantages of traditional breeding together with CRISPR/Cas9 methodologies could potentially expedite the production of crops that are high yielding which will enhance food security (Abdullah et al., 2023). The concentration of bioactive compounds in rye like essential fatty acids and antioxidants that are of benefit for human health can be increased through genetic engineering. In particular, genome editing approaches have great potential toward increasing yield by producing rye varieties that are more stress-tolerant and optimal for yielding in stressful conditions (Tang et al., 2023). Rye varieties that are biofortified could fulfill the growing appeal for nutraceuticals and food, hence encouraging improved health parameters (Kiran et al. 2024).

7.9.7 Improving By-product Processing Techniques

Bioactive compounds like bioactive peptides or polyphenolic compounds made by enzymatic bioprocesses can be used to enrich the food items with health benefits (Chourasia et al., 2020). Techniques like hydrothermal and enzymatic extraction produce functional ingredients from by-products of rye that have antioxidant properties (Bityukova et al., 2019). The nutritional composition of the by-products of rye can be improved by the process of fermentation, supporting the growth of good microbes and enhancing nutrient bioavailability (Saleh et al., 2019). There are many hydrolysis processes for the dissolution of lignocellulosic biomass, allowing rye straw to be economically viable for the production of biofuel. Eco-friendly extraction processes can be developed through the adoption of green technologies in the processing promoting waste reduction and sustainability (Fierascu et al., 2021).

7.9.8 Sustainability and Environmental Impact

Making good use of rye by-products could lead to the reduction of the cost and creation of new job opportunities while tackling the environmental concerns regarding the disposal of waste (Ajayi et al.,2024). However, the biorefinery concept offers an opportunity for extracting important components from the by-products of rye like soluble dietary fibers and proteins, which can be used in different food applications (Galanakis, 2022). Valuing the by-products of rye can decrease the dependance on reliance on virgin resources of raw materials, hence aiding in reducing the environmental impact and supporting economic stability in rural groups (Danciu et al.,

2023). Product innovation using waste can meet the consumer's demand for healthy, environmentally friendly foods as it helps to enhance the food system's sustainability (Vardakas et al., 2024).

7.9.9 Expansion of Rye's Role in Functional Foods

Among the commonly consumed cereals, rye has the highest fiber content, which promotes the health of the digestive tract and the likelihood of preventing obesity (Iversen et al., 2022). Additionally, it actually contains bioactive compounds, like flavonoids and pentosans that lower inflammation and have favorable metabolic responses. Hydrolyzed rye protein has antioxidant, antihypertensive and anti-diabetic activities, thus demonstrating its multiple health benefits (Islas-Martínez et al., 2023). An increasing trend has been observed in the adoption of rye in breakfast cereals, snack foods and convenience foods, depicting the shift toward healthier option. Although global rye production has reduced, there seems to be growing interest in its health advantages, which indicate a possibility of revived demand (Németh & Tömösközi, 2021). Current breeding strategies have concentrated on improving traits associated with quality, which will also enhance the usage of rye in a wider array of food industries (Brzozowski et al., 2023).

7.9.10 Future Trends in Rye and Its By-Products

Rye is able to adapt to most soil types especially less fertile soils, and therefore, it can be grown in a variety of farming systems. This crop seems to have good ability to adapt to climate change which is of high importance given the growing global climatic alterations. The development of hybrid rye cultivars has enhanced the yield and stress resistance, which further supports the case for sustainable agriculture (Korzun et al., 2021). Rye, as a winter cover crop, has been shown to be effective for weed suppression, thereby reducing reliance on herbicides. A number of studies highlight the reduction of weed biomass when rye is used as a mulch material. The high biomass production of rye enables nutrient cycling and improves soil structures, which leads to better soil health (Boselli et al., 2021). In addition, rye grain can serve as a source of animal feed and bioenergy, which are new sources of revenue as well as the solution of green gas emission (Herbstritt et al., 2022).

Biopolymers can be manufactured by using rye, which is important for environment-friendly packaging solutions. Agricultural by-products such as rye straw can be converted into valuable products like bioplastics, which are better than conventional plastics and cut down the over-dependence on fossil fuels, which is detrimental to the environment (Patel & Blumberga, 2024). Biopolymers coming from rye are not only biodegradable but also produced from renewable resources, making them a better sustainable substitute than petroleum-based plastics (Sinha, 2024). These biopolymers might be modified for strengthening characteristics of the barriers to moisture and oxygen, the key ingredients needed in maintaining the quality of food products (Gigante et al., 2023). There remain numerous avenues of research to cover to enhance the properties of biopolymers obtained from rye or seek new uses in food, cosmetics and other sectors (Perera et al., 2023).

Rye is known for its high concentration of dietary fibers, which is greater than that of other cereals (Dziki, 2022). It possesses bioactive compounds, which have several health benefits, including better digestion and lower risk of chronic diseases. The trend of rye-fortified convenience foods is growing, presumably because they meet the needs of contemporary consumers (Kamal-Eldin et al., 2008). Traditionally, rye is mainly used in bread production, but now its range of uses encompasses breakfast cereals, noodles and snack products. The development of wholemeal and dark rye flours is gaining traction, enhancing the nutritional profile of various food items (Ikram et al., 2023a).

Systems driven by AI utilize remote sensing and various data sources to continually observe the status of soil and plants, enabling interventions when needed. With precision agriculture technologies like GPS and UAVs, field mapping can be done in great detail, which enhances the application of resources while minimizing wastage. Artificial intelligence helps in early pest and disease detection by ensuring timely and appropriate intervention, hence lesser dependence on pesticides (Jeyalakshmi et al., 2024). Machine learning models can also predict the outbreak of diseases and other harmful events, thereby allowing preparatory measures to be put in place (Prabha et al., 2024). AI models work on lots of data to make predictions on the crop yield and the right time to plant; this ensures optimal use of key resources. Efficiency in labor, mainly for sowing and harvesting, can also be improved by the automation of AI-powered robotics applications, which would ultimately improve production (Polwaththa et al., 2024).

7.10 CONCLUSION

In advancing sustainable food systems, rye plays a crucial role due to its exceptional adaptability to diverse growing environments, resistance to climate variability and capacity to enhance soil health. To maximize its potential, the food industry should explore broader applications of rye, including its use in bio-based materials, animal feed and processed food products, while also improving processing methods to boost its sensory properties and nutritional value, thus increasing consumer acceptance. For optimizing rye cultivation techniques, developing high-yielding, disease-resistant varieties, and investigating its potential for climate resilience, collaborative efforts between farmers, researchers and industry stakeholders are essential to further integrate rye into global food systems, enhancing food security, promoting biodiversity and advancing environmental sustainability through more sustainable agricultural practices.

REFERENCES

Abdullah, M. A., Fatima, S., Khaliq, F., Mehmood, M., Yousaf, T., Mustafa, M. B., … Hassan, A. (2023). Using CRISPR/Cas9 technology to improve crops and address the global food crisis: A review. *Asian Journal of Research in Crop Science*, *8*(4), 454–464.

Åberg, S. (2024). *Metabolic effects of whole grains: Emphasis on glycemic control, appetite, and body weight*. Chalmers Tekniska Hogskola.

Ahmad, R., Srivastava, S., Ghosh, S., Khare, S. K. J. C., & Biointerfaces, S. B. (2021). Phytochemical delivery through nanocarriers: A review. *Biointerfaces*, *197*, 111389.

Ahmar, S., Hensel, G., & Gruszka, D. (2023). CRISPR/Cas9-mediated genome editing techniques and new breeding strategies in cereals–current status, improvements, and perspectives. *Biotechnology Advances*, *70*, 108248.

Ajayi, O. O., Toromade, A. S., & Olagoke, A. (2024). Circular agro-economies (CAE): Reducing waste and increasing profitability in agriculture. *International Journal of Advanced Economics*, *6*(11), 598–611.

Alijošius, S., Švirmickas, G. J., Bliznikas, S., Gružauskas, R., Šašytė, V., RACEVIČIŪTĖ-STUPELIENĖ, A., … Daukšienė, A. (2016). Grain chemical composition of different varieties of winter cereals. *Zemdirbyste-Agriculture*, *103*(3).

Almonajjed, M. B., Wardeh, M., Atlagh, A., Ismaiel, A., Popa, S.-L., Rusu, F., & Dumitrascu, D. L. (2025). Impact of microbiota on irritable bowel syndrome pathogenesis and management: A narrative review. *Medicina*, *61*(1), 109.

Amanov, B. N., Rakhmonov, K. S., Isabaev, I. B., Atamuratova, T. I., Oltiyev, A. T., & Nikolaevna, M. E. (2021). Applications of natural additives-acidiments and probiotic bacterial preparations for prevention of rye bread cretal disease. *Natural Volatiles & Essential Oils*, *8*(4), 5976–5988.

Aslaner, G., Sumnu, G., & Sahin, S. (2021). Encapsulation of grape seed extract in rye flour and whey protein–based electrospun nanofibers. *Food and Bioprocess Technology*, *14*(6), 1118–1131.

Auerbach, H., Theobald, P., Kroschewski, B., & Weiss, K. (2020). Effects of various additives on fermentation, aerobic stability and volatile organic compounds in whole-crop rye silage. *Agronomy*, *10*(12), 1873.

Baeyens, J., Kang, Q., Appels, L., Dewil, R., Lv, Y., & Tan, T. (2015). Challenges and opportunities in improving the production of bio-ethanol. Progress in Energy and Combustion Science, *47*, 60–88.

Bağcı, A., Gecgel, Ü., Dursun, N., Özcan, M. M., Tamkoç, A., Özer, İ., & Özcan, M. M. (2019). The oil yields, mineral contents and fatty acid compositions of some rye (*Secale cereale*) grains. *Iranian Journal of Chemistry and Chemical Engineering (IJCCE)*, *38*(5), 285–292.

Banu, I., Neagu, C., & Aprodu, I. (2021). Effect of milling and sourdough fermentation on bioactive compounds of selected multigrain blends. *Romanian Biotechnological Letters*, *26*(6), 3022–3030.

Banu, Z., Somraj, B., Geetha, A., Khan, A. A., Srinu, B., Vamshi, J., … Saidaiah, P. (2024). A comprehensive review of biochemical, functional, and dietary implications of gluten. *Annals of Phytomedicine*, *13*(1), 196–208.

Barak, S. (2018). Cereal based beverages. In D. Mudgil & S. Barak (Eds.), *Beverages: Processing and technology* (pp. 73–90). Scientific Publishers.

Benito, J., Marques, G., Barro, F., Gutiérrez, A., Del Río, J. C., & Rencoret, J. (2025). Comprehensive study of lipophilic compounds from various cereal straws (wheat, triticale, rye, and tritordeum)—A promising source of valuable phytochemicals. *Journal of Agricultural and Food Chemistry*, *73*(12), 7282–7297.

Biel, W., Kazimierska, K., & Bashutska, U. (2020). Nutritional value of wheat, triticale, barley and oat grains. *Acta Scientiarum Polonorum Zootechnica*, *19*(2), 19–28.

Bityukova, A., Amelkina, A., Evteev, A., & Bannikova, A. (2019). *New biotechnology for the production of phytocompounds from secondary products of grain processing.*

Boselli, R., Anders, N., Fiorini, A., Ganimede, C., Faccini, N., Marocco, A., … Tabaglio, V. (2021). Improving weed control in sustainable agro-ecosystems: Role of cultivar and termination timing of rye cover crop. Italian Journal of Agronomy, *16*(4).

Boskov Hansen, H., Andreasen, M., Nielsen, M., Larsen, L., Knudsen, B. K., Meyer, A., … Technology. (2002). Changes in dietary fibre, phenolic acids and activity of endogenous enzymes during rye bread-making. *Food Research and Technology*, *214*, 33–42.

Brzozowski, L. J., Szuleta, E., Phillips, T. D., Van Sanford, D. A., & Clark, A. J. (2023). Breeding cereal rye (*Secale cereale*) for quality traits. *Crop Science*, *63*(4), 1964–1987.
Сафонова, И., Аниськов, Н., & Кобылянский, В. (2019). База данных генетических ресурсов коллекции озимой ржи ВИР как средство классификации генетического разнообразия, анализа истории коллекции и эффективного изучения и сохранения. *Вавиловский журнал генетики и селекции*, *23*(6), 780–786.
Carter, S. (2025). *Fart science*. Publifye AS.
Čertík, M., Adamechová, Z., & Guothová, L. (2013). Simultaneous enrichment of cereals with polyunsaturated fatty acids and pigments by fungal solid state fermentations. *Journal of biotechnology*, *168*(2), 130–134.
Chiş, M. S., & Fărcaş, A. C. (2025). Cereal by-products valorization in bakery, pastry, and gastronomy products manufacturing. In *Exploring the world of cereal crops*. IntechOpen.
Chourasia, R., Phukon, L. C., Singh, S. P., Rai, A. K., & Sahoo, D. (2020). Role of enzymatic bioprocesses for the production of functional food and nutraceuticals. In: *Biomass, biofuels, biochemicals* (pp. 309–334). Elsevier.
Chudzik, G., Nunes, J. J., Arneson, N. J., Stoltenberg, D. E., & Werle, R. (2025). Cereal rye biomass effects on giant ragweed suppression inform management decisions. *Agrosystems, Geosciences & Environment*, *8*(1), e70023.
Chusak, C., Wang, C.-K., & Adisakwattana, S. (2021). RD43 rice flour: The effect on starch digestibility and quality of noodles, glycemic response, short-acting satiety hormones and appetite control in humans. *Food & Function*, *12*(17), 7975–7985.
Coda, R., Katina, K., & Rizzello, C. G. (2015). Bran bioprocessing for enhanced functional properties. *Current Opinion in Food Science*, *1*, 50–55.
Cullinan, P., Sigsgaard, T., Jeebhay, M. F., Raulf, M., & Tarlo, S. M (2021). Occupational asthma in the baking industry. In: *Asthma in the workplace* (pp. 153–164). CRC Press.
Danciu, C.-A., Tulbure, A., Stanciu, M.-A., Antonie, I., & Capatana, C. (2023). *Opportunities in grain processing, through the use of waste and by-products: An overview of sustainable valorization in the food industry.*
Deleu, L. J., Lemmens, E., Redant, L., & Delcour, J. A. (2020). The major constituents of rye (Secale cereale L.) flour and their role in the production of rye bread, a food product to which a multitude of health aspects are ascribed. *Cereal Chemistry*, *97*(4), 739–754.
Demirel, F., Germec, M., & Turhan, I. J. E. T. (2022). Fermentable sugars production from wheat bran and rye bran: Response surface model optimization of dilute sulfuric acid hydrolysis. *Environmental technology*, *43*(24), 3779–3800.
Din, A. M. U., Mao, H.-T., Khan, A., Raza, M. A., Ahmed, M., Yuan, M., … Liu, Z.-H (2023). Photosystems and antioxidative system of rye, wheat and triticale under Pb stress. *Ecotoxicology and Environmental Safety*, *249*, 114356.
Drăgan, M., Tătărîngă, G., Mircea, C., Cioancă, O., Dragostin, O., Iacob, A. T., & Stan, C. D. (2018). Ferulic acid—A versatile molecule. *Acta Biologica Marisiensis*, *1*(2), 53–60.
Drzymała, K., Mirończuk, A. M., Pietrzak, W., & Dobrowolski, A. (2020). Rye and oat agricultural wastes as substrate candidates for biomass production of the non-conventional yeast *Yarrowia lipolytica*. *Sustainability*, *12*(18), 7704.
Dutta, M., Das, P., & Bhattacharyya, S (2025). Understanding prebiotics: A review on their role, benefits, and applications. *European Journal of Pharmaceutical and Medical Research*, *1*(6), 7–8.
Dynkowska, W. M. (2019). Rye (*Secale cereale* L.) Phenolic compounds as health-related factors. *Plant Breeding and Seed Science*, *79*, 9–24.
Dziki, D. (2022). Rye flour and rye bran: New perspectives for use. *Processes*, *10*(2), 293.
Edelmann, H., Thieme, N., Ehrenreich, A., Zverlov, V., & Liebl, W. (2024). Valorization of milling byproducts and ergot-sclerotia-contaminated rye via clostridial ABE fermentation. *Biotechnology for Biofuels and Bioproducts*, *17*(1), 139.

Eisenhuber, K., Jäger, A., Wimberger, J., & Kahr, H. (2013). *Comparison of different pretreatment methods for straw for lignocellulosic bioethanol production.*

El-Mahis, A., Baky, M. H., & Farag, M. A. (2023). How does rye compare to other cereals? A comprehensive review of its potential nutritional value and better opportunities for its processing as a food-based cereal. *Food Reviews International, 39*(7), 4288–4311. doi: 10.1080/87559129.2021.2023817

Fallah, N., Marvizadeh, M. M., Jahangiri, R., Zeinalzadeh, A., & Nafchi, A. M. (2023). High-barrier and light-protective bionanocomposite film based on rye starch/nanorod-ZnO for food packaging applications. *Journal of Chemical Health Risks, 13.*

Fattakhova, Z. F., Shakirov, S. K., Krupin, E. O., Bikchantaev, I. T., Shayakhmetova, L. N., & Askarova, A. A. (2020). The chemical composition, nutrition and fractional composition of winter rye grain proteins after various methods of exposure. *Carpathian Journal of Food Science & Technology, 12*(1).

Faubel, N., Blanco-Morales, V., Sentandreu, V., Barberá, R., & Garcia-Llatas, G. (2025). Modulation of microbiota composition and markers of gut health after in vitro dynamic colonic fermentation of plant sterol-enriched wholemeal rye bread. *Food Research International, 201*, 115570.

Feng, G. (2019). Rye. *Bioactive Factors and Processing Technology for Cereal Foods.* 151–169.

Fierascu, I., Fierascu, R. C., & Sieniawska, E (2021). Advanced techniques for recovery of active compounds from food by-products. In: *Valorization of agri-food wastes and by-products* (pp. 693–710). Elsevier.

Galanakis, C. M. (2022). Sustainable applications for the valorization of cereal processing by-products. *Foods, 11*(2), 241.

Geiger, H., & Miedaner, T. (2009). Rye (*Secale cereale* L.). *Cereals*, 157–181.

Ghafoor, A., Wijata, M., Rozbicki, J., Krysztofik, R., Banaszak, K., Karim, H., …, Studnicki, M (2024). Influence of crop management on stability rye yield and some grain quality traits. *Agronomy Journal, 116*(5), 2263–2274.

Ghafoor, A. Z., Karim, H., Studnicki, M., Raza, A., Javed, H. H., & Asghar, M. A. (2024). Climate change and rye (*Secale cereale* L.) Production: Challenges, opportunities and adaptations. *Journal of Agronomy and Crop Science, 210*(4), e12725.

Ghazvini, M., Ghanbari-Gohari, F., Foshati, S., & Akhlaghi, M. (2025). Effect of rye consumption on markers of glycemic control: Evidence on the "rye factor":. *Nutrition & Metabolism, 22*(1), 27.27.

Gigante, V., Aliotta, L., Ascrizzi, R., Pistelli, L., Zinnai, A., Batoni, G., … Lazzeri, A. (2023). Innovative biobased and sustainable polymer packaging solutions for extending bread shelf life: A review. *Polymers, 15*(24), 4700.

Go, E. J., & Song, K. B. (2019). Antioxidant properties of rye starch films containing rosehip extract and their application in packaging of chicken breast. *Starch-Stärke, 71*(11–12), 1900116.

Goodman, R. E., Baumert, J. L., & Amnuaycheewa, P. (2025). Allergens and gluten associated with cereal grains. In: *Food safety* (pp. 145–161). Elsevier.

Hansen, H. B., Møller, B., Andersen, S. B., Jørgensen, J. R., Hansen, Å. (2004). Grain characteristics, chemical composition, and functional properties of rye (*Secale cereale* L.) as influenced by genotype and harvest year. *Journal of Agricultural and Food Chemistry, 52*(8), 2282–2291.

Herbstritt, S., Richard, T. L., Lence, S. H., Wu, H., O'Brien, P. L., Emmett, B. D., … Malone, R. W. (2022). Rye as an energy cover crop: Management, forage quality, and revenue opportunities for feed and bioenergy. *Agriculture, 12*(10), 1691.

Ikram, A., Saeed, F., Afzaal, M., Iqbal, J., & Hassan Mohamed, M (2023a). Vitamin B complex, amino acid profile and bioactive properties of newly developed rye variants. *International Journal of Food Properties, 26*(2), 3556–3566.

Ikram, A., Saeed, F., Noor, R. A., Imran, A., Afzaal, M., Rasheed, A., … Naz, S. (2023b). A comprehensive review on biochemical and technological properties of rye (*Secale cereale* L.). *International Journal of Food Properties*, *26*(1), 2212–2228.

Islas-Martínez, D., Ávila-Vargas, Y. N., Rodríguez-Serrano, G. M., González-Olivares, L. G., Pérez-Flores, J. G., Contreras-López, E., … Pérez-Escalante, E. (2023). *Multi-bioactive potential of a rye protein isolate hydrolysate by enzymatic processes.* Paper presented at the Biology and Life Sciences Forum.

Iversen, K. N., Dicksved, J., Zoki, C., Fristedt, R., Pelve, E. A., Langton, M., & Landberg, R. (2022). The effects of high fiber rye, compared to refined wheat, on gut microbiota composition, plasma short chain fatty acids, and implications for weight loss and metabolic risk factors (the RyeWeight study). *Nutrients*, *14*(8), 1669.

Iversen, K. N., Jonsson, K., & Landberg, R. (2022). The effect of rye-based foods on postprandial plasma insulin concentration: The rye factor. *Frontiers in Nutrition*, *9*. doi:10.3389/fnut.2022.868938

Jaksics, E., Németh, R., Farkas, A., Horváth, R., Dúzs, D., Drozdik, Á. A., ... & Tömösközi, S. (2022). Comparative compositional and functional characterisation of rye varieties and novel industrial milling fractions. *International Journal of Food Science and Technology*, *57*(7), 4463–4472.

Jansson, A., Harris, P., Davey, S. L., Luthersson, N., Ragnarsson, S., & Ringmark, S. (2021). Straw as an alternative to grass forage in horses—Effects on post-prandial metabolic profile, energy intake, behaviour and gastric ulceration. *Animals*, *11*(8), 2197.

Jeyalakshmi, J., Vijay, K., Saiganesh, B., & Manohar, E (2024). AI deployment models in precision agriculture. In *Artificial intelligence for precision agriculture* (pp. 125–142). Auerbach Publications.

Jonsson, K., Andersson, R., Bach Knudsen, K. E., Hallmans, G., Hanhineva, K., Katina, K., … Landberg, R. (2018). Rye and health – Where do we stand and where do we go? *Trends in Food Science & Technology*, *79*, 78–87. https://doi.org/10.1016/j.tifs.2018.06.018

Kamal-Eldin, A., Aman, P., Zhang, J.-X., Knudsen, K.-E. B., & Poutanen, K. (2008). Rye bread and other rye products. In Technology of functional cereal products (pp. 233–260). Woodhead Publishing Limited.

Kan, A. (2015). Characterization of the fatty acid and mineral compositions of selected cereal cultivars from Turkey. *Records of Natural Products*, *9*(1), 124.

Karasu, G. K., Rogers, C. W., & Gee, E. K. (2023). Dietary transitions toward sustainable horse feeding. *Journal of Equine Veterinary Science*, 104880.

Kärkönen, A., Korpinen, R., Järvenpää, E., Aalto, A., & Saranpää, P. (2022). Properties of oat and barley hulls and suitability for food packaging materials. *Journal of Natural Fibers*, *19*(16), 13326–13336.

Kataria, D., Kaur, J., Singh, J., Rasane, P., Bhadariya, V., & Kaur, S. (2024). Rye. In: *Cereals and nutraceuticals* (pp. 107–119). Springer.

Kaur, P., Sandhu, K. S., Bangar, S. P., Purewal, S. S., Kaur, M., Ilyas, R. A., … Razman, M. R (2021a). Unraveling the bioactive profile, antioxidant and DNA damage protection potential of rye (*Secale cereale*) flour. *Antioxidants*, *10*(8), 1214.

Kaur, P., Singh Sandhu, K., Singh Purewal, S., Kaur, M., & Kumar Singh, S. (2021b). Rye: A wonder crop with industrially important macromolecules and health benefits. *Food Research International*, *150*, 110769. https://doi.org/10.1016/j.foodres.2021.110769

Kaushal, P., & Kumar, N. (2022). Processing of cereals. In *Agro-processing and food engineering: Operational and application aspects* (pp. 415–454). Springer.

Khiaosa-ard, R., Czermak, S., Hollmann, M., Penagos-Tabares, F., Sulyok, M., Krska, R., & Zebeli, Q. (2024). Changes in nutritional and hygienic quality due to storage of common native and processed grain cereals intended for horse feeding. *Journal of Stored Products Research*, *106*, 102310.

Kim, J., Kim, J., Kim, G., Kim, S., & Shin, H. (2025). Prebiotic potential of wheat and dark rye: Modulation of gut microbiota and short-chain fatty acid-producing bacteria. *Food Science and Biotechnology*, *34*(11), 2563–2570.

Kiran, G. D., Tiwari, U., Pandey, P. K., Devi, O. R., Gireesha, D., Laishram, B., & Patel, A. K. (2024). Discovering new frontiers in plant breeding: The fascinating world of advancements shaping future growth. *International Journal of Research in Agronomy*, *7*(1), 441–445.

Kołodziejczyk, P., Michniewicz, J., Buchowski, M. S., & Paschke, H. (2020). Effects of fibre-rich rye milling fraction on the functional properties and nutritional quality of wholemeal rye bread. *Journal of Food Science and Technology*, *57*, 222–232.

Korzun, V., Ponomareva, M. L., & Sorrells, M. E. (2021). Economic and academic importance of rye. *The rye genome*, 1–12.

Kumari, A., Price, A. J., Li, S., Gamble, A., & Jacobson, A. (2025). Effects of cereal rye residue biomass and preemergence herbicide on the emergence of troublesome southeastern weed species. *Frontiers in Agronomy*, *6*, 1502864.

Kyrylchuk, A., Bezprozvana, I., Chukhlieb, S., Liashenko, S., & Kulyk, T. (2024). Variability of grain yield and quality of new varieties of winter rye. *Scientific Collection «InterConf»* (198), 413–416.

Lajvardi, M., Moghimi-Rad, J., Hadi, I., Gavili, A., Isfahani, T. D., Zabihi, F., … Materials, M. (2010). Experimental investigation for enhanced ferrofluid heat transfer under magnetic field effect. *Journal of Magnetism and Magnetic Materials*, *322*(21), 3508–3513.

Laskowski, W., Górska-Warsewicz, H., Rejman, K., Czeczotko, M., & Zwolińska, J. (2019). How Important are Cereals and Cereal Products in the Average Polish Diet?. Nutrients, 11(3), 679. https://doi.org/10.3390/nu11030679

Li, B., Zhou, Y., Wen, L., Yang, B., Farag, M. A., & Jiang, Y. (2024). The occurrence, role, and management strategies for phytic acid in foods. *Comprehensive Reviews in Food Science and Food Safety*, *23*(5), e13416.

Lu, J., Hansson, H., Johansson, D. P., Landberg, R., & Langton, M. (2024). Microstructure and viscosity of in vitro-digested rye and wheat food products. *Food Hydrocolloids*, *154*, 109990.

Luithui, Y., Baghya Nisha, R., Meera, M. J. (2019). Cereal by-products as an important functional ingredient: Effect of processing. *Journal of Food Science and Technology*, *56*, 1–11.

Luthria, D. L., Lu, Y., & John, K. M (2015). Bioactive phytochemicals in wheat: Extraction, analysis, processing, and functional properties. *Journal of Functional Foods*, *18*, 910–925.

Magdziarz, A., Jerzak, W., Wądrzyk, M., & Sieradzka, M. (2024). Benefits from co-pyrolysis of biomass and refuse derived fuel for biofuels production: Experimental investigations. *Renewable Energy*, *230*, 120808.

Maguire, R. (2024). Hay is for horses: Making and using a traditional-style Irish straw harness. *EXARC Journal*, (2024/2). https://exarc.net/ark:/88735/10779

Mihajlovski, K., Buntić, A., Milić, M., Rajilić-Stojanović, M., & Dimitrijević-Branković, S. (2021). From agricultural waste to biofuel: Enzymatic potential of a bacterial isolate *Streptomyces fulvissimus* CKS7 for bioethanol production. *Waste and Biomass Valorization*, *12*, 165–174.

Mirski, R., Banaszak, A., & Bekhta, P. (2021). Selected properties of formaldehyde-free polymer-straw boards made from different types of thermoplastics and different kinds of straw. *Materials*, *14*(5), 1216.

Mishra, L. K., Sarkar, D., Zwinger, S., & Shetty, K. (2017). Phenolic antioxidant-linked antihyperglycemic properties of rye cultivars grown under conventional and organic production systems. *Journal of cereal science*, *76*, 108–115.

Nadiya, P., Dmytro, S., Iryna, I., Oleksiy, Y., & Maryana, R. (2026). Forecasting the implementation of innovative processes by food industry enterprises. *Management Theory and Studies for Rural Business and Infrastructure Development*, *48*(1), 109–116.

Németh, R., & Tömösközi, S (2021). Rye: Current state and future trends in research and applications. *Acta Alimentaria*, *50*(4), 620–640.

Oest, M., Bindrich, U., Voß, A., Kaiser, H., & Rohn, S (2020). Rye bread defects: Analysis of composition and further influence factors as determinants of dry-baking. *Foods*, *9*(12), 1900.

Öncü Öner, T. (2023). Health effects of plant-based foods and their components. In *Plant-Based foods: ingredients, technology and health aspects* (pp. 137–178). Springer International Publishing.

Ozdemir, S., Seydosoglu, S., Kokten, K., & Cil, A. (2024). Lipid contents and fatty acid compositions of grains of Italian ryegrass varieties. *Chemistry of Natural Compounds*, *60*(6), 1126–1129.

Pandey, V., Singh, A., Patwa, N., Ankush, Gupta, O. P., Gopalareddy, K., Kumar, S., Kumar, A., Ram, S., & Singh, G. P. (2023). Wheat-based anti-nutritional factors and their reduction strategies: An overview. In O. P. Gupta, S. Kumar, A. Pandey, M. K. Khan, S. K. Singh, & G. P. Singh (Eds.), *Wheat science: Nutritional and anti-nutritional properties, processing, storage, bioactivity, and product development* (pp. 183–218). CRC Press.

Papageorgiou, M., & Skendi, A. (2018). Introduction to cereal processing and by-products. In *Sustainable recovery and reutilization of cereal processing by-products* (pp. 1–25). Woodhead Publishing.

Parrinello, G., Fontana, D. E., & Villalta, D. (2025). An overview of hidden food allergens: Need for change to the priority food allergen lists? *European Annals of Allergy and Clinical Immunology*, *57*, 51–67.

Pastrana-Pastrana, Á. J., Rodríguez-Herrera, R., Solanilla-Duque, J. F., & Flores-Gallegos, A. C. (2025). Plant proteins, insects, edible mushrooms and algae: More sustainable alternatives to conventional animal protein. *Journal of Future Foods*, *5*(3), 248–256.

Patel, N., & Blumberga, D. (2024). *A Novel GE-Mackinsey Market Approach: Investment Opportunity for the Biopolymer Packaging Materials*. Paper presented at the CONECT. International Scientific Conference of Environmental and Climate Technologies.

Patel, S., Singh, A., Sawyer, J. E., & Lundvall, J. P. (2024). Root and shoot biomass and nutrient composition of winter rye cover crop following corn and soybean. *Agrosystems, Geosciences & Environment*, *7*(4), e70017.

Perera, K. Y., Jaiswal, A. K., & Jaiswal, S (2023). Biopolymer-based sustainable food packaging materials: Challenges, solutions, and applications. *Foods*, *12*(12), 2422.

Pihlava, J.-M., Hellström, J., Kurtelius, T., & Mattila, P. (2018). Flavonoids, anthocyanins, phenolamides, benzoxazinoids, lignans and alkylresorcinols in rye (*Secale cereale*) and some rye products. *Journal of Cereal Science*, *79*, 183–192.

Pihlava, J.-M., Nordlund, E., Heiniö, R.-L., Hietaniemi, V., Lehtinen, P., & Poutanen, K. (2015). Phenolic compounds in wholegrain rye and its fractions. *Journal of Food Composition and Analysis*, *38*, 89–97.

Plaha, N. S., Awasthi, S., Sharma, A., & Kaushik, N. (2022). Distribution, biosynthesis and therapeutic potential of lignans. *3 Biotech*, *12*(10), 255.

Polwaththa, K., Amarasinghe, S., Amarasinghe, A., & Amarasinghe, A. (2024). Exploring artificial intelligence and machine learning in precision agriculture: A pathway to improved efficiency and economic outcomes in crop production. *American Journal of Agricultural Science, Engineering, and Technology*, *8*(3), 50–59.

Prabha, D., Subramanian, R. S., Dinesh, M., & Girija, P. (2024). Sustainable farming through AI-enabled precision agriculture. In *Artificial intelligence for precision agriculture* (pp. 159–182). Auerbach Publications.

Pržulj, N., Grujić, R., Velimirović, A., Jovović, Z., Brenjo, D., Trkulja, V., ... Govedar, Z.(2024). Nutritional properties of whole small grains and potential benefits for improving human health. *SANUS 2024 Knjiga 3: Zbornik radova*, 351.

Purohit, N. N., Ghosh, R. K., Price, A. J., & Maity, A. (2025). Potential ecological implications of extensive cereal rye cover cropping in the United States. *Crop Science*, *65*(3), e70056.

Qian, M., Liu, D., Zhang, X., Yin, Z., Ismail, B. B., Ye, X., & Guo, M. (2021). A review of active packaging in bakery products: Applications and future trends. *Trends in Food Science & Technology*, *114*, 459–471.

Ran, X.-l., Zhang, M., Wang, Y., & Adhikari, B. (2019). Novel technologies applied for recovery and value addition of high value compounds from plant byproducts: A review. *Critical Reviews in Food Science and Nutrition*, *59*(3), 450–461.

Rani, M., Singh, G., Siddiqi, R. A., Gill, B. S., Sogi, D. S., & Bhat, M. A. (2021). Comparative quality evaluation of physicochemical, technological, and protein profiling of wheat, rye, and barley cereals. *Frontiers in Nutrition*, *8*, 694679.

Ravisankar, S., Queiroz, V. A., & Awika, J. M. (2020). Rye flavonoids—Structural profile of the flavones in diverse varieties and effect of fermentation and heat on their structure and antioxidant properties. *Food Chemistry*, *324*, 126871.

Redant, L., Buggenhout, J., Brijs, K., & Delcour, J. A. (2017). Extractability and chromatographic separation of rye (*Secale cereale* L.) flour proteins. *Journal of Cereal Science*, *73*, 68–75.

Saitov, A., Kurbanov, R., & Saitov, V. (2020). *Taring the unloading of grain material from the loading hopper.* Paper presented at the International Scientific Conference on Architecture and Construction.

Saleh, A. S., Wang, P., Wang, N., Yang, S., & Xiao, Z. (2019). Technologies for enhancement of bioactive components and potential health benefits of cereal and cereal-based foods: Research advances and application challenges. *Critical Reviews in Food Science and Nutrition*, *59*(2), 207–227.

Segura, T., Zanoni, P., Brémond, U., Lucet, C., Pradel, A., Escudie, R., & Steyer, J. P. (2025). Modelling anaerobic digestion of agricultural waste: From lab to full scale. *Waste Management*, *200*, 114739.

Sinha, S. (2024). An overview of biopolymer-derived packaging material. *Polymers from Renewable Resources*, *15*(2), 193–209.

Skřivan, P., Sluková, M., Jurkaninová, L., & Švec, I (2021). Preliminary investigations on the use of a new milling technology for obtaining wholemeal flours. *Applied Sciences*, *11*(13), 6138.

Sluková, M., Jurkaninová, L., Švec, I., & Skřivan, P. (2021). Rye–the nutritional and technological evaluation in Czech cereal technology—A review: Grain and flours. *Czech Journal of Food Sciences*, *39*(1).

Stępniewska, S., Hassoon, W. H., Szafrańska, A., Cacak-Pietrzak, G., & Dziki, D. (2019). Procedures for breadmaking quality assessment of rye wholemeal flour. *Foods*, *8*(8), 331.

Studnicki, M., Macholdt, J., Macdonald, A., & Stępień, W. (2021). Effects of fertilizers and manures on temporal yield variability of winter rye. *Agronomy*, *11*(3), 519.

Surówka, K., & Rzepka, M. (2024). Biogenic amines as an indicator of rye leaven quality during production and storage. *Food Chemistry*, *452*, 139523.

Swamy, C. T., Sen, A., Maramreddy, P. R., Chelladurai, P. K., Sivadurga, K., Pandey, A., & Singh Purewal, S. (2025). Rye Fiber: Extraction, Identification Methods and Health Benefiting Features. In *Rye: Processing, Nutritional Profile and Commercial Uses* (pp. 95–116). Springer Nature Switzerland.

Talebnia, F., Karakashev, D., & Angelidaki, I. J. B. t. (2010). Production of bioethanol from wheat straw: An overview on pretreatment, hydrolysis and fermentation. *Bioresource Technology, 101*(13), 4744–4753.

Tang, Q., Wang, X., Jin, X., Peng, J., Zhang, H., & Wang, Y. (2023). CRISPR/Cas technology revolutionizes crop breeding. *Plants, 12*(17), 3119.

Tsykhanovska, I., Stabnikova, O., Oliinyk, S., Lazarieva, T., & Gubsky, S. (2024). Combined food additive based on iron oxide nanoparticles and кombu in a rye-wheat bread technology. *Ukrainian Food Journal, 13*(3).

Torres Rivera, F. A. (2022). *Effects on the combustion properties of wheat straw after different thermobiological pretreatments* (Doctoral dissertation, Hochschulbibliothek der Technischen Hochschule Köln).

Valli, L (2024). *The meaning of rye*. Washington State University.

Vardakas, A., Giannakas, A. E., Dimitrakouda, M., Ndreka, A., Chaintari, C., Iordanidis, G., … Potsaki, P. (2024). Sustainable development of an innovative spreadable plant-based product of high added value through the valorization of an agro-food by-product. *Applied Sciences, 14*(15), 6525.

Wang, N., Zhang, Q., Zhang, L., Lu, W., Xiao, Y., Wang, H., … Zhao, J. (2024). Characteristics of dietary patterns, and serum metabolites during the low-gluten diet among Chinese young adults. *Food Bioscience, 61*, 104829.

Warechowska, M., Warechowski, J., Tyburski, J., Siemianowska, E., Nawrocka, A., Miś, A., & Skrajda-Brdak, M. (2019). Evaluation of physicochemical properties, antioxidant potential and baking quality of grain and flour of primitive rye (*Secale cereale* var. Multicaule). *Journal of Food Science and Technology, 56*, 3422–3430.

Wojciechowski, K. (2022). Surface tension of native and modified plant seed proteins. *Advances in Colloid and Interface Science, 302*, 102641.

Wrigley, C (2019). Rye. In *Whole grains* (pp. 173–196). CRC Press.

Wrigley, C., & Bushuk, W. (2017). Rye: Grain-quality characteristics and management of quality requirements. In *Cereal grains* (pp. 153–178). Elsevier.

Xhaferaj, M., Muskovics, G., Schall, E., Bugyi, Z., Tömösközi, S., & Scherf, K. A. (2023). Characterization of rye flours and their potential as reference material for gluten analysis. *Food Chemistry, 408*, 135148.

Yalcin, S. (2024). Comparison of sensory analysis results of gluten-free hosmerim samples with hosmerim sample made from wheat flour. *İçindekiler Tablosu*, 30.

Yılmaz, B., Sırbu, A., Altıntaş Başar, H. B., Goksen, G., Chabı, I. B., Kumagaı, H., & Ozogul, F. (2025). Potential roles of cereal bioactive compounds in the prevention and treatment of type 2 diabetes: A review of the current knowledge. *Critical Reviews in Food Science and Nutrition, 65*(7), 1326–1343.

Ying, T., Zheng, J., Kan, J., Li, W., Xue, K., Du, J., … He, G. (2024). Effects of whole grains on glycemic control: A systematic review and dose-response meta-analysis of prospective cohort studies and randomized controlled trials. *Nutrition Journal, 23*(1), 47.

Yu, J., Wu, Y., Zhu, Z., & Lu, H. (2025). The impact of dietary patterns on gut microbiota for the primary and secondary prevention of cardiovascular disease: A systematic review. *Nutrition Journal, 24*(1), 17.

Zanella, I., Biasiotto, G., Holm, F., & di Lorenzo, D. (2017). Cereal lignans, natural compounds of interest for human health? *Natural Product Communications, 12*(1), 1934578X1701200139.

8 Health Endorsing Properties of Rye

Muhammad Shahbaz, Ushna Momal, Muhammad Hammad ul Hassan, Asfa Perween, Hammad Naeem and Nighat Raza

8.1 INTRODUCTION

Rye (*Secale cereale* L.) originated as a weedy grass in wheat/barley agriculture systems and spread throughout Europe. Rye is classified botanically with other cereals like barley and wheat in the tribe *Triticea*, the subfamily Pooideae, and the grass family Poaceae (Sapirstein & Bushuk, 2016). Rye cultivation peaked in the Middle Ages, and its importance as a staple grain greatly aided post-Roman European communities (Hackauf et al., 2022). Since rye requires colder growing temperatures and there are significant geographical variations in the taste for rye-based items, its production and distribution are different from wheat (Wrigley & Bushuk, 2017). Over 85% of the global production of rye (13 mt) comes from Europe, which includes the top rye-producing states: Belarus, Poland, Germany, Russia and Denmark. The world's rye production fell dramatically in the previous ten years. Simultaneously, rye's increasing amount of research emphasizes both its value as a basic ingredient in nutritious foods and its key function in the manufacture of innovative goods (Németh & Tömösközi, 2021).

Similar to other cereals like wheat, barley and triticale, the chemical composition of rye is known for its increased fiber content, particularly in pentosan. Rye is an excellent source of vitamins, proteins, fiber, antioxidants and carbohydrates. One of the main carbohydrates about 65% found in rye and other cereal grains is starch. In addition to carbohydrates, dietary fibers are important because they help to regulate weight management and keep the digestive tract active. The dietary fiber composition of rye shows that arabinoxylan is the primary component, followed by lignin, fructan and beta-glucan (Biel et al., 2020). A variety of bioactive chemicals found in rye grains, including sinapic acid, catechol, ferulic acid, vanillic acid and vanillin, have antioxidant protentional and support immune system functions and helps in preventing age-related problems. Bioactive substances with antioxidant qualities are essential for maintaining a healthy lifestyle because they aid in processes that reduce the adverse effects caused by free radicals and oxidative stress (Jonsson et al., 2018). Specifically, the main phenolic components present in rye and wheat grains are alkylresorcinols. They serve as biomarkers to determine how much whole grain is included in different cereal products (Dziki, 2022). According to analysis, macronutrients and

DOI: 10.1201/9781003585817-8

micronutrients are equally crucial for sustaining the body's diverse biochemical processes. Rye is naturally enhanced with an exclusive combination of minerals (potassium, magnesium, iron, calcium, zinc and zinc) and vitamins (B1, B2, B3, B6, B9, choline and tocopherol) (Bağcı et al., 2019). One of the major aspects of rye is that it contains a high level of fiber, which is soluble fiber such as fructans and arabinoxylans. These fibers have been proven to boost colon health, prevent constipation and help maintain a healthy microflora in the gut. It has been found to increase the production of beneficial short-chain fatty acids (SCFAs) in the human body associated with better metabolism profiles and reduced inflammation when one consumes more rye foods (Iversen et al., 2022). There are several health benefits associated with eating items made from processed rye grains. Eating food items made with rye raises feelings of satisfaction and reduces inflammation and blood cholesterol levels. Additionally, rye bran can be a useful raw material for the synthesis of biofuels (Joye, 2020).

Rye meals have a beneficial effect on the metabolism of insulin compared to other cereal grains when consumed at standardized doses of readily accessible carbohydrates under isocaloric conditions. These advantages could have a substantial impact on the prevention of diabetes (Aune et al., 2016). The most popular rye products include morning cereals, hot bread, rye flour bread, black and sour bread, porridge and flakes. Rye is becoming a more and more popular component for snacking. Consumers' preference for rye is growing because of its nutritional value as well as its therapeutic advantages, such as its gut-friendly metabolism and health-promoting qualities due to bioactive components. However, careful consideration and investigation are required to determine how different qualities of rye affect health at the cellular level and how they affect biological activities (Ikram et al., 2023).

8.2 NUTRITIONAL PROFILE OF RYE

Consumers demand nutrient-rich diets; thus, researchers and the industrial sectors are paying attention to the nutritional contents of rye. Cereals are recommended over junk food and fried meals since they include a significant quantity of dietary fiber, which can improve overall health. In general, rye is an excellent source of proteins, carbohydrates, minerals and dietary fiber. A wide range of bioactive substances found in rye grains have potentially complementary or independent impacts on health (Kaur et al., 2021).

8.2.1 Macro and Micronutrients Composition

Rye has a very variable protein concentration (10–15%), mostly based on growing circumstances. The main protein present in rye is albumin, which makes about 30–40% of the grain's total protein composition (Laidig et al., 2017). Globulins comprise 8–11% of the overall protein composition, whereas prolamins and glutelins, the rye storage proteins also known as secalins, make about 17–19 and 9–15%, respectively. Based on their high glutamine and proline concentrations and solubility in a solution of alcohol and water, prolamins comprise the majority of storage protein components found in cereal grains. Up to 80% of the total protein content found in barley, rye and wheat is made up of these proteins. The solubility of cereal storage seed protein prolamins in aqueous alcohol mixtures enables the differentiation of these prolamins into alcohol-soluble monomeric glutenin and alcohol insoluble polymeric glutenin. These components of the polymeric fractions may be classified as high-molecular-weight (HMW) or low-molecular-weight (LMW) groups following reduction, while based on their electrophoretic mobility monomeric prolamins can be further divided into α-γ- and ω-types (Bauer et al., 2017).

The protein that stores prolamins is significant because it triggers responses mediated by the non-Immunoglobulin E (IgE) and IgE immune system in certain people. About 1% of people worldwide suffer from celiac disease, a non-IgE immune-mediated food intolerance that is caused by protein prolamin found in several cereal grains, including barley, wheat and rye (Sharma et al., 2020). Rye proteins have a somewhat better amino acid composition than wheat proteins because of their greater lysine level, but their tryptophan and isoleucine contents are limited (Redant et al., 2017).

Unlike other grains, starch is the most prominent carbohydrate in rye. Rye grain starch content is between 55 and 65%: this is somewhat less than wheat starch content (65–75%) and more than barley starch content (50–64%) (Grossmann & Koehler, 2016). Similar to other cereal grain starches, rye starches are made up of large (A-type; 60–90%) and microscopic (B-type; 10–40%) granules with dimensions of 23–40 and less than 10 μm, respectively. Rye starches often have an amylose concentration of 22–26% (Kan, 2015).

Rye has a lipid concentration of 2–3% similar to wheat, and its fatty acid composition is similar as well, with linoleic acid (19–60%) being a prominent component (Sapirstein & Bushuk, 2016). Micronutrients known as vitamins are necessary for general health, development, metabolism and reproduction. Based on their solubility, vitamins are divided into two groups: water-soluble (B and C) and fat-soluble (A, D, E and K). Because the human body is unable to synthesize all vitamins except D, so consumption of vitamins through diet is essential. The deficiency of vitamins can cause a number of diseases (Yasin et al., 2017). Cereal grains consist of the germ, endosperm and the bran are edible seeds of grasses. These essential foods retain their nutritious content even when kept for a long time. Cereal grains include carbohydrates, proteins, dietary fiber, vitamins and minerals. The ratio of nutrients varies between cereal crops and depends on the techniques of processing and cooking. The process of grinding grains to produce refined flour lowers the nutritional content since the number of nutrients in the seed covering is higher than in the endosperm (Guo et al., 2019). Rye is similar to other cereals in terms of its mineral and vitamin content. Vitamin B complex (B1, B2, B3, B5, B6 and B9) are all found in rye cereal grain, along with phosphorus, potassium and magnesium. Usually, it has higher levels of iron, zinc, manganese and copper than other grains. Moreover, rye and wheat are also excellent sources of alpha tocopherol (Bağcı et al., 2019).

8.2.1.1 Dietary Fibers

Based on its solubility, dietary fiber can be divided into two main categories: soluble dietary fiber (SDF) and insoluble dietary fiber (IDF). Beta-glucan, arabinoxylans, insulin, pectin, gums and non-cellulosic polysaccharides such as non-digestible oligosaccharides make SDF, whereas lignin and resistant starch (RS) make IDF. The high dietary fiber content of rye is found throughout the rye kernel, and the dietary fiber level of rye endosperm is greater than that of wheat endosperm. Rye contains arabinoxylan, cellulose, fructans, beta-glucan and lignin as dietary fiber (Dahl et al., 2017). Rye and wheat are comparable in this regard; however, rye has a greater fiber content and arabinoxylan solubility than wheat. The most common fiber in rye is arabinoxylan, which makes about 6–12% of the dry grain weight and can be found in variable proportions and quantities in the endosperm, aleurone and pericarp tissues of the grain. Numerous xylose units that have been substituted with arabinose residues form the backbone of arabinoxylan. Furthermore, ferulic acid groups are esterified to a large number of arabinose residues, allowing polysaccharide cross-linking. For example, the water solubility of arabinoxylan varies between fractions; in endosperm, it is around 70%, but in the aleurone and pericarp, it is nearly zero (Holscher, 2017). The substitution pattern (arabinoxylan/xylose) also varies for various milling fractions. The molecular weight of rye arabinoxylan varies depending on the method of analysis used and the source of material (bran or whole grain). Rye's viscous SDF is attributed to soluble beta-glucan and arabinoxylan that can be extracted in water, which may also play a role in the numerous health benefits and technical capabilities. For the bacteria present in the large intestine, the most fermentable substrate is provided by SDF and fructans, which may positively impact both gut health and general health (Vuholm et al., 2017). The dietary fiber that is insoluble has an impact on intestinal transit time and fecal bulk, which can alleviate constipation symptoms and lower the risk of colon cancer that has been linked to large intakes of whole grains (De Vries et al., 2016). Rye has the highest fructan content. Fructan is a SDF that can include either a terminal glucose residue or beta-d-fructofuranosyl units. Rye fructans can have a branching or linear structure. Fructan polymerization in rye typically varies from 2 to 60 degrees (Meija & Krams, 2019). Types of dietary fibers present in rye are shown in Table 8.1.

TABLE 8.1
Type of Dietary Fiber Present in Rye

Dietary Fiber	Solubility	Health Benefits	References
Fructans	Soluble fiber concentrations of 4.6–6.6% present in rye	Promote regular bowel movement Improves mineral absorption	Peicz et al. (2020)
Lignin	Insoluble fiber present in significant amount	Add to the volume of the feces Have antioxidant characteristics	Ikram et al. (2023)
B-glucan	Soluble fiber present in concentrations of 1.3–2.2%	Enhances immune system activity Improves blood sugar regulation Reduces low-density lipoprotein (LDL) cholesterol	Kaur et al. (2021)
Arabinoxylan	Non-starch polysaccharide makes up to 6–12% of the weight of the dry grains of rye	Reduces cholesterol level increases satiety and decreases appetite Improve gut health	Jonsson et al. (2018)
Cellulose	Insoluble fiber present in concentration of 1–1.7%	Lower the chances of colorectal cancer support digestive health	Németh and Tömösközi (2021)

8.2.1.2 Biological Active Substances

The rye grains are known to have high fiber content as the main factor contributing to positive health impacts. However, dietary fiber is not sufficient to account for the effective biological benefits of rye. Additional elements, such as bioactive chemicals, are proposed to play a role in improving body functions (Bach Knudsen et al., 2017). Plant-produced substances known as phytochemicals serve a variety of purposes in plants, including defense and signaling. Most phytochemicals in grains are found in bran, especially in the aleurone cells. Rye grains are rich in bioactive chemicals (alkylresorcinols, ferulic acid, catechol, sinapic acid, vanillin and vanillic acid) that have the ability to increase immunity and prevent from age-related problems (Koistinen & Hanhineva, 2017). There are a number of other bioactive substances found in rye, including polyphenols, phenolamides, flavonoids, anthocynanidins and lignins (Kulichová et al., 2019). Natural polyphenolic substances called flavonoids are frequently present in seeds, grains and plant-based foods. Anthocyanins, flavonol, flavones, flavanones and flavane are the six primary subclasses of flavonoids (Panche et al., 2016). The class of flavonoids that occurs most frequently in cereal grains is called flavones. The structure of the flavones, the conjugated B and C rings and the oxo-group at the C-4 position is linked to increase anti-inflammatory effects, increase insulin sensitivity and improve interaction with intestinal membrane transporters, which makes them particularly significant (Suchowilska et al., 2020). These have positive impacts on human health, which are related to their anti-inflammatory and anticarcinogenic effects. According to the researchers, the number of flavones varied from 58 to 137 μg/g o-glycosides, which are derived from apigenin, chrysoeriol and tricin, were found to be the predominant flavones, ranging from 50 to 68% of the total flavone content. Rye has a greater concentration of flavone than rice (≤20 μg/g) and barley (6–30 μg/g), but it still has a similar quantity of flavones to wheat. Additionally, researchers revealed that rye flour and rye bran contain anthocyanidins and flavonoid C-glycosides. Whereas wholemeal rye flour had the lowest concentration, while rye bran had the greatest concentration of it (Martínez et al., 2018). Phytoestrogens called lignins can be found in vegetables, seeds and grains. Numerous hundred lignin compound structures were found in the grains of various cereals. These substances were discovered to be esterified to the cereal matrix and to exist as free aglycones coated with sugar. According to several studies, eating a diet high in lignins may improve human health and provide protection against a number of illnesses linked to an unhealthy lifestyle (Bolvig et al., 2016). Gut bacteria transform plant lignins into enterolactone, enterodiol and enterolignins. The small intestine is where these substances are absorbed, and they are good for human health. Specifically, enterolignins are important in lowering the probability of developing of cancer and cardiovascular disorders. Rye has the highest content of the lignin than other cereal grains. However, rye grains vary in their concentration of these compounds from 12 to 68 mg/kg (Zanella et al., 2017). Lignins are mostly concentrated in the bran of the rye, which is similar to the distribution of other bioactive component in the grain. Many types of dietary lignins (secoisolariciresinol, matairesinol and pinresinol) have been found in rye grains. Secoisolariciresinol constituted the majority of the lignin content in rye, making up around 80% of

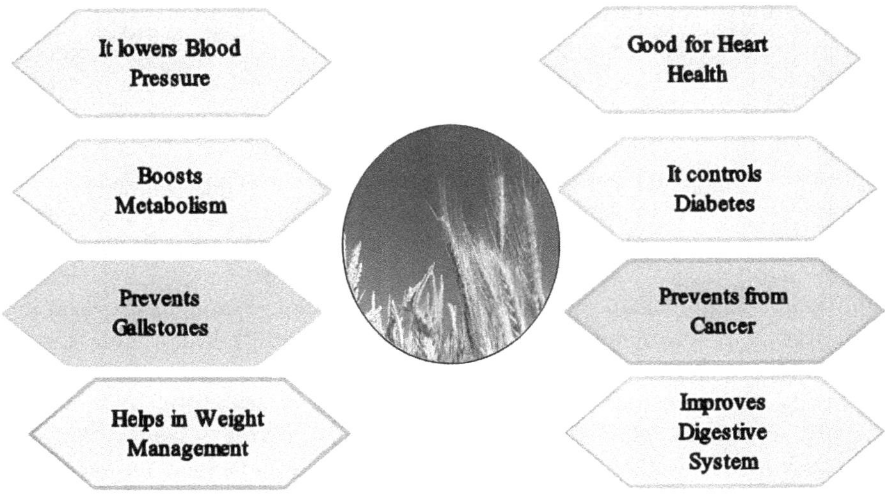

FIGURE 8.1 Health benefits of rye.

the grain. The percentages of medioresinol and pinresinol are around 10% each, while rye grains also contain trace levels of secoisolariciresinol, matairesinol and 7-hydroxymatairesinol (Eriksen et al., 2020). A class of chemicals known as benzoxazinoids was discovered in rye recently and gained attention because of its highly bioactive structure and health benefits. Although there have been a few research specifically focused on benzoxazinoids in humans, few is known from animal and *in vitro* analysis indicated that these compounds have immunoregulatory, appetite and weight reducing and anticancer properties (Adhikari et al., 2015). Benzoxazinoids have been identified in prostate tissue, supporting their potential to play a role in prostate health. It is proposed that these compounds have particularly strong impacts on prostate cancer. It is concluded that further research is required to determine the health benefits of bioactive substances obtained from rye grains. Health benefits of rye are mentioned in Figure 8.1. However, the variety of phytochemical identified and bioactive chemical properties and their role in improving health status (Steffensen et al., 2016). Bioactive substances present in rye grains are shown in Table 8.2.

8.3 HEALTH BENEFITS OF RYE

Rich in essential minerals and vitamins, rye is a good grain for dieting, weight loss, heart patients, diabetes and immune system support. Its potent antioxidant qualities also make it a good grain for cancer prevention, immune system stimulation and improving cell metabolism. Rye is beneficial for the digestive tract. It not only relieves gout, ulcerative colitis and cancer but also prevents gallstones formation in the body. Rye with high fiber content bulk up feces and helps it pass through the

TABLE 8.2
Bioactive Substances Present in Rye Grain

Rye Grains	Bioactive Substances in Rye	Main Components	Present in	Content	References
Phenolic Compounds	• Ferulic • Ferulic dimer • Sinapic • p-Coumaric • Vanillin • p-Hydrobenzoic acids • Salicyclic • Flavonoids	Sinapic p-Coumarin Ferulic	Bran	1300–1500 mg/kg	Gupta et al. (2019)
Phytosterol	• Sitosterol • Sitistanol • Stanol • Campesterol • Campestanol Stigmasterol	Sitosterol	Germ and aleurone layer	761–1000 mg/kg	Feng (2019)
Phenolic Lipids	• Alkylresorcinol • Alkenylresorcinol homologous C17:0, C19:0, C19:1, C21:0, C23:0, C25: 0	Majorly alkenylresorcinol C 19:0 and C 17:0	Bran	500–1000 mg/kg	Kulichová et al. (2019)
Tacols	• α-Tocopherols • α-Tocoterinol • β-Tocoterinol	α-Tocotrienol	Aleurone layer and germ	20–55 mg/kg	Nascimento et al. (2023)

digestive tract more quickly, resulting in a smoother digestive process. Consequently, stomach secretes less bile to aid in digestion.

So, gallstone development is facilitated by bile, hence reducing the secretion completely prevents gallstones formation (Bach Knudsen et al., 2017). For those who have diabetes, rye is an excellent grain. As its molecules are larger than wheat, it does not digest quickly. This may assist in avoiding blood sugar level's abnormal an increase and falls. The body's first defense against all threats is its immune system. An immune system that is functioning well protect against diseases, including autoimmune disorders. The repeated depletion of immune system resources over little renders the immune system incapable of handling larger issues and causes infection. Antioxidants are essential for strong immunity, and rye is high in them. These antioxidants support body in eradicating free radicals, enabling immune system to function normally. Additionally, rye is proven to reduce blood pressure. This is mostly due to the grain's high fiber vitamin and mineral content (Aune et al., 2016).

All these health endorsing benefits are briefly explained here:

8.3.1 Fiber Rich Rye: Digestive Health

Multiple research investigations have demonstrated that SDF and IDF can both enhance the human gastrointestinal systems and work in different ways. The three main processes of bulking, viscosity and fermentation are responsible for beneficial effects. Certain fibers, mostly IDF, have a bulking effect that improves regularity, reduces constipation and increases stool mass. Both the actual presence of DF and the water retained in fiber matrix are responsible for the higher stool weight. Because of their ability to form particles and retain water, IDF's constituents are cellulose and lignin and can effectively increase fecal volume even they are usually not fermentable in the colon. Additionally, IDFs have been linked to a reduction in intestinal transit time, which helps in both prevention and treat constipation (Siddiqui et al., 2023). On the other hand, easily fermented SDFs have the potential to increase stool size via encouraging the growth of intestinal and fecal bacteria as well as their byproducts, such as gas, SCFA. By firming loose or liquid stool in diarrhea and softening hard stool in constipation, these qualities may help to restore normal shape (Nadar et al., 2018). The whole digestive system, from mouth to anus, is impacted by dietary fibers. Eating meals high in fiber typically requires more time and has a lower energy density. Gastric emptying is often delayed by soluble fibers. Food ingredients may pass through the small intestine more slowly when soluble fibers are present, but insoluble fibers typically cause intestinal discomfort. Dietary fibers have the ability to detect reactions of broad range of gastrointestinal hormones in the small intestine. These hormones act as incretins, inducing insulin and influencing hunger (Ötles & Ozgoz, 2014). Certain soluble fibers bind bile acids and prevent the production of micelles, which increases the amount of cholesterol and bile acids excreted in the feces. There are various physiological benefits that arises from eating diets high in RS and beta-glucan present in rye grains. Consequently, in healthy individual's beta-glucan reduces postprandial blood glucose levels and insulin reactions. For both people and animals, beta-glucan derived from rye, oats as well as cereal products can lower serum cholesterol and lipoprotein concentrations (Kieffer et al., 2016). It becomes healthier and more advantageous when bran is added to the rye and wheat flour. As an excellent laxative, it is well regarded. Due to quickly breakdown by intestinal microflora, cereals bran has considerably better laxative properties than cellulose found in fruits and green vegetables. Because of the high cellulose content of the fiber, constipation may be prevented and treated. Due to enhanced peristalsis, its ingestion provides the intestine with bulk, which has a smoothing effect and facilitates comfortable transit. Moreover, it has shown that prevention not only facilitates bowel movement but also protects against related problems like hemorrhoids (Sana Afzal et al., 2013).

8.3.2 Rye Fiber Helps Reduce LDL (Bad) Cholesterol and Its Implications for Heart Disease Prevention

The other major carbohydrates found in rye grains are the dietary fibers followed by the starch. These are primarily occurring in outer bran layer of seeds; it has been noted that shell is really full of these granules. Dietary fibers are present with varying percentage in rye grains; it is found to be between 18 and 22%. The fiber found in the cereal grains is in soluble and insoluble fibers and its total quantity. For instance, rye contains 14.41% IDF and 4.59% SDF, making up a total dietary fiber content of 19%. Arabinoxylan (8–12%) and fructan (4.5–6%) are found in rye grains followed by β-glucan (2–3%). Lignin and cellulose are two examples of insoluble fiber that may also be found in rye grains in the amount of between 1 and 3%. It is also well known that food containing rye have an impact on feces' transit time and movement of stools in the digestive system. Cereal grains which contain an appreciable amount of β glucan appear to have cholesterol reducing capacity. The popular scientists have examined that consumption of rye breads reduces the cholesterol level by 8%. They were able to show that the intake of rye breads helped to decrease quantity of LDL in diet. The study further showed that by consuming rye-based diets, there are negative alterations in amount and proportion of LDL cholesterol as compared to HDL cholesterol (Kaur et al., 2021).

8.3.3 The Role of Magnesium and Potassium in Rye for Maintaining Healthy Blood Pressure Levels

In rye, potassium content was found to be 3 87 and 4.48 g/kg, respectively. Rye flour also contains magnesium at concentrations of 1.19–1.30 g/kg (Kaur et al., 2021). Another condition that is dangerous for development of heart disease or stroke is high arterial pressure. Lastly, there are the substance that contains magnesium keeps the blood pressure in check. Any shift of the endogenous magnesium position results to alterations in tone of arteries which in turn, affect the arterial blood pressure. Mg deficiency enhances hypertensive angiotensin II–stimulated aldosterone secretion and thromboxane as well as vasoconstrictor prostaglandins formation. In addition, the etiology of primary hypertension has been linked to alternation of magnesium and calcium metabolism. The researchers demonstrated an analysis of 22 experiments showed that 1.104 adults with normotensive and hypertensive that taking magnesium supplements for the duration of 3–24 weeks of follow up significantly reduced systolic blood pressure by 3–4 mmHg and diastolic blood pressure by 2–3 mmHg. The additional magnesium level was between 120 and 973 mg daily. The changes were somewhat larger when the participant's magnesium supplementation was above 370 mg/day. Dietary magnesium is useful in managing hypertension and diminution of the cardiovascular risk factors such as atherosclerosis in hypertensive patients, especially those on chronic diuretics or with low dietary intake of magnesium (Gröber et al., 2015). The 'American Heart Association' indicated that having high consumption of potassium would reduce the occurrence of high blood pressure and that life expectancy would be extended. On the other hand, potassium loss through the diet in normotensive human beings can actually increase blood pressure. Potential relationship between blood pressure and potassium was first proposed for the first time in 1928 by W. L. Addison. New meta-analysis revealed that the potassium intake in the level of 90–120 mmol/day decrease the blood pressure and produce positive link with stroke occurrence rate. There are so many ways through which potassium may cause a reduction in blood pressure. For instance, potassium

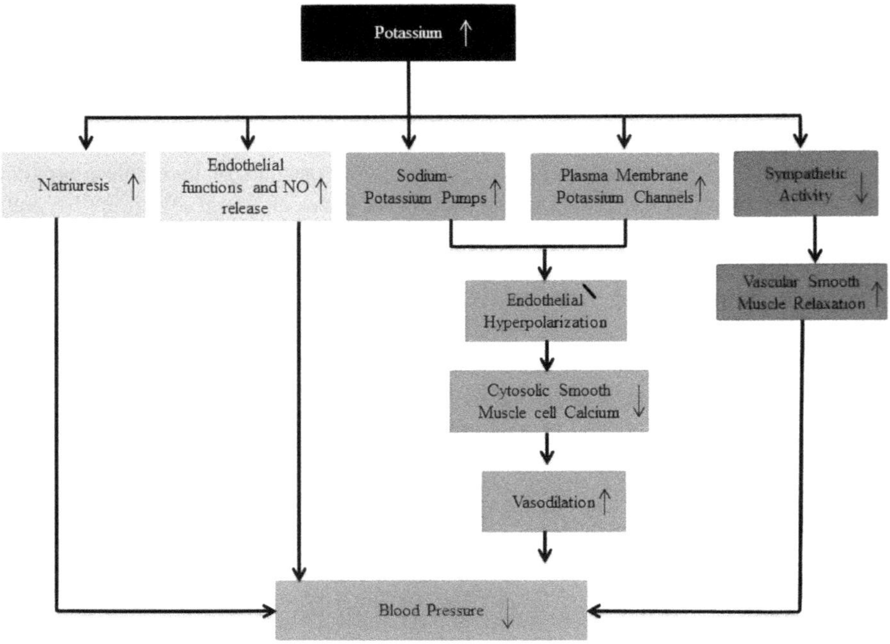

FIGURE 8.2 Potassium role in maintaining blood pressure.

supplementation tends to promote sodium excretion within four days. Different pathways are involved in the stimulation of natriuresis such as the decrease in active reabsorption of sodium in the proximal tubule possibly through inhibition of renin secretion and the increase in GFR by vasodilation of renal afferent arteriole. The endothelium-dependent vasodilation can be attributed to potassium, which hyperpolarizes the endothelial cell through activation of activated sodium potassium pump and plasma membrane potassium channel. Endothelial products cause changes of membrane potential that leads to reduction of cytoplasmic calcium in smooth muscle cells hence vasodilation.

Potassium may also possess vasodilation properties due to relaxation of smooth muscles within the blood vessels and enhanced blood flow due to the prevention of over stimulation of the sympathetic nervous system owing to increased NA + uptake at the pre-synaptic terminal. This way, potassium may exhibit some kind of sympatholytic effect, which should, in turn, help preserve the cardiovascular system from prolonged deterioration (Ekmekcioglu et al., 2016). Potassium role in maintaining blood pressure is shown in Figure 8.2.

8.3.4 Cardiovascular Disease

Systemic inflammation as indicated by the C-reactive protein (CRP) test is a continuous variable that has causal relevance for the onset and development of chronic disorders, including metabolic syndrome, CVDs and diabetes (Omar et al., 2023).

Another pilot study about the effect of high-fiber rye foods on prostate cancer and value of CRP composed that this inflammatory marker was lowered in men with this disease after intake of fiber rich rye meals in comparing with refined cereal grain products. Some blood glucose levels associated with reduced insulin and reduced the release of free fatty acids in a later phase after ingesting whole grain rye. On the other hand, other refined grains products may reduce inflammatory response. Reduced permeability of the gastrointestinal tract after a diet high in rye may also be one of the factors for the reduced the pro-inflammatory level of R-GAL residues getting into the blood circulation and triggering inflammation. Although the results seem promising and represent a possible avenue of intervention, more research has to be done before any conclusions can be made and the processes that operate behind the identified effects have to be explored. This makes a scene that whole grain rye products can reduce the chance of getting cardiovascular diseases because it reduces low-grade inflammation to some degree (Jonsson et al., 2018). Likewise, as with 3-indolepropionic acid, trimethylamine N-oxide (TMAO) level was increased compared to baseline only with the higher whole grain rye intervention. It is the liver metabolite of trimethylamine which is a microbial metabolite of several dietary molecules such as glycine betaine, carnitine and choline. Whole grains are food group linked with reduced cardiovascular disease risk, and the rise in TMAO levels under the whole grain rye intervention may imply that TMAO is a biomarker of the increased activity of gut microbiota and a player with an innocent role in cardiovascular disease development (Koistinen & Hanhineva, 2017).

8.3.5 Stroke

Whole-grain rye intake is evaluated by level of 'plasma 3-(3,5-dihydroxyphenyl)-1-propanoic acid (DHPPA)'—bio marker of grain consumption, and that DHPPA level is related to occurrence of ischemic stroke (IS). The results available from a hospital-based case–control study suggested that raised plasma level of DHPPA, being a marker for whole-grain rye, was connected with reduced chance of the IS, thus recommended the high consumption of whole grains. More studies have to be conducted to establish the given conclusions in the prospective cohort populations (Suhr et al., 2017). New epidemiological studies have revealed that high whole grain intake, but especially rye, help decrease the rate of IS. Scientists identified recently that consumption of whole-grain rye enhances the metabolic values and decreases the content of pro-inflammatory compounds, both of which are associated with stroke prevention. The study established how the rye's dietary fiber, particularly the arabinoxylans and lignin, protects against inflammation and improves vascular health (Hanhineva et al., 2021).

Besides, the same study conducted in 2022 linked intake of rye to this positive impact, reduced blood pressure and improved endothelial function, which are both key factors for decreased risk of stroke. The flavones and ferulic acid in rye that has been analyzed by the researchers are potent antioxidants that may protect against inflammation and oxidative stress, the two vices that cause stroke (Ikram et al., 2023).

8.3.6 Anti-Inflammatory Properties

Inflammation is an immune response to negative stimuli such as radiation, poison or injury within the human body. It is associated with redness, pain, heat, increased soft tissue bulk and loss of tissue function due to activation of specific chemical mediators like CRP, interleukins and tumor necrosis factor. Although there is some evidence that acute inflammation when left uncontrolled can facilitate tissue repair and minimize tissue injury by releasing cellular and molecular mediators, it can also lead to chronic inflammation and severe, non-reversible damage to DNA and tissue (Chen et al., 2019). Among the many chronic diseases, obesity, metabolic syndrome, type 2 diabetes (T2D), cardiovascular disease and some malignancies are associated with chronic inflammation (Cicero et al., 2017). Many studies have showed that high-fiber cereal diets have a positive impact on low-grade and chronic inflammation by reducing circulating inflammation-related lipid profile and inflammation in adipose tissues (Roager et al., 2017). There have been a small number of researches done on whole grain rye foods. When high-fiber rye foods were consumed instead of refined wheat, a short trial on males with prostate cancer revealed a substantial decrease in CRP an indicator of inflammation. Furthermore, when grain bran enriched with rye was consumed 28% of total caloric consumption for a 12-week trial of Chinese participants, the results showed a substantial decrease in CRP when compared to other refined grains. However, in another research when cereal grain rye and fortified bran were consumed 25% of total caloric intake for a 12-week trial of Chinese participants, the results showed a substantial decrease in CRP when compared to refined wheat. After a meal of whole grain rye, as compared to other refined products, advantageous blood glucose level linked to reduce the insulin reactions and a lower later phase reaction of free fatty acids may help reduce inflammation (Sandberg et al., 2016).

8.3.7 Weight Management

Whole grain porridge made from rye, soft bread and crispbread have been shown to enhance feelings of fullness, reduce feelings of hungriness and decrease the cravings to eat when compared to refined wheat bread, which has the same caloric content (Lee et al., 2016). The amount and type of fiber in rye meals impact physiological qualities like level of encapsulation and viscosity; the enhanced satiety may be corelated with the nutrient availability and digestion (Johansson et al., 2017). Moreover, it has been reported that rye bread contains compounds that resemble serotonin and may boost satiety. Additionally, one research demonstrated a correlation between enhanced satiety and SCFAs, specifically butyrate, which is produced during fermentation of dietary fiber in rye (Sandberg et al., 2017). When rye crispbread (fermentation or non-fermentation) was compared in studies, the results showed that the former had a great effect on satiety while the latter had no little impact on appetite. Over-time managing weight may benefit from more satiety. After consuming whole grain rye, wheat and refined wheat for six weeks, the effects on body weight and fat mass were measured in 60 persons who were at the risk of developing metabolic syndrome. The analysis showed that people who ate whole grain rye lost more weight than those who ingested whole grain or refined wheat (Suhr et al., 2017).

8.3.8 Diabetes

Diabetes and its types; type 1 diabetes (T1D) and T2D are prevalent health issues that impact millions of people globally and are on rise. Of all diabetic patients, over 90% have T2D. T2D is causing an increasing number of deaths and disabilities every day. Diabetes mellitus is a metabolic illness with several physiologies characterized by persistent hyperglycemia, protein, lipid and carbohydrate metabolism abnormalities and deficiencies in either insulin action or secretion or both. A variety of causes, including genetic and environmental that impact beta cell function and the tissue of the pancreas, muscles and liver, can result in T2D (Panhwar et al., 2018). Diabetes develops when the body's cells are unable to properly use the insulin that the pancreases produce or when the pancreas is unable to synthesize enough insulin. Acute or long-term hyperglycemia can result from dysregulation in the glucose homeostasis system. The chronic condition of diabetes mellitus results in adverse health effects from impaired insulin secretion. These include elevated levels of triglycerides in the liver, increased lipolysis that produces fatty acids, increased cholesterol level, increased rate of gluconeogenesis and decreased protein synthesis (Hong et al., 2018). A variety of nutritional factors have been linked to glycemic metabolism, and their effects on insulin secretion, insulin sensitivity and blood glucose levels have been studied. These factors also vary in concentration across a range of food sources. Reduced body weight is correlated with modified intake of fat, carbohydrates and cereal fiber along with low GI food consumption. Maintaining a healthy lifestyle and proper bodily metabolism depend heavily on proper nutrition, which also plays a major role in disease prevention. As compared to genetic, epigenetic or environmental variables, dietary causes of T2D are thought to be more significant and dependable risk factors (Lee & Halter, 2017). Non-starchy polysaccharide present in rye cereal is called arabinofuranosyl. It is structurally similar polymer of xylose that has been substituted with arabinofuranosyl. Consumption of arabinoxylan has been linked to decreased serum glucose level, especially in those with T2D who have poorer metabolic blood glucose control and impaired glucose tolerance. In small intestine, arabinoxylans inhibit glucose absorption and restrict access to enzymes. A reduction in alpha glucosidase activity has resulted in antiglycemic characteristics in water extracts of arabinoxylan from rye bran and aleurone layers. Production of SCFAs like acetate and butyrate is stimulated by arabinoxylan. Blood sugar is regulated by SCFAs because they inhibit the liver's gluconeogenesis. Consuming arabinoxylan crackers has been demonstrated in clinical trials to have a positive impact on the regulation of blood sugar in overweight individuals (Coe & Ryan, 2016). In humans, long-term consumption of long-chain arabinoxylans has been shown the ability to restore insulin and glucose metabolic functions in people with T2D; similarly, beta-glucan has been found to improve insulin sensitivity. Of the cereals containing arabinoxylans, rye has the lowest glycemic index of 29, which can significantly help to improve the glycemic control. Supplementation with arabinoxylan rich fiber at a rate of 15/day dosage may help to maintain blood glucose level. Rye contains phenolic compounds called alkyresorcinols, which are catabolized into two distinct metabolites in the liver. Researchers have shown that among Chinese individuals, T2DM and poor glucose metabolism were inversely correlated with plasma alkylresorcinol metabolites 3-(3,5-dihydroxyphenyl)-1-propanoic acid level. Alpha glucosidase and alpha amylase can attach to alkylresorcinols, which reduces their activity. Researchers also have examined the relationship between plasma alkylresorcinol and T2D risk and found that rye fiber intake is directly correlated with alkylresorcinol metabolites in plasma, and that whole grain consumption which includes a significant amount of rye is linked to a lower risk of T2DM in the Scandinavian population (VSS et al., 2018).

8.3.9 Cancer

All around the world, cancer is still a leading cause of both mortality and morbidity. Among women, breast cancer is one of the most common cancers which are diagnosed. According to GLOBOCAN report, 2.1 million new cases of breast cancer were diagnosed in 2018 (Nicholls, 2018). In addition, breast cancer ranks sixth among Chinese women's causes of death and is the cancer that is diagnosed with the highest frequency. Different histological, clinical and molecular characteristics are present in breast cancer, making it complicated disease. Numerous variables can influence the development of breast cancer including extrinsic factors like radiations, lifestyle choices, family history and reproductive status as well as internal factors like genetic mutation and environment (Rakha & Green, 2017). Many foods are regarded as essential components in the prevention of cancer, with whole grains being particularly rich in phytochemicals. It has been demonstrated that rye grain consumption reduces the risk of non-communicable disorders. Researchers have demonstrated that consuming a high-quality whole grain product can help to prevent cancer. Bioactive nutrients present in plant food are called phytochemicals. Antioxidants, phytosterols, lignins, phenolic acids, vitamin E and polysaccharides are among the many phytochemicals found in whole grain rye. Numerous experimental investigations have demonstrated that the bioactive elements present in whole grain rye exhibit anti breast cancer properties by inhibiting the proliferation, regulating the immune system and preventing the spread of breast tumor cells (Li et al., 2017). Furthermore, whole grain rye may influence hormone levels, which may have an impact on breast cancer. Foods high in whole grain contain phytoestrogen like iso-flavones, which may have an impact on hormone level and functions. Both phytoestrogens and dietary fiber can reduce inner estrogen concentrations, inhibit tumor growth and make early-stage cancer risk marker expression less prominent. Moreover, a variety of bioactive phytochemicals that are present in rye have a potent ability to alter signaling pathways linked to the development and progression of breast cancer (Makarem et al., 2016). One of the most common disorders is colon cancer. The main cause of morbidity is colon cancer. Numerous investigations and analysis have proposed a negative correlation between high fiber intake and colon cancer. It is more significant that phytate inhibits the development of colon cancer. Animal modeling has been used to explain how the type and sources of dietary fiber affects its ability to suppress the growth of colon cancer (Ahmed, 2020). As compared to wheat and rye bran, oats and corn bran appear to be less efficient in reducing the occurrence of malignant growth in the colon in human. The scientist demonstrated in a study that the phytate slowed the growth of human colon cancer

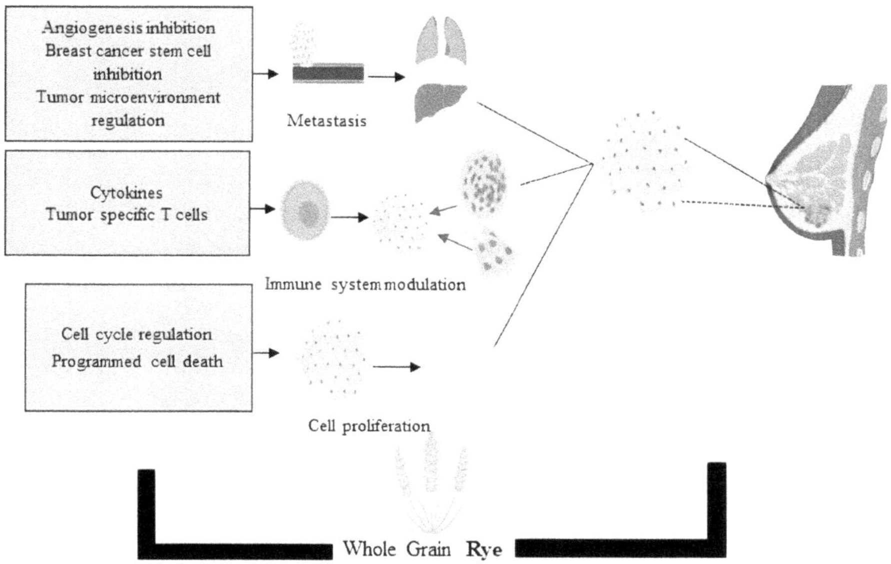

FIGURE 8.3 Overview of rye whole grain phytochemical's molecular targeted mechanism of action against breast cancer.

cells, HT-29. There was also a down parameter of Proliferating Cell Nuclear Antigen (PCNA), a tumor growth marker. Additionally, using phytate as a colon biomarker decreased the incidence of aberrant crypts (Cicero et al., 2017). Overview of rye whole grain phytochemical's molecular targeted mechanism of action against breast cancer as shown in Figure 8.3.

8.4 EFFECTS OF RYE-BASED FOOD ON METABOLISM

The benefits of rye and whole grains on health have been studied in a variety of analysis. The relationship between whole grains and health has been subject of several observational studies. Eating diets high in whole grains may help prevent heart disease, T2D and several malignancies, most notably colon cancer. On the other hand, the evidence on illness risk factors derived from diet related changes in human is not clear. However, it's evident from researches that whole grain foods are most important dietary components for the protection of non-communicable illness (Zhang et al., 2018a, b). Researchers have indicated the potential for whole grain meals, such as whole grain barley or rye products rich in intact kernels, to prevent diabetes and obesity. These foods are also rich in dietary fiber, especially non-starch polysaccharides (NSP) and RS. Moreover, eating whole grain rye bread, which is made of flour and kernels in a 1:1 ratio and contains 'RS type 2 (RS2) from high amylose maize (HAMRS2)', reduced insulin and glucose responses. In order to compensate for the loss of RS during the milling process of the kernels into flour, and to raise the RS level above that of intact kernels, HAMRS2 was added to the bread mix. Consequently, these results support the anti-diabetic properties of whole grain rye and suggest that

TABLE 8.3
Effects of Rye-Based Food on Metabolism

Rye-Based Food	Method	Effect on Metabolism	Results	Citations
Rye crispbread	Overweight women's randomized crossover trial	Increased satiety and reduced energy consumption	Participants reported feeling better and having less of an appetite, which could aid in controlling their weight.	Lee et al. (2016)
Rye-based dietary pattern	Human (n = 500) observational study	Improved lipid profile and a decreased risk of type 2 diabetes	Frequent rye meal consumption has been associated with decreased triglycerides, LDL cholesterol and fasting glucose.	Iversen et al. (2022)
Rye pasta	Crossover trial conducted at random in older individuals	Decreased postprandial blood sugar and increased production of insulin	Compared to white pasta, rye pasta produced better insulin secretion and a significantly reduced postprandial glucose response.	Juntunen et al. (2020)
Whole rye bread	Healthy adults (n = 30)	Increased insulin sensitivity after meals	When rye bread was consumed instead of wheat bread, blood glucose and insulin levels were lower, suggesting better glycemic control.	Jonsson et al. (2018)
Whole grain rye flour	Randomized controlled trial in men with metabolic syndrome	Decreased inflammatory markers and enhanced insulin sensitivity	Eating rye flour was linked to decreased C-reactive protein (CRP) levels and enhanced insulin sensitivity.	Iversen et al. (2022)

RS is crucial for the cereals' ability to improve metabolism. As compared to white wheat–based bread, whole grain rye with additional HAMRS2 also lowered levels of free fatty acid and raised peptide YY (PYY), the gut hormone that induces satiety. Products made from whole grain rye boost gut hormones like PYY and GLP-1 and enhance the level of metabolic control in healthy individuals (Sandberg et al., 2016). Effects of rye-based food on metabolism are shown in Table 8.3.

8.5 FUTURE PROSPECTIVE AND CONCLUSION

Rye grains have been shown to have positive impacts on human health due to their high presence of chemicals with antioxidant qualities. Numerous substances, including dietary fiber, phenolic acids, tocols, alkylresorcinols and sterols are found in significant amount in rye grains. Rye's health-promoting qualities are increasingly recognized as being essential for treating a number of chronic illnesses, including T2D, cancer and obesity. Its high dietary fiber content is essential for the metabolism of glucose. Research suggested that eating rye produces a more stable blood glucose response than eating refined grains, which is advantageous for people who have T2D or are at risk for the disease. Soluble fibers and bioactive substances in rye's unique makeup increase insulin sensitivity while decreasing the requirement for insulin secretion. This impact aids in blood sugar regulation. Rye's strong antioxidant content and anti-inflammatory qualities may help prevent cancer by reducing the risk of colon cancer in particular. SCFAs are produced when rye ferments in the stomach. These fatty acids have been connected to decreased inflammation and better gut health. This implies that rye might be beneficial to digestive health in general and cancer prevention in particular. Eating rye has been connected to better metabolic profiles, which include decreased levels of LDL cholesterol and better gut health. Due to their ability to maintain healthy lipid profiles and aid in digestion, lignins and phytosterols found in rye help provide these advantages. Another important area where rye shows potential benefits is in weight management. Its capacity to enhance satiety by inducing the release of gut hormones such as peptide aids in appetite regulation and lowers total calorie consumption. This can be especially helpful in the fight against obesity, which is a significant risk factor for a number of chronic illnesses. Cereals grains especially rye grain and food product are becoming more and more popular around the world. As they promote satiety and enhance the efficiency of digestive and metabolic systems, cereal grains high in protein and fiber can be used to satisfy hunger. It has been recommended that rye and products made with rye are included in a regular diet plan in order to achieve a disease-free healthy lifestyle. The well-established benefits of rye intake for insulin metabolism need further investigation to fill knowledge gaps about the mechanism and long-term health implications, including increased risk of diabetes and problems. The production of several tasty, inventive and novel rye products together with related health claims, effective marketing and effective communication are essential to raising consumer knowledge and consumption of rye meals. In summary, rye is a promising dietary ingredient with a variety of health benefits that merit more research. By concentrating on these goals, future studies will be able to offer more profound understandings of how rye might be used as a preventive strategy against chronic illnesses, which will ultimately lead to better public health outcomes.

REFERENCES

Adhikari, K. B., Tanwir, F., Gregersen, P. L., Steffensen, S. K., Jensen, B. M., Poulsen, L. K., & Fomsgaard, I. S. (2015). Benzoxazinoids: Cereal phytochemicals with putative therapeutic and health-protecting properties. *Molecular Nutrition & Food Research*, *59*(7), 1324–1338.

Ahmed, M. (2020). Colon cancer: A clinician's perspective in 2019. *Gastroenterology Research*, *13*(1), 1.

Aune, D., Keum, N., Giovannucci, E., Fadnes, L. T., Boffetta, P., Greenwood, D. C., ..., & Norat, T. (2016). Whole grain consumption and risk of cardiovascular disease, cancer, and all cause and cause specific mortality: Systematic review and dose-response meta-analysis of prospective studies. *BMJ*, *353*.

Bach Knudsen, K. E., Nørskov, N. P., Bolvig, A. K., Hedemann, M. S., & Laerke, H. N. (2017). Dietary fibers and associated phytochemicals in cereals. *Molecular Nutrition & Food Research*, *61*(7), 1600518.

Bağcı, A., Gecgel, Ü., Dursun, N., Özcan, M. M., Tamkoç, A., Özer, İ., & Özcan, M. M. (2019). The oil yields, mineral contents and fatty acid compositions of some rye (*Secale cereale*) grains. *Iranian Journal of Chemistry and Chemical Engineering (IJCCE)*, *38*(5), 285–292.

Bauer, E., Schmutzer, T., Barilar, I., Mascher, M., Gundlach, H., Martis, M. M., …, & Scholz, U. (2017). Towards a whole-genome sequence for rye (*Secale cereale* L.). *The Plant Journal*, *89*(5), 853–869.

Biel, W., Kazimierska, K., & Bashutska, U. (2020). Nutritional value of wheat, triticale, barley and oat grains. *Acta Scientiarum Polonorum Zootechnica*, *19*(2), 19–28.

Bolvig, A. K., Adlercreutz, H., Theil, P. K., Jørgensen, H., & Knudsen, K. E. B (2016). Absorption of plant lignans from cereals in an experimental pig model. *British Journal of Nutrition*, *115*(10), 1711–1720.

Chen, H., Chen, Z., Fu, Y., Liu, J., Lin, S., Zhang, Q., …, & Qin, W. (2019). Structure, antioxidant, and hypoglycemic activities of arabinoxylans extracted by multiple methods from triticale. *Antioxidants*, *8*(12), 584.

Cicero, A. F., Fogacci, F., & Colletti, A. (2017). Potential role of bioactive peptides in prevention and treatment of chronic diseases: A narrative review. *British Journal of Pharmacology*, *174*(11), 1378–1394.

Coe, S., & Ryan, L. (2016). Impact of polyphenol-rich sources on acute postprandial glycaemia: A systematic review. *Journal of Nutritional Science*, *5*, e24.

Dahl, W. J., Agro, N. C., Eliasson, Å. M., Mialki, K. L., Olivera, J. D., Rusch, C. T., & Young, C. N. (2017). Health benefits of fiber fermentation. *Journal of the American College of Nutrition*, *36*(2), 127–136.

De Vries, J., Birkett, A., Hulshof, T., Verbeke, K., & Gibes, K. (2016). Effects of cereal, fruit and vegetable fibers on human fecal weight and transit time: A comprehensive review of intervention trials. *Nutrients*, *8*(3), 130.

Dziki, D. (2022). Rye flour and rye bran: New perspectives for use. *Processes*, *10*(2), 293.

Eriksen, A. K., Brunius, C., Mazidi, M., Hellström, P. M., Risérus, U., Iversen, K. N., …, & Landberg, R. (2020). Effects of whole-grain wheat, rye, and lignan supplementation on cardiometabolic risk factors in men with metabolic syndrome: A randomized crossover trial. *The American Journal of Clinical Nutrition*, *111*(4), 864–876.

Ekmekcioglu, C., Elmadfa, I., Meyer, A. L., & Moeslinger, T. (2016). The role of dietary potassium in hypertension and diabetes. *Journal of physiology and biochemistry*, *72*(1), 93–106.

Feng, G. (2019). Rye. *Bioactive Factors and Processing Technology for Cereal Foods*, 151–169.

Gröber, U., Schmidt, J., & Kisters, K. (2015). Magnesium in prevention and therapy. *Nutrients*, *7*(9), 8199–8226.

Grossmann, I., & Koehler, P. (2016). Fractionation-reconstitution studies to determine the functional properties of rye flour constituents. *Journal of Cereal Science*, *70*, 1–8.

Guo, J., Lovegrove, J. A., & Givens, D. I. (2019). Food fortification and biofortification as potential strategies for prevention of vitamin D deficiency. *Nutrition Bulletin*, *44*(1), 36–42.

Gupta, A., Sharma, S., & Singh, B. (2019). *Processing effect on the bioactive compounds of cereals*. ResearchGate; unknown. https://www.researchgate.net/publication/337561635_Processing_Effect_on_the_Bioactive_Compounds_of_Cereals#pf7

Hackauf, B., Siekmann, D., & Fromme, F. J. (2022). Improving yield and yield stability in winter rye by hybrid breeding. *Plants*, *11*(19), 2666.

Hanhineva, K., Katina, K., Kolehmainen, M., Kyrø, C., Langton, M., Nordlund, E., & Poutanen, K. (2021). Whole grain rye intake improves metabolic health markers in humans: A systematic review. *American Journal of Clinical Nutrition*, *113*(5), 1235–1250.

Holscher, H. D. (2017). Dietary fiber and prebiotics and the gastrointestinal microbiota. *Gut Microbes*, *8*(2), 172–184.

Hong, K. W., Kim, S. H., Zhang, X., & Park, S. (2018). Interactions among the variants of insulin-related genes and nutrients increase the risk of type 2 diabetes. *Nutrition Research*, *51*, 82–92.

Ikram, A., Saeed, F., Noor, R. A., Imran, A., Afzaal, M., Rasheed, A., …, & Kinki, A. B. (2023). A comprehensive review on biochemical and technological properties of rye (*Secale cereale* L.). *International Journal of Food Properties*, *26*(1), 2212–2228.

Iversen, K. N., Jonsson, K., & Landberg, R. (2022). The effect of rye-based foods on postprandial plasma insulin concentration: The rye factor. *Frontiers in Nutrition*, *9*, 868938.

Johansson, D. P., Gutierrez, J. L. V., Landberg, R., Alminger, M., & Langton, M. (2017). Impact of food processing on rye product properties and their in vitro digestion. *European Journal of Nutrition*, *57*(4) 1651–1666.

Jonsson, K., Andersson, R., Knudsen, K. E. B., Hallmans, G., Hanhineva, K., Katina, K., …, & Landberg, R. (2018). Rye and health-where do we stand and where do we go? *Trends in Food Science & Technology*, *79*, 78–87.

Joye, I. J. (2020). Dietary fibre from whole grains and their benefits on metabolic health. *Nutrients*, *12*(10), 3045.

Juntunen, K. S., Niskanen, L. K., Liukkonen, K. H., Poutanen, K. S., Holst, J. J., & Mykkänen, H. M. (2020). Postprandial glucose, insulin, and incretin responses to grain products in healthy subjects. *The American Journal of Clinical Nutrition*, *75*(2), 254–262.

Kan, A. (2015). Characterization of the fatty acid and mineral compositions of selected cereal cultivars from Turkey. *Records of Natural Products*, *9*(1), 124.

Kaur, P., Sandhu, K. S., Purewal, S. S., Kaur, M., & Singh, S. K. (2021). Rye: A wonder crop with industrially important macromolecules and health benefits. *Food Research International*, *150*, 110769.

Kieffer, D. A., Martin, R. J., & Adams, S. H. (2016). Impact of dietary fibers on nutrient management and detoxification organs: Gut, liver, and kidneys. *Advances in Nutrition*, *7*(6), 1111–1121.

Koistinen, V. M., & Hanhineva, K. (2017). Microbial and endogenous metabolic conversions of rye phytochemicals. *Molecular Nutrition & Food Research*, *61*(7), 1600627.

Kulichová, K., Sokol, J., Nemeček, P., Maliarová, M., Maliar, T., Havrlentová, M., & Kraic, J. (2019). Phenolic compounds and biological activities of rye (*Secale cereale* L.) grains. *Open Chemistry*, *17*(1), 988–999.

Laidig, F., Piepho, H. P., Rentel, D., Drobek, T., Meyer, U., & Huesken, A. (2017). Breeding progress, variation, and correlation of grain and quality traits in winter rye hybrid and population varieties and national on-farm progress in Germany over 26 years. *Theoretical and Applied Genetics*, *130*, 981–998.

Lee, I., Shi, L., Webb, D. L., Hellström, P. M., Risérus, U., & Landberg, R. (2016). Effects of whole-grain rye porridge with added inulin and wheat gluten on appetite, gut fermentation and postprandial glucose metabolism: A randomised, cross-over, breakfast study. *British Journal of Nutrition*, *116*(12), 2139–2149.

Lee, P. G., & Halter, J. B. (2017). The pathophysiology of hyperglycemia in older adults: Clinical considerations. *Diabetes Care*, *40*, 444–452.

Li, Y., Li, S., Meng, X., Gan, R. Y., Zhang, J. J., & Li, H. B. (2017). Dietary natural products for prevention and treatment of breast cancer. *Nutrients*, *9*(7), 728.

Makarem, N., Nicholson, J. M., Bandera, E. V., McKeown, N. M., & Parekh, N. (2016). Consumption of whole grains and cereal fiber in relation to cancer risk: A systematic review of longitudinal studies. *Nutrition Reviews*, *74*(6), 353–373.

Martínez, M., Motilva, M. J., de Las Hazas, M. C. L., Romero, M. P., Vaculova, K., & Ludwig, I. A. (2018). Phytochemical composition and β-glucan content of barley genotypes from two different geographic origins for human health food production. *Food Chemistry*, *245*, 61–70.

Meija, L., & Krams, I. (2019). Rye. *Whole grains and their bioactives: Composition and health*, 169–208.

Nadar, S. S., Rao, P., & Rathod, V. K. (2018). Enzyme assisted extraction of biomolecules as an approach to novel extraction technology: A review. *Food Research International*, *108*, 309–330.

Nascimento, R. D. P. D., da Rocha Alves, M., Noguera, N. H., Lima, D. C., & Junior, M. R. M (2023). Cereal grains and vegetables. In *Natural plant products in inflammatory bowel diseases* (pp. 103–172). Academic Press.

Németh, R., & Tömösközi, S. (2021). Rye: Current state and future trends in research and applications. *Acta Alimentaria*, *50*(4), 620–640.

Nicholls, E. (2018). Breast cancer in a Renaissance book of the dead. *The Lancet Oncology*, *19*(8), 1023–1024.

Omar, N., Yeoh, B. S., Chellappan, K., Chui, S. Z., Salamt, N., & Aminuddin, A. (2023). The effects of pedometer-based exercise on central and peripheral vascular functions among young sedentary men with CVD risk factors. *Frontiers in physiology*, *14*, 1062751.

Ötles, S., & Ozgoz, S. (2014). Health effects of dietary fiber. *Acta scientiarum polonorum Technologia alimentaria*, *13*(2), 191–202.

Panche, A. N., Diwan, A. D., & Chandra, S. R. (2016). Flavonoids: An overview. *Journal of Nutritional Science*, *5*, e47.

Panhwar, R. B., Akbar, A., Ali, M. F., Yang, Q., & Feng, B. (2018). Phytochemical components of some minor cereals associated with diabetes prevention and management. *Journal of Biosciences and Medicines*, *6*(02), 9.

Rakha, E. A., & Green, A. R. (2017). Molecular classification of breast cancer: What the pathologist needs to know. *Pathology*, *49*(2), 111–119.

Redant, L., Buggenhout, J., Brijs, K., & Delcour, J. A. (2017). Extractability and chromatographic separation of rye (*Secale cereale* L.) flour proteins. *Journal of Cereal Science*, *73*, 68–75.

Roager, H. M., Vogt, J. K., Kristensen, M., Hansen, L. B. S., Ibrugger, S., Maerkedahl, R. B., …, & Licht, T. R. (2017). Whole grain-rich diet reduces body weight and systemic lowgrade inflammation without inducing major changes of the gut microbiome: A randomised cross-over trial. *Gut*, *68*(1), 83–93.

Sana Afzal, S. A., Aamir Shehzad, A. S., Randhawa, M. A., Ali Asghar, A. A., Muhammad Shoaib, M. S., & Jahangir, M. A. (2013). Health benefits and importance of utilizing wheat and rye. *Pakistan Journal of Food Science.*, *23*(4), 212–222.

Sandberg, J. C., Bjorck, I. M. E., & Nilsson, A. C. (2017). Effects of whole grain rye, with and without resistant starch type 2 supplementation, on glucose tolerance, gut hormones, inflammation and appetite regulation in an 11-14.5-hour perspective; a randomized controlled study in healthy subjects. *Nutrition Journal*, *16*(1), 25.

Sandberg, J. C., Björck, I. M., & Nilsson, A. C. (2016). Rye-based evening meals favorably affected glucose regulation and appetite variables at the following breakfast; a randomized controlled study in healthy subjects. *PLoS One*, *11*(3), e0151985.

Sapirstein, H. D., & Bushuk, W. (2016). Rye grain: Its genetics, production, and utilization. *Encyclopedia of Food Grains*, *1*, 159–167.

Sharma N, Bhatia S, Chunduri V, Kaur S, Sharma S, Kapoor P, Kumari A, Garg M. Pathogenesis of Celiac Disease and Other Gluten Related Disorders in Wheat and Strategies for Mitigating Them. Front Nutr. 2020 Feb 7;7:6. doi: 10.3389/fnut.2020.00006. PMID: 32118025; PMCID: PMC7020197.

Siddiqui, H., Sultan, Z., Yousuf, O., Malik, M., & Younis, K. (2023). A review of the health benefits, functional properties, and ultrasound-assisted dietary fiber extraction. *Bioactive Carbohydrates and Dietary Fibre, 30*, 100356.

Steffensen, S. K., Pedersen, H. A., Adhikari, K. B., Laursen, B. B., Jensen, C., Høyer, S., ..., & Fomsgaard, I. S. (2016). Benzoxazinoids in prostate cancer patients after a rye-intensive diet: Methods and initial results. *Journal of Agricultural and Food Chemistry, 64*(43), 8235–8245.

Suchowilska, E., Bieńkowska, T., Stuper-Szablewska, K., & Wiwart, M. (2020). Concentrations of phenolic acids, flavonoids and carotenoids and the antioxidant activity of the grain, flour and bran of *Triticum polonicum* as compared with three cultivated wheat species. *Agriculture, 10*(12), 591.

Suhr, J., Vuholm, S., Iversen, K. N., Landberg, R., & Kristensen, M. (2017). Wholegrain rye, but not wholegrain wheat, lowers body weight and fat mass compared with refined wheat: A 6-week randomized study. *European Journal of Clinical Nutrition, 71*(8), 959–967.

VSS, P., Adapa, D., Vana, D. R., Choudhury, A., Asadullah, J., & Chatterjee, A. (2018). Nutritional components relevant to type-2-diabetes: Dietary sources, metabolic functions and glycaemic effects. *Journal of Research in Medical and Dental Science, 6*(5), 52–75.

Vuholm, S., Nielsen, D. S., Iversen, K. N., Suhr, J., Westermann, P., Krych, L., ..., & Kristensen, M. (2017). Whole-grain rye and wheat affect some markers of gut health without altering the fecal microbiota in healthy overweight adults: A 6-week randomized trial. *The Journal of Nutrition, 147*(11), 2067–2075.

Wrigley, C., & Bushuk, W. (2017). Rye: Grain-quality characteristics and management of quality requirements. In *Cereal grains*, 153–178.

Yasin, M., Butt, M. S., & Zeb, A. (2017). Vitamin K2 rich food products. *Vitamin K2-Vital for Health and Wellbeing; InTechOpen: London, UK*, 35–61.

Zanella, I., Biasiotto, G., Holm, F., & di Lorenzo, D. (2017). Cereal lignans, natural compounds of interest for human health? *Natural Product Communications, 12*(1) 139–146.

Zhang, B., Zhao, Q., Guo, W., Bao, W., & Wang, X. (2018a). Association of whole grain intake with all-cause, cardiovascular, and cancer mortality: A systematic review and dose-response meta-analysis from prospective cohort studies. *European Journal of Clinical Nutrition, 72*(1), 57–65.

Zhang, Y., Cao, X., Juhasz, A., Islam, S., Qi, P., She, M., ..., & Ma, W. (2018b). Wheat avenin-like protein and its significant Fusarium head blight resistant functions. *bioRxiv*, 406694.

9 Allergens and Toxicity in Rye

Hamna Zafar, Mariam Islam, Farhan Saeed, Amara Rasheed, Aliza Naeem, Bushra Niaz, Muhammad Afzaal, Hareem Fatima Gulzar Ahmad Nayik, Aanchal Sharma and Tuba Mustafa

9.1 INTRODUCTION

A common grass that is grown for grain, cover and fodder is cereal rye (*Secale cereale* L.). It is linked to barley (*Hordeum vulgare* L.) and wheat (*Triticum aestivum*), delegate of the wheat family (*Triticeae*). Originally cultivated in Northern Europe, rye is a relatively recent cereal. It is believed to have originated from wild rye variants that spread like weeds in Asian grain fields (Grabiński et al., 2021). Due to their physical characteristics, rye grains are smaller, darker and more effective. Hectoliter (hL) weight is normally between 70 and 75 kg for marketing purposes, with a minimum of 70 kg/hL and a maximum moisture level of 12%. The protein fulfilled by rye grains is lower than that of wheat grains. The dough lacks the wheaten dough's elastic properties. According to reports, rye bread has a mild flavor and a firm texture (Gumul & Berski, 2021). Rye demonstrated a variety of nutritional and technical potentials despite its intricate chemical structure. Though their species diverge from an ideal milieu, rye kernels physiologically resemble wheat grain (Schwingshackl et al., 2019). The testa and pericarp, which surround the endosperm and germ which include the starchy endosperm and aleurone, are the outermost layers of the rye seed.

According to Food and Agriculture Organization Corporate Statistical Database (FAOSTAT) (2006), rye is grown and consumed across Russia, Poland, Northern Europe, Germany, Belarus and Ukraine being the top producers (Bondia-Pons et al., 2009). Both the growing conditions and genetic factors affect the rye's yield and quality. Different rye varieties have different chemical compositions, end-use values and agricultural traits like yields and resistance to lodging, diseases and pests. The population cultivars of rye are the most widely grown. They are grown and reproduced in a conventional manner, meaning that every stage of reproduction is the same (Buksa et al., 2012). Rye hybrids have become more and more common in recent years. These cultivars exhibit a 20–30% increase in yield in comparison to population types, as well as greater resistance to lodging and diseases, excellent adaptability and resilience to severe environmental conditions and soil quality (Grabiński et al., 2021).

DOI: 10.1201/9781003585817-9

Ergot, also known as *Claviceps purpurea*, is a disease that gradually affects wheat and rye crops. The ergot alkaloids (EAs) have the potential to seriously harm both people and animals. Non-celiac gluten sensitivity (NCGS), celiac disease (CD) and wheat allergy are among the negative reactions to wheat that can result from consuming wheat and the closely related grains rye and barley (Catassi et al., 2017). Storage proteins (gluten) are the primary triggers, although non-gluten proteins that are immunoreactive include puroindolines, β-amylases, α-amylase/trypsin-inhibitors (ATIs) and lipid transfer proteins (Scherf, 2019). The complex protein mixture of wheat, rye or barley that is insoluble in water or salt solution is commonly referred to as gluten (Lexhaller et al., 2019). Rye bread consumption has been linked to favorable physiological effects in a number of studies, particularly in relation to inflammation and glucose metabolism (Kallio et al., 2007). The therapeutic properties of whole grain diets proposed attributed to a variety of phytochemicals, vitamins and minerals in addition to dietary fiber (Fardet, 2010).

9.2 CHEMICAL COMPOSITION OF RYE

As a proportion of dry matter, whole grain rye flour's composition consists of total dietary fiber (15–21%), ash (2%), proteins (8–13%), lipids (2–3%) and carbs (56–70%). Compared to wheat, rye has a higher dietary fiber content (15–21%), with about 20% of the fiber being soluble. Arabinoxylans (8–12%), ß-glucan (1.3–2.2%) and cellulose (1–1.7%) are the main fibers found in rye (Brownlee et al., 2017). Ingesting rye cereals benefits human health since they have a high concentration of chemicals with antioxidant properties. Whereas rye bran is mostly composed of cellulose and insoluble arabinoxylans, oat bran has a higher concentration of dietary fiber, primarily β-glucans. Numerous substances, including folate, tocols, phenolic acids, alkylresorcinols, dietary fiber and sterols, are found in significant amounts in rye grains. Individual rye cultivars did, however, differ significantly in the amounts of these chemicals (Nyström et al., 2008).

Additionally, phenolic acids liberated from glycosides and soluble esters (vanillic, caffeic, p-coumaric and sinapic) as well as free phenolic acids (p-coumaric, ferulic) were discovered. The common prevalent phenolic acid found in rye cereals and rye-derived products was ferulic acid. The ABTS (2,2′-azino-bis(3-ethylbenzothiazoline-6-sulfonic acid)) and DPPH (1,1-diphenyl-2-picrylhydrazyl radical) tests have been employed in numerous investigations to determine the antioxidant capacity of rye extracts (Bondia-Pons et al., 2009). Aqueous concentrates from grains of all researched rye cultivars showed the best antioxidant effect (up to 98%) based on the ABTS free radical scavenging (Mishra et al., 2017). Moreover, the antioxidant capacity of rye grain concentrates was notably impacted by the enzymatic treatment of rye flour with tannase from *Paecilomyces variotii*; the DPPH and FRAP ratios rose by 1150 and 61%, respectively (de Lima et al., 2018).

Rye was placed among the cereals with the greatest antioxidant activity, as assessed by FRAP and DPPH, in flour milling by-products (Ivanišová et al., 2012). When compared the Trolox equivalent antioxidant capacity (TEAC) of wheat bread, rye crops, which is measured by their ability to scavenge free radicals against DPPH and ABTS radicals, respectively, revealed more antioxidant content and superior antioxidative qualities. The increased amount of alkylresorcinols in the bran fraction of rye grains compared to wheat, oat and triticale cereals may be the cause of their high antioxidant capacity. Protease inhibition, including thrombin and trypsin, was associated with greater levels of phenolic acids. Extracts of a whole meal of rye additionally showed additional aminopeptidase, exoproteolytic, endoproteolytic, carboxypeptidase and N-α-benzoyl-arginine-p-nitroanilide hydrolyzing actions. According to numerous research, rye (*S. cereale* L.) contains enzyme inhibitors, namely those that inhibit α-amylase, trypsin and bi-functional proteinase/α-amylase (Kulichová et al., 2019).

Even though a significant amount of research has been done on the dispersal and amounts of specific phytochemicals in rye kernels, there are still some substances, like flavonoids and benzoxazinoids, about which little is known. While wheat and maize include additional kinds of benzoxazinoid compounds, the main benzoxazinoids in rye are the hexose and mono-conjugates of cyclic hydroxamic acid DIBOA (2,4-dihydroxy-1,4-benzoxazin-3-one) and lactam HBOA (2-hydroxy-1,4-benzoxazin-3-one) (Pihlava et al., 2015). According to a recent study by Adhikari et al. (2013), people absorb and metabolize dietary benzoxazinoids.

9.3 TOXIC COMPOUNDS IN RYE

Ergot often does not result in a significant decrease in yield; nonetheless, EAs can contaminate grain (Florea et al., 2017) that is dangerous as they are poisonous to both people and animals. Research conducted in various countries showed that EAs were often found in all popular grains (Schwake-Anduschus et al., 2020). The European Union (EU) has established maximum limits of 500 μg/kg for the six most common EAs, including ergosine and ergocryptine and their corresponding -inine epimers, in rye milling products intended for human consumption. In addition, a limit of 0.5 g/kg has been set for ergot sclerotia and sclerotial fragments in unprocessed cereals intended for human use (European Union, 2021).

In rye, hybrid cultivars are in light of cytoplasmic male sterility (CMS), which is due to a unique cytoplasm in the female parent. The male stepparent provides nuclear-encoded DNA of restorer-to-fertility (Rf), which should completely restore pollen shedding in the resultant hybrids (Miedaner & Laidig, 2019). However, the Rf genes employed result in restoration capacity and, consequently, pollen shedding that varies from poor to exceptional.

In terms of food safety, agriculture and animal nutrition, rye (*S. cereale*) has a number of hazardous substances. These include mycotoxins, particularly T-2/HT-2 and deoxynivalenol (DON), which are generated by *Fusarium* species that frequently infect rye under humid environments. In cattle, these poisons have been linked to immunological suppression, gastrointestinal issues and low weight gain (Uhlig et al., 2016). EAs, which are produced when *C. purpurea* contaminates the body with fungi, are still a serious problem. Hallucinations, gangrene, formerly known as ergotism, vasoconstriction and reproductive problems can result from long-term exposure to these substances in tainted grains (Crews, 2015). Additionally, rye contains naturally occurring benzoxazinoids (BXs), including DIBOA and DIMBOA, which have dual functions as defense chemicals and phytotoxicity. Some research showed that these substances may have cytotoxic or genotoxic effects on mammalian cells. They can also seep into the soil from rye leftovers and prevent the germination of seeds or the establishment of roots in further crops (Schulz 2013). The major toxic compounds present in rye are summarized in Table 9.1.

Rye bran contains phenolic lipids called alkylresorcinols, which have also garnered interest in recent studies. High doses may damage cell membranes and alter the gut microbiota, even though they have antibacterial action and may be beneficial to health when used in moderation (Ross et al., 2017). Because of these chemicals, rye has both agronomic and toxicological value, underscoring the significance of post-harvest monitoring and regulated production. In cross-pollinated crops such as rye, susceptibility to ergot infection is influenced by multiple factors, including

TABLE 9.1
Toxic Compounds in Rye

Compound Type	Specific Compounds	Source/Cause	Health Effects	References
Mycotoxins	Deoxynivalenol (DON)	*Fusarium* spp. during humid conditions	Vomiting, immunosuppression, growth retardation	Qu et al. (2024)
	Zearalenone (ZEN)	*Fusarium* spp. contamination	Endocrine disruption, reproductive toxicity	Qu et al. (2024)
	Ochratoxin A (OTA)	*Aspergillus*, *Penicillium* spp.	Nephrotoxicity, carcinogenicity	EFSA (European Food Safety Authority) (2020)
	Trichothecenes (T-2, HT-2)	*Fusarium* spp.	Cytotoxicity, immunosuppression	EFSA (2011)
	Fusarin C	*Fusarium moniliforme*	Mutagenicity	Conti Taguali et al. (2023)
Heavy metals	Cadmium (Cd)	Soil contamination, fertilizers	Nephrotoxicity, bone demineralization	WHO (2024)
	Lead (Pb)	Atmospheric deposition, water contamination	Neurotoxicity, hypertension	WHO (2024)
Phytochemicals	Benzoxazinoids (DIBOA, DIMBOA)	Natural plant metabolites	Allelopathic effects, toxicity to non-target organisms	Hussain et al. (2022)
	Mercury (Hg)	Industrial pollution	Neurotoxicity, kidney damage	WHO (2024)
	Phenolic acids (ferulic, p-coumaric acid)	Naturally occurring in rye grain	Antioxidants; high doses may have anti-nutritional roles	Wiczkowski et al. (2019)
	Alkylresorcinols	Rye bran	Whole grain biomarkers, possible anti-nutrients	Landberg et al. (2018)

pollen fertility, morphological disease-escape characteristics, and maternal genetic resistance. Strong pollen fertility limits pathogen establishment by ensuring rapid ovary fertilization, whereas certain floral traits decrease the plant's physical susceptibility to infection (Menzies & Turkington, 2015). Plant resistance is initiated immediately after the conidia reach the pistil, according to recent transcriptome investigations in wheat (Boyd et al., 2020). For totally male-sterile entry under pollen isolation in rye and pearl millet, slight but noteworthy variations in ergot severity were previously observed. In order to determine if genetic variations result from post-host refusal, we injected CMS rye without pollen availability, meaning that none of the flowers were fertilized.

9.4 ALLERGENIC COMPOUNDS IN RYE

An allergen is a material, frequently a dietary protein, that causes sensitized people to have an undesirable reaction after being exposed to a particular meal due to antigen-specific immune systems (Ebisawa et al., 2020). The 'Big Eight' food allergies include gluten (such as wheat, barley and rye), crustaceans, eggs, fish, milk, tree nuts (such as cashews, walnuts and almonds), peanuts and soyabeans (WHO/FAO, 2020); they are responsible for about 90% of food allergies (Teuber et al., 2015). According to reports, 8% of children globally and 3–11% of adults, depending on the location, suffer from food allergies (Gupta et al., 2019). Allergy and asthma have been linked to inhaling flour made from wheat, rye, barley, soy, *Lathyrus sativus* and buckwheat; most of these instances have been linked to wheat, rye or barley flour.

Because of its flour and pollen, rye (*S. cereale*) is a major source of allergens that can cause respiratory allergies and occupational asthma. Research has demonstrated a substantial correlation between baker's asthma and certain IgE antibodies to rye flour, allowing for a precise diagnosis without the need for inhalation challenge testing (Fernández-Nieto et al., 2024). Furthermore, rye pollen has a significant degree of cross-reactivity with the pollens of other grasses, including perennial ryegrass and timothy, which makes immunotherapy and allergy testing more difficult (Aleksić et al., 2014). Temperature variations and ozone pollution are two environmental elements that have been demonstrated to affect rye pollen's allergenic potency, increasing its clinical significance in seasonal allergies (Venkatesan et al., 2024). Additionally, there have been isolated reports of rye-related food-dependent exercise-induced anaphylaxis (FDEIA), particularly when consumed before physical activity (García-Menaya et al., 2016). Due to variations in commercial rye extract formulations, diagnostic techniques like skin prick testing (SPT) exhibit variability, suggesting the need for standardization (van Kampen., 2009).

Strong clinical data has shown that allergen immunotherapy, especially sublingual immunotherapy (SLIT), is a safer and more effective treatment for rye and grass pollen allergies (Dhami et al., 2017). Among occupational asthma, baker's asthma remains one of the most prevalent types. A number of cereal antigens, flour additives and storage mites cause an IgE-mediated allergic reaction, which is part of its pathophysiology. Wheat flour is thought to be the most significant allergen source among these causative factors. Nonetheless, bakers and farmers are using rye flour more and more. According to reports, the most well-characterized allergens from

TABLE 9.2
Allergenic Compounds in Rye

Allergen Type	Specific Allergens	Source/Protein Family	Clinical Relevance	References
Seed storage proteins	Secalin (γ-secalin, ω-secalin)	Prolamins (gluten-like proteins)	Triggers celiac disease and non-celiac gluten sensitivity	EFSA (2014)
Flour protein allergen	Sec c 38	α-Amylase/trypsin inhibitor	Major allergen in baker's asthma; positive in 71% of cases	Thermo Fisher (2024)
Pollen allergen	Sec c 5	Group 5 grass pollen allergen	Causes seasonal allergic rhinitis and asthma	Thermo Fisher (2024)
Pollen allergen	Sec c 20	γ-Secalin (70 kDa)	Cross-reactive with other grass pollens; contributes to respiratory allergies	Thermo Fisher (2024)
Cross-reactive proteins	Profilins, pathogenesis-related proteins	Conserved plant proteins	Cross reactivity with other cereals, nuts and pollens; complicates diagnosis	Castor et al. (2024)
Environmental influence	Increased allergen expression	Elevated ozone exposure	Higher allergen content in pollen under ozone stress; exacerbates allergic responses	Eckl-Dorna et al. (2010)

cereal flour are the trypsin and α-amylase inhibitors, which make up the majority of the salt-soluble protein family. These endosperm proteins have molecular weights between 12 and 16 kDa and are soluble in salt.

The majority of these family members have the ability to bind IgE in vitro and produce favorable results in skin prick tests. The allergic significance of rye flour enzymatic inhibitors in baker's asthma is, however, poorly understood. At the genetic level predisposed people who are treated with gluten and other environmental variables are at risk for developing CD, a systematic illness mediated by T helper type 1 (Th1) cells. This combination of variables results in an inflammatory response that damages the intestinal mucosa. Eliminating wheat, rye, barley and malt, a derivative of barley, from the diet is the sole proven remedy (Hill et al., 2016). Allergies can be brought on by eating, skin contact and inhalation of wheat, rye and barley (Cianferoni, 2016). Some allergenic compounds reported in rye are summarized in Table 9.2.

9.5 HEALTH IMPACTS OF RYE TOXINS AND ALLERGENS

9.5.1 Respiratory Allergies (Pollen)

Rye cereals are notable for major peak of fiber content relative to the other seeds often consumed by people. Rye is one of the best nutritional sources and bioactive substances, and significant raw material for bread and bakery goods after wheat (Ikram et al., 2023). However, it can also pose various health risks due to its allergenic proteins, gluten content and potential contamination with natural toxins. Rye grass are significant allergen-producing grasses, common in the Midwest from May to July. Southern areas like Florida and California have grass seasons lasting 10–11 months, affecting the air quality (Lei & Grammer, 2019). Pollens are generally the main cause of pollen allergy in many people and are emitted by a hundred distinct species of plants for fertilization. Rye-grass pollen sensitization is worldwide; patients were affected with various allergic diseases, like allergic rhinitis (Vusthepalli et al., 2022).

9.5.2 Food Allergy

Food allergy symptoms can result from a range of processes and tend to be classified as unfavorable reactions brought on by the consumption and ingestion of a nutritious component. They are generally reported in cross-sectional studies; between 20 and 25% of those polled had food opinions (Warren et al., 2020). Presenting with symptoms that affect both children and adults from mild itching, stomach pain and rash to severe anaphylaxis, it leads to anxiety and dietary restrictions for many patients (Tedner et al., 2022). Anaphylaxis is a critical allergic reaction that is fast in onset and may be life-threatening. It can be regulated by IgE, or by other immunological or non-immunological mechanisms. The NIAID criteria suggest that anaphylaxis is highly probable when an acute illness onset occurs with either cutaneous, respiratory or cardiovascular system symptoms or at least two or more manifestations in cutaneous, respiratory, gastrointestinal and cardiovascular systems (Cardona et al., 2020). A novel disorder with IgE-mediated mechanisms (like OAS), FDEIA takes place if the raising meal arrives within 4 hours of carrying out physical activity (dancing, running, long walks etc.) or within the hour after. Urticaria, angioedema, pulmonary, gastrointestinal and anaphylactic shock make up the indications (Gargano et al., 2021).

9.5.3 Rye Toxins: Ergot Alkaloids

A recurring possibility to the total rye value chain is ergot contamination brought on by the phytopathogenic fungus *C. purpurea*. The fungus sclerotia develop EAs, which are poisonous to both humans and animals and are not allowed in human food (Miedaner et al., 2021). Ergotism, a severe health condition caused by the consumption of ergot sclerotia, can lead to seizures, psychosis, stomach pain, nausea/vomiting, reduced milk production and blood circulation. The disease's occurrence in grain can vary due to pathogen variation, host specificities and crop origin (Walkowiak et al., 2022). The entangled EAs and ergotamine cause dysregulation of biological functions, defined by vasoconstriction as well as thermoregulation. EAs interact as agonists or antagonists at the various monoamine neurotransmitter receptor sites with no specificity (Reddy et al., 2020). According to the EFSA et al. (2017), studies found that rye and rye-containing commodities had the highest levels of EAs, with the highest exposure in humans being infants and children. Animal feed exposure varied between 0.31 and 0.46 μg/kg bw per day in beef cattle and 6.82–8.07 μg/kg bw per day in piglets.

9.5.4 Celiac Disease and Gluten Sensitivity

Cereal dietary consumption is induced by gluten-linked disarray, CD, NCGS (Roszkowska et al., 2019). Rye, wheat and barley contain gluten proteins that trigger immune-mediated inflammation of the small intestine in people with CD (Pasquali et al., 2019). Symptoms profile can be gastrointestinal, as shown in Figure 9.1. It is an unusually small bowel chronic inflammatory disorder with an occurrence of 1% worldwide and an increasing occurrence globally (Lebwohl & Rubio-Tapia, 2021). Signs of CD can vary significantly between individuals and are primarily manifested with many gastrointestinal problems, including diarrhea, abdominal pain and constipation (Tye-Din et al., 2018). Diagnosing CD is based on several tests, which consist of examining medical history, physical examination, serological tests to find antibodies to CD and small bowel biopsy for confirmation (Kaswala et al., 2015). To avoid false test results, patients with CD must maintain their usual gluten-containing diet before small bowel biopsy. The treatment involves a lifetime gluten-free diet, eliminating wheat, barley and rye sources and products (Sharma et al., 2015).

In the last 20 years, the incidence of diseases believed to be induced by the ingestion of wheat and related gluten-containing cereals, including rye and barley, has

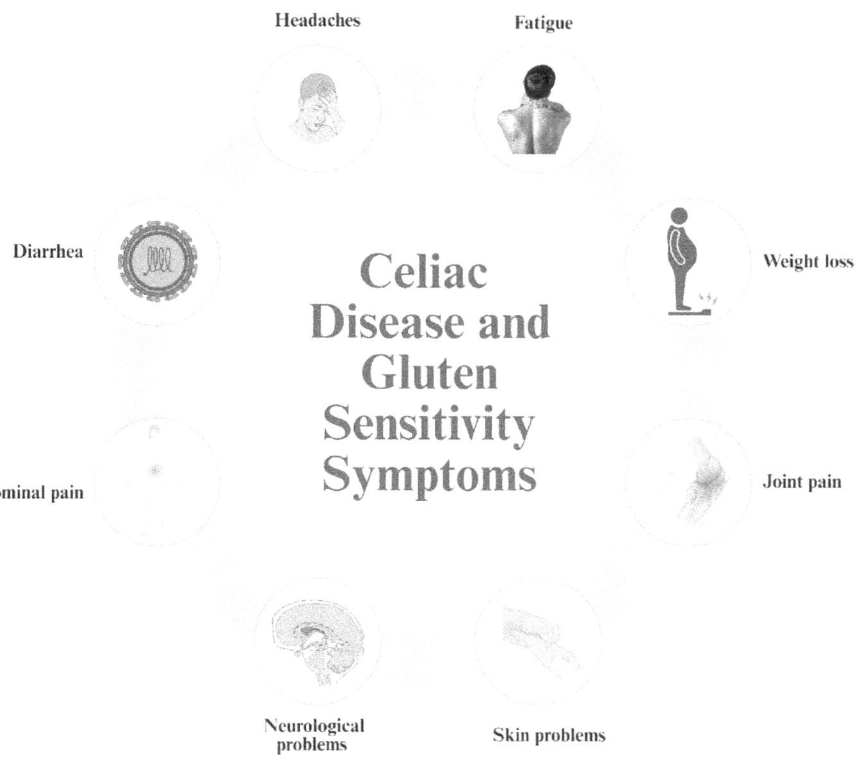

FIGURE 9.1 Symptoms of celiac disease and gluten sensitivity.

increased (Leonard et al., 2017). NCGS is a novel syndrome of gluten intolerance. In the chapter, some other names have been suggested for this disarray, such as gluten sensitivity (GS), gluten hypersensitivity and non-celiac gluten intolerance (Pieczyńska, 2018). The extent of the spread of this disorder is poorly defined, that 13% of the citizens had a self-reported sensitivity to gluten, and that it is more prevalent in women (Miranda et al., 2023).

9.5.5 Detection and Quantification Techniques of Allergens and Toxins in Rye

Rye (*S. cereale*) is a food cereal, but its safety is compromised by allergens, like gluten, and mycotoxins (EAs and *Alternaria* toxins). The report delivers new analytical methods for their determination and quantitation, like immunoassays (enzyme-linked immunosorbent assay (ELISA), lateral flow assays), molecular biology (polymerase chain reaction (PCR)) and novel chromatographic methods, as shown in Figure 9.2 (Kononenko et al., 2021). Rye (*S. cereale*) is a common cereal that is a food staple worldwide. Allergens, immune-stimulating proteins inducing mild to severe anaphylaxis, and mycotoxins, toxic fungal secondary metabolites such as those from *Aspergillus*, *Penicillium* and *Fusarium*, are threats to its safety. However, detection of allergens depends on the presence of certain proteins or DNA (Yu & Pedroso, 2023). The commonly employed methods for the detection of rye allergens are summarized in Table 9.3.

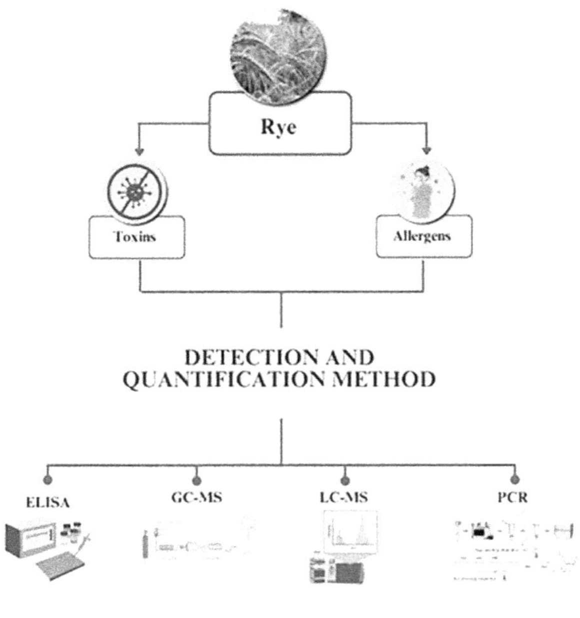

FIGURE 9.2 Detection and quantification methods of rye toxins and allergens.

TABLE 9.3
Rye Allergen Detection Methods

Method	Principle	Target Analyte	Specificity	Advantages	Limitations	Typical Applications
ELISA	Antibody–antigen interaction, enzyme-linked colorimetric detection	Protein	Specific but susceptible to matrix cross-reactions	Widely used, quantitative, routine analysis, cleaning validation	Matrix interference, cross-reactivity (false positives/negatives), limited selectivity, affected by protein denaturation/processing	Routine allergen detection, raw/semi-processed foods, cleaning validation
Lateral flow immunoassay (LFA)	Immunochromatographic, antibody–antigen reaction on strip, visual/reader detection	Protein (or mycotoxin)	Highly specific	Fast (minutes), on-site, easy handling, low cost, no lab equipment, qualitative/semi-quantitative/quantitative with reader	Visual evaluation can be subjective for semi-quantitative, requires reader for quantitative	Rapid on-site screening, hygiene control (swab tests)
Polymerase chain reaction (PCR)	DNA amplification using specific primers	DNA	Highly specific	Highly sensitive, specific, good for processed foods (DNA stability)	Identifies DNA, not immunogenicity (cannot assess allergic reaction potential), low sensitivity for some heat-processed foods (DNA extraction issues)	Allergen detection in processed foods, contamination identification
Liquid chromatography–mass spectrometry (LC-MS/MS)	Protein extraction, enzymatic digestion to peptides, LC separation, MS/MS detection of specific peptides	Peptides (derived from proteins)	High specificity, can differentiate closely related allergens	High sensitivity, specificity, accuracy, multiplexing, robust for processed foods (peptides), overcomes ELISA/PCR limitations	Lack of comprehensive protein sequence repositories, genome format issues, need for robust species-specific peptide markers, intact protein separation challenges, variability based on allergen/heat treatment/sample prep, high cost, complex operation	Quantitative analysis of contaminants/allergens, multi-allergen detection, gluten quantification, confirmatory analysis

9.5.6 Techniques of Detection and Quantitation of Rye Allergens

ELISA is widely applied in specific protein detection; it is very sensitive but susceptible to matrix interference and cross-reactivity and can produce false-positive results, particularly in processed food. ELISA is also a commonly used method known for its sensitivity and specificity in identifying allergenic proteins (Smits et al., 2024). A secondary antibody, which has been conjugated with an enzyme, is added, which attaches to the complex and enhances the signal. Monoclonal antibodies, which bind to a single epitope, and polyclonal antibodies, which bind to multiple epitopes, are used (Radomirović et al., 2025).

Lateral flow immunoassay (LFA) is a quick on-site test strip for qualitative or semi-quantitative screening and provides ease of use and mobility for rapid checks. LFAs are quick, immunochromatographic test strips that combine an entire antibody-based test system on one membrane. Visual inspection is suitable for qualitative or semi-quantitative analysis. For quantitative analysis, a digital reader is available. LFAs find application in quick, in situ testing of mycotoxins and allergens (Zhan et al., 2018). They are also utilized for hygiene control through swab tests in food processing. Key benefits are quick results (usually within 5 minutes), ease of use, low expense, portability (no laboratory equipment needed) and not requiring hazardous material, and hence they are ideal to be performed by untrained staff (Sadeghi et al., 2024).

PCR is applied to detect specific DNA sequences, in processed foods where DNA remains stable. It identifies allergenic material (DNA) rather than immunogenic characteristics, and it generally identifies single analytes (Dettmann et al., 2025). PCR is an analytical technique applied for the detection of allergens by recognizing specific sequences of DNA that are unique to a specific allergen. The process includes homogenizing the food sample, DNA extraction and purification to eliminate inhibitors, followed by target DNA sequence amplification by specific primers. Real-time PCR is commonly used to detect amplified DNA, allowing for quick identification (Liu et al., 2022). PCR is very specific and sensitive. But it has a major drawback that it detects allergenic material (DNA) but not its immunogenic properties. Also, with certain heat-processed foods, sensitivity of PCR can become affected because DNA gets extracted poorly (Niedziela et al., 2021).

Liquid chromatography–mass spectrometry (LC-MS/MS): It is a sensitive technique that directly measures and detects species- and allergen-specific marker peptides that have been digested. Careful sample preparation (digestion, extraction) is necessary, marker selection is best and MS/MS analysis is necessary for high specificity and multiplexing (Radomirović et al., 2025). High-performance liquid chromatography (HPLC) is applied for disjunction, typically with UV/Vis, fluorescence or MS. Gas chromatography (GC) is applied primarily for volatile analytes, typically in conjunction with MS. No technique is superior to all; they supplement MS-based techniques, providing definitive analysis (Qu et al., 2025).The technique entails the extraction of proteins from the food source, enzyme digestion to peptides (e.g., with trypsin), liquid chromatographic separation of the peptides and subsequent analysis with high delicacy and specificity by tandem mass spectrometry (MS/MS) (Benedé et al., 2022).

9.5.7 Detection and Quantification Methods for Rye Toxins

Mycotoxins are secondary, toxic metabolites of filamentous fungi (*Penicillium*, *Aspergillus*, *Fusarium*) that spoil crops such as rye and carry serious food safety hazards. Mycotoxins need sensitive and quick analytical techniques for their detection, raising stringent demands for adequate sample treatment (Camus-Ela et al., 2025). Table 9.4 summarizes the major analytical approaches used for the detection of mycotoxins in rye.

MS is very selective and discriminative for the identification and quantification of toxins. HPLC-MS/MS is stable for non-volatile toxins, whereas GC-MS is used for volatile toxins. LC-MS/MS and LC-FLD are used for EAs. HPLC-UV/FLD and IAC cleanup are common for deoxynivalenol. GC-MS, HPLC-FD and HPLC-MS/MS are used for T-2/HT-2 toxins. 15 HPLC-MS/MS is the most preferred for *Alternaria* toxins (Mekonnen & Labuschagne, 2025).

Multiple reaction monitoring (MRM) mode of the mass spectrometer allows for the simultaneous analysis of peptides from more than one allergen protein with high specificity. Chromatographic methods, including HPLC and GC, when combined with MS or MS/MS, are very effective for the analysis and espial of principal mycotoxins. HPLC-MS/MS is also very strong and sensitive for regular analysis and can identify non-volatile compounds like mycotoxins with high accuracy and precision (Zhang et al., 2025). GC-MS is normally used for volatile compounds and provides quicker analysis and high selectivity. External standard calibration is not typically precise due to matrix effects. IDMS is more precise, adjusting for matrix effects by using isotopically enriched internal standards. IDMS forms (ID1MS, ID2MS, ID5MS) are precise, with ID5MS sufficient for high-throughput analysis (Chelladurai et al., 2025).

ELISA is used to screen and detect mycotoxins. Lateral flow assays offer rapid on-site screening, with digital reader quantification. Analytical innovations are initiated by new or reduced maximum levels of mycotoxins, which demand more sensitive analysis (Cardona-Alzate et al., 2025).

High-resolution mass spectrometry (HRMS) is useful for the attribution of unknown materials and effective in non-targeted analyses, detecting even slight mass changes with high precision. This never-ending loop of regulation-facilitated innovation is essential for continually advancing food safety standards that detection capabilities keep abreast of advancing health protection objectives (Guardia et al., 2025).

Food safety requirements are a top priority for rye. A strong declaration of gluten-containing rye is required under new allergen labeling (February 25, 2024). New/lowest maximum limits for DON (deoxynivalenol) and lowered EA limits for rye (to 250 µg/kg by July 1, 2028) drive analytical innovation. FDA leads on allergen management and labeling. 'Big 9' allergens are declared, but the declaration of rye/barley is voluntary except for gluten-containing, a potential loophole. Increased

TABLE 9.4
Mycotoxin Detection Methods in Rye

Mycotoxin	Primary Methods	Key Principles	Specificity	Advantages	Limitations
Ergot alkaloids	LC-MS/MS, LC-FLD, ELISA	LC separation, MS/MS detection of specific ions; LC separation, fluorescence detection; antibody–antigen reaction	High (LC-MS/MS)	High sensitivity, selectivity, handles complex mixtures (LC-MS/MS); rapid (ELISA)	High cost, complex operation (LC-MS/MS)
Deoxynivalenol (DON)	HPLC-UV, HPLC-FLD, GC-MS (confirmation), ELISA	HPLC separation, UV/fluorescence detection; GC separation, MS detection for volatile compounds; antibody–antigen reaction	Can be good, but matrix interference can be an issue	Quick, simple, common equipment (HPLC); rapid (ELISA)	Selectivity not always sufficient for quantification without cleanup (HPLC); high cost/complexity (GC-MS)
T-2 and HT-2 toxins	GC-FID/ECD/MS, HPLC-FD, UPLC/FLD, HPLC-MS/MS, ELISA	GC separation, various detectors; HPLC/UPLC separation, fluorescence detection; LC separation, MS/MS detection; antibody–antigen reaction	Good, but needs cleanup	High sensitivity (HPLC-MS/MS); rapid screening (ELISA/UPLC)	Destructive (GC); can have cross-reactivity (ELISA)
***Alternaria* toxins**	HPLC-MS/MS	LC separation, MS/MS detection of specific ions	High selectivity	High sensitivity, selectivity, handles complex mixtures, multiplexing	High cost, complex operation
Ochratoxin A (OTA)	LC-MS/MS, HPLC-UV/Vis/FLD, FPIA, LFA	LC separation, MS/MS detection; HPLC separation, UV/Vis/fluorescence detection; fluorescence polarization immunoassay; immunochromatographic strip	High selectivity (LC-MS/MS); specific (FPIA/LFA)	High sensitivity, selectivity, accuracy (LC-MS/MS); rapid, on-site, cost-effective (LFA)	Matrix effects (external calibration); can have cross-reactivity (LFA)

mycotoxin prevalence (climate change) and 'modified mycotoxins' necessitate adaptive strategies (Sørensen et al., 2024).

LC-MS/MS is still needed because it is more specific and multiplexing, which is required for confirmatory analysis and attaining the stringent limits. Good sample preparation and IDMS are necessary for accuracy. Merging new technologies with risk management is the key to safeguarding public health (Salvador et al., 2023). Detection and quantitation of allergens and toxins in rye are essential for food safety. Rye is a hazard from allergens (secalins, gluten) and mycotoxins (EAs, *Alternaria* toxins), with impact ranging from allergy to extreme toxicity (Das et al., 2025). LC-MS/MS enhanced sensitivity, specificity and multiplexing for precise quantitation in complex matrices, underlined by stringent sample preparation and isotope dilution. The dynamic regulatory landscape, driven by health risk and climate change, demands ongoing analytical innovation (Pompeia et al., 2022).

9.6 CONCLUSION

A nutrient-dense grain rich in fiber, antioxidants and phytochemicals, rye is essential for human nutrition and the creation of functional foods, but it also presents toxicological and allergic problems. Because of their gluten-like storage proteins, rye flour and pollen can cause non-celiac gluten-related disorder and CD, and occupational exposure can aggravate respiratory disorders. A staple food, rye is susceptible to contamination by harmful substances such as mycotoxins and EAs, which can lead to serious illnesses that include immunotoxicity and neurotoxicity. Strict food safety procedures, such as controlled cultivation, post-harvest monitoring and sophisticated analytical detection techniques, are necessary for rye, a functional food and possible source of allergies and toxins.

REFERENCES

Adhikari, K. B., Laursen, B. B., Gregersen, P. L., Schnoor, H. J., Witten, M., Poulsen, L. K., Jensen, B. M., & Fomsgaard, I. S. (2013). Absorption and metabolic fate of bioactive dietary benzoxazinoids in humans. *Molecular Nutrition & Food Research*, *57*(10), 1847–1858.

Benedé, S., Lozano-Ojalvo, D., Cristobal, S., Costa, J., D'Auria, E., Velickovic, T. C., Garrido-Arandia, M., Karakaya, S., Mafra, I., Mazzucchelli, G., Picariello, G., Romero-Sahagun, A., Villa, C., Roncada, P., & Molina, E. (2022). New applications of advanced instrumental techniques for the characterization of food allergenic proteins. *Critical Reviews in Food Science and Nutrition*, *62*(31), 8686–8702.

Bondia-Pons, I., Aura, A. M., Vuorela, S., Kolehmainen, M., Mykkänen, H., & Poutanen, K. (2009). Rye phenolics in nutrition and health. *Journal of Cereal Science*, *49*(3), 323–336.

Boyd, L. A., Tente, E., Ereful, N., Rodriguez, A. C., Grant, P., O'Sullivan, D. M. & Gordon, A. (2020). *Reprogramming of the wheat transcriptome in response to infection with Claviceps purpurea, the causal agent of ergot*, 16 December 2020, PREPRINT (Version 1) Retrieved February 11, 2021, from Research Square. https://doi.org/10.21203/rs.3.rs-126182/v1

Brownlee, I. A., Chater, P. I., Pearson, J. P., & Wilcox, M. D. (2017). Dietary fibre and weight loss: Where are we now? *Food Hydrocolloids*, *68*, 186–191.

Buksa, K., Nowotna, A., Gambuś, H., Krawontka, J., Sabat, R., & Noga, M. (2012). Technological evaluation and chemical composition of rye grain of selected varieties cultivated over 3 consecutive years. *Acta Agrophysica*, *19*(2), 265–276.

Camus-Ela, M., Wang, Y., Rennie, G. H., Raghavan, V., & Wang, J. (2025). Update on hazelnut allergy: Allergen characterization, epidemiology, food processing technique and detecting strategy. *Comprehensive Reviews in Food Science and Food Safety*, *24*(2), e70098.

Cardona, V., Ansotegui, I. J., Ebisawa, M., El-Gamal, Y., Rivas, M. F., Fineman, S., Geller, M., Gonzalez-Estrada, A., Greenberger, P. A., Borges,M. S., Senna, G., Sheikh, A., Tanno, L. K., Thong, B. Y., Turner, P. J., & Worm, M. (2020). World allergy organization anaphylaxis guidance 2020. *World Allergy Organization Journal*, *13*(10), 100472.

Cardona-Alzate, L., Preciado-Rojo, L. M., Barahona-Sanchez, I., Cardona-Cartagena, V., & Pereañez, J. A. (2025). Inhibition of a myotoxic phospholipase A2 isolated from *Crotalus durissus* cumanensis by pentacyclic triterpenes. *Vitae*, *32*(1).

Castor, M. A. R., Cruz, M. K. D. M., Balanag, G. A. M., Hate, K. M., Reyes, R. D. C., Agcaoili-De Jesus, M. S., Ocampo-Cervantes, C. C., & Dalmacio, L. M. M. (2024). Identification of cross-reactive IgE-binding proteins from Philippine grass pollen extracts. *Asia Pacific Allergy*, *14*(3), e30. https://journals.lww.com/apallergy/fulltext/2024/09000/identification_of_cross_reactive_ige_binding.3.aspx

Catassi, C., Alaedini, A., Bojarski, C., Bonaz, B., Bouma, G., Carroccio, A., Castillejo, G., De Magistris, L., Dieterich, W., Di Liberto, D., Elli, L., Fasano, A., Hadjivassiliou, M., Kurien, M., Lionetti, E., Mulder, C. J., Rostami, K., Sapone, A., Scherf, K., Schuppan, D., … Sanders, D. S. (2017). The overlapping area of non-celiac gluten sensitivity (NCGS) and wheat-sensitive irritable bowel syndrome (IBS): An update. Nutrients, 9(11), 1268.

Chelladurai, P. K., Pandey, A., Swamy, C. T., Nartunai, G., & Purewal, S. S. (2025). Rye phenolics: Extraction, identification, structure and health benefits. In S. S. Purewal (Ed.), *Rye: Processing, nutritional profile and commercial uses* (pp. 117–156). Springer.

Cianferoni, A. (2016, Jan 29). Wheat allergy: Diagnosis and treatment. *Journal of Asthma and Allergy*, *9*, 13–25.

Crews, C. (2015). Analysis of ergot alkaloids. *Toxins*, *7*(6), 2024–2050. https://doi.org/10.3390/toxins7062024

Das, P., Chatterjee, A., Bose, S., & Mondal, S. (2025). Emerging applications of biotoxin detection in food. *Biotoxins in Food*, 344–374.

de Lima, F. A., Martins, I. M., Faria, A., Calhau, C., Azevedo, J., Fernandes, I., Mateus, N., & Macedo, G. A. (2018). Influence of rye flour enzymatic biotransformation on the antioxidant capacity and transepithelial transport of phenolic acids. *Food & Function*, *9*(3), 1889–1898.

Dettmann, P., Skiba, M., Stern, D., Weisemann, J., Mages, H. W., Krez, N., Dorner, M. B., Schaarschmidt, S., Avondet, M. A., Fulde, M., Rummel, A., Strommenger, B., Maurischat, S., & Dorner, B. G.(2025). Highly sensitive suspension immunoassay for multiplex detection, differentiation, and quantification of eight *Staphylococcus aureus* enterotoxins (SEA to SEI). *Toxins*, *17*(6), 265.

Dhami, S., Nurmatov, U., Arasi, S., Khan, T., Asaria, M., Zaman, H., Agarwal, A., Netuveli, G., Roberts, G., Pfaar, O., Muraro, A., Ansotegui, I. J., Calderon, M., Cingi, C., Durham, S., Gerth van Wijk, R., Halken, S., Hamelmann, E., Hellings, P., … Sheikh, A. (2017). Allergen immunotherapy for allergic rhinoconjunctivitis: A systematic review and meta-analysis. *Allergy*, *72*(11), 1597–1631. https://doi.org/10.1111/all.13201

van Kampen, V., Merget, R., Rabstein, S., Sander, I., Bruening, T., Broding, H. C., Keller, C., Muesken, H., Overlack, A., Schultze-Werninghaus, G., Walusiak, J., & Raulf-Heimsoth, M. (2009). Comparison of wheat and rye flour solutions for skin prick testing: a multi-centre study (Stad 1). *Clinical and experimental allergy : journal of the British Society for Allergy and Clinical Immunology*, *39*(12), 1896–1902. https://doi.org/10.1111/j.1365-2222.2009.03342.x

Ebisawa, M., Ito, K., Fujisawa, T., Aihara, Y., Ito, S., Imai, T., Ohshima, Y., Ohya, Y., Kaneko, H., Kondo, Y., Shimojo, N., Nagao, M., Ito, Y., Inoue, Y., Okafuji, I., Sato, S., Nakajima, Y., Nishimoto, H., Fukuie, T., Futamura, M., Manabe, T., Yanagida, N., Yamada, Y., Urisu, & A. (2020). Japanese guidelines for food allergy 2020. *Allergology International*, *69*(3), 370–386. https://doi.org/10.1016/j.alit.2020.03.004

Eckl-Dorna, J., Klein, B., Reichenauer, T. G., Niederberger, V., & Valenta, R. (2010). Exposure of rye (*Secale cereale*) cultivars to elevated ozone levels increases the allergen content in pollen. *Journal of Allergy and Clinical Immunology*, 126(6), 1315–1317. https://doi.org/10.1016/j.jaci.2010.07.031

Venkatesan, S., Zare, A., & Stevanovic, S. (2024). Pollen and sub-pollen particles: External interactions shaping the allergic potential of pollen. *Science of The Total Environment*, *926*, 171593. https://doi.org/10.1016/j.scitotenv.2024.171593.

EFSA (European Food Safety Authority). (2011). Scientific opinion on the risks related to the presence of T-2 and HT-2 toxin in food and feed. *EFSA Journal*, *9*(12), 2481. https://doi.org/10.2903/j.efsa.2011.2481

EFSA (European Food Safety Authority). (2014). Scientific opinion on the evaluation of allergenic foods and food ingredients for labelling purposes. *EFSA Journal*, *12*(11), 3894. https://www.efsa.europa.eu/en/efsajournal/pub/3894

EFSA (European Food Safety Authority). (2020). Risk assessment of ochratoxin A in food. *EFSA Journal*, *18*(5), e06113. https://doi.org/10.2903/j.efsa.2020.6113

EFSA (European Food Safety Authority), Arcella, D., Gómez Ruiz, J. Á., Innocenti, M. L., & Roldán, R. (2017). Human and animal dietary exposure to ergot alkaloids. *EFSA Journal*, *15*(7), e04902.

European Union. (2021, Aug 25). Commission Regulation (EU) 2021/1399 of 24 August 2021 amending Regulation (EC) No 1881/2006 as regards maximum levels of ergot sclerotia and ergot alkaloids in certain foodstuffs. *Official Journal of the European Union*, *301*, p. 1–5. Retrieved January 18, 2022, from http://data.europa.eu/eli/reg/2021/1399/oj

Fardet, A. (2010). New hypotheses for the health-protective mechanisms of whole-grain cereals: What is beyond fibre. *Nutrition Research Reviews*, *23*(1), 65–134.

Fernández-Nieto, M., Sastre, J., & Quirce, S. (2024). Diagnostic accuracy of specific IgE against wheat and rye in flour-induced occupational asthma. *Journal of Allergy and Clinical Immunology: In Practice*. https://doi.org/10.1016/j.jaip.2024.02.031

Florea, S., Panaccione, D. G., & Schardl, C. L. (2017). Ergot alkaloids of the family Clavicipitaceae. *Phytopathology*, *107*, 504–518. https://doi.org/10.1094/PHYTO-12-16-0435-RVW

Hussain, M. I., Araniti, F., Schulz, M., Baerson, S., Vieites-Álvarez, Y., Rempelos, L., Bilsborrow, P., Chinchilla, N., Macías, F. A., Weston, L.A. and Reigosa, M. J. (2022). Benzoxazinoids in wheat allelopathy–From discovery to application for sustainable weed management. *Environmental and Experimental Botany*, *202*, p.104997.

Gargano, D., Appanna, R., Santonicola, A., De Bartolomeis, F., Stellato, C., Cianferoni, A., Casolaro, V., & Iovino, P. (2021). Food allergy and intolerance: A narrative review on nutritional concerns. *Nutrients*, *13*(5), 1638.

Conti Taguali, S., Riolo, M., Dopazo, V., Meca, G., & Cacciola, S. O. (2024). Characterization of mycotoxins produced by two *Fusarium* species responsible for postharvest rot of banana fruit. *Journal of Plant Pathology*, *106*, 1785–1800. https://doi.org/10.1007/s42161-024-01751-8

Grabiński, J., Sułek, A., Wyzińska, M., Stuper-Szablewska, K., Cacak-Pietrzak, G., Nieróbca, A., & Dziki, D. (2021). Impact of genotype, weather conditions, and production technology on the quantitative profile of anti-nutritive compounds in rye grains. *Agronomy*, *11*(1), 151.

Guardia, G., Abalos, D., Ribatti, E., Ardenti, F., Capra, F., Mortella, G., Tabaglio, V., Ibáñez, M. Á., Chen, J., & Fiorini, A. (2025). Balancing legume-cereal proportions in cover crop mixtures to minimize N2O emissions. *Geoderma*, *454*, 117195.

Gumul, D., & Berski, W. (2021). The polyphenol profile and antioxidant potential of irradiated rye grains. *International Journal of Food Science*, *2021*(1), 8870754.

Gupta, R. S., Warren, C. M., Smith, B. M., Jiang, J., Blumenstock, J. A., Davis, M. M., Schleimer, R. P., & Nadeau, K. C. (2019). Prevalence and severity of food allergies among US adults. *JAMA Network Open*, *2*, e185630.

Hill, I. D., Fasano, A., Guandalini, S., Hoffenberg, E., Levy, J., Reilly, N., & Verma, R. (2016). NASPGHAN clinical report on the diagnosis and treatment of gluten-related disorders. *Journal of Pediatric Gastroenterology and Nutrition*, *63*(1), 156–65.

Ikram, A., Saeed, F., Noor, R. A., Imran, A., Afzaal, M., Rasheed, A., Islam, F., Iqbal, A., Zahoor, T., Naz, S., & Waheed, W. (2023). A comprehensive review on biochemical and technological properties of rye (*Secale cereale* L.). *International Journal of Food Properties*, *26*(1), 2212–2228.

Ivanišová, E., Ondrejovič, M., & Šilhár, S. (2012). Antioxidant activity of milling fractions of selected cereals. *Nova Biotechnologica et Chimica*, *11*(1), 45–56.

Kallio, P., Kolehmainen, M., Laaksonen, D. E., Kekäläinen, J., Salopuro, T., Sivenius, K., Pulkkinen, L., Mykkänen, H. M., Niskanen, L., Uusitupa, M., & Poutanen, K. S. (2007). Dietary carbohydrate modification induces alterations in gene expression in abdominal subcutaneous adipose tissue in persons with the metabolic syndrome: The FUNGENUT Study. *The American Journal of Clinical Nutrition*, *85*(5), 1417–1427.

Kaswala, D. H., Veeraraghavan, G., Kelly, C. P., & Leffler, D. A. (2015). Celiac disease: Diagnostic standards and dilemmas. *Diseases*, *3*(2), 86–101.

Aleksić, I., Vučković, O., Smiljanić, K., Gavrović-Jankulović, M., Krsmanovic, V. and Burazer, L.M., 2014. The importance of cross-reactivity in grass pollen allergy. *Archives of biological sciences*, *66*(3), pp.1149–1155

Kononenko, G. P., Zotova, E. V., & Burkin, A. A. (2021). Advances in mycotoxicological research of forage grain crops. *Sel'skokhozyaistvennaya biologiya [Agricultural Biology]*, *56*(5), 958–967.

Kulichová, K., Sokol, J., Nemeček, P., Maliarová, M., Maliar, T., Havrlentová, M., & Kraic, J. (2019). Phenolic compounds and biological activities of rye (*Secale cereale* L.) grains. *Open Chemistry*, *17*(1), 988–999.

Landberg, R., et al. (2018). Alkylresorcinols as biomarkers of whole grain wheat and rye intake. *Nutrition Reviews*, *76*(10), 659–678. https://doi.org/10.1093/nutrit/nuy030

Lebwohl, B., & Rubio-Tapia, A. (2021). Epidemiology, presentation, and diagnosis of celiac disease. *Gastroenterology*, *160*(1), 63–75.

Lei, D. K., & Grammer, L. C. (2019, November). An overview of allergens. *Allergy & Asthma Proceedings*, *40*(6).

Leonard, M. M., Sapone, A., Catassi, C., & Fasano, A. (2017). Celiac disease and nonceliac gluten sensitivity: A review. *Jama*, *318*(7), 647–656.

Lexhaller, B., Colgrave, M. L., & Scherf, K. A. (2019). Characterization and relative quantitation of wheat, rye, and barley gluten protein types by liquid chromatography–tandem mass spectrometry. *Frontiers in Plant Science*, *10*, 1530.

Liu, C., Wang, J., Fu, S., Wang, L., Li, H., Wang, M., Huang, Y., Shi, Q., Zhou, Y., Guo, X., Zhu, C., Zhang, J., & Han, F. (2022). Establishment of a set of wheat-rye addition lines with resistance to stem rust. *Theoretical and Applied Genetics*, *135*(7), 2469–2480.

Mekonnen, T. W., & Labuschagne, M. (2025). Rye protein: Extraction, identification and properties. In *Rye: Processing, nutritional profile and commercial uses* (pp. 77–94). Springer.

Menzies, J. G., & Turkington, T. K. (2015). An overview of the ergot (*Claviceps purpurea*) issue in western Canada: Challenges and solutions. *Canadian Journal of Plant Pathology*, *37*, 40–51. https://doi.org/10.1080/07060661.2014.986527

Miedaner, T., Kodisch, A., Raditschnig, A., & Eifler, J. (2021). Ergot alkaloid contents in hybrid rye are reduced by breeding. *Agriculture*, *11*(6), 526.

Miedaner, T. & Laidig, F. (2019). Hybrid breeding in rye (*Secale cereale* L.). In J. Al-Khayri, S. Jain, & D. Johnson (Eds.), *Advances in plant breeding strategies: Cereals* (Vol. 5, pp. 343–372). Springer. https://doi.org/10.1007/978-3-030-23108-8_9

Miranda, M. E., García-Valdés, L., Espigares-Rodríguez, E., Leno-Durán, E., & Requena, P. (2023). Non-celiac gluten sensitivity: Clinical presentation, etiology and differential diagnosis. *Gastroenterología y Hepatología (English Edition)*, *46*(7), 562–571.

Mishra, L. K., Sarkar, D., Zwinger, S., & Shetty, K. (2017). Phenolic antioxidant-linked anti-hyperglycemic properties of rye cultivars grown under conventional and organic production systems. *Journal of Cereal Science*, *76*, 108–115.

Niedziela, A., Wojciechowska, M., & Bednarek, P. T. (2021). New PCR-specific markers for pollen fertility restoration QRfp-4R in rye (*Secale cereale* L.) with Pampa sterilizing cytoplasm. *Journal of Applied Genetics*, *62*(4), 545–557.

Schulz, M., Marocco, A., Tabaglio, V., Macias, F. A., & Molinillo, J. M. (2013). Benzoxazinoids in rye allelopathy-from discovery to application in sustainable weed control and organic farming. *Journal of chemical ecology*, *39*(2), pp.154–174.

Nyström, L., Lampi, A.-M., Andersson, A. A. M., Kamal-Eldin, A., Gebruers, K., Courtin, C. M., Delcour, J. A., Li, L., Ward, J. L., Fraś, A., Boros, D., Rakszegi, M., Bedő, Z., Shewry, P. R., & Piironen, V. (2008). Phytochemicals and dietary fiber components in rye varieties in the HEALTHGRAIN diversity screen. *Journal of Agricultural and Food Chemistry*, *56*(21), 9758–9766.

García-Menaya, J. M., Chiarella, G. M., Cordobés-Durán, C., Mahecha, A. C., & Bobadilla-González, P. (2016). Rye-dependent exercise-induced anaphylaxis. *Annals of Allergy, Asthma & Immunology*, *117*(5), 566–568. https://doi.org/10.1016/j.anai.2016.08.032

Pasquali, D., Blundell, M., Howitt, C. A., & Colgrave, M. L (2019). Catcher of the rye: Detection of rye, a gluten-containing grain, by LC–MS/MS. *Journal of Proteome Research*, *18*(9), 3394–3403.

Pieczyńska, J. (2018). Do celiac disease and non-celiac gluten sensitivity have the same effects on reproductive disorders? *Nutrition*, *48*, 18–23.

Pihlava, J. M., Nordlund, E., Heiniö, R. L., Hietaniemi, V., Lehtinen, P., & Poutanen, K. (2015). Phenolic compounds in wholegrain rye and its fractions. *Journal of Food Composition and Analysis*, *38*, 89–97.

Pompeia, C., Frare, E. O., Peigneur, S., Tytgat, J., da Silva, Á. P., de Oliveira, E. B., Pereira, A., Kerkis, I., & Kolonin, M. G. (2022). Synthetic polypeptide crotamine: Characterization as a myotoxin and as a target of combinatorial peptides. *Journal of Molecular Medicine*, *100*(1), 65–76. https://doi.org/10.1007/s00109-021-02140-9

Qu, B., et al. (2024). Fusarium mycotoxins: The major food contaminants. *mLife*, *3*(1), 1–10. https://doi.org/10.1002/mlf2.12112

Qu, X., Ma, Z., Wu, X., & Lv, L. (2025). Recent advances of processing and detection techniques on crustacean allergens: A review. *Foods*, *14*(2), 285.

Radomirović, M., Gligorijević, N., & Rajković, A. (2025). Immuno-PCR in the analysis of food contaminants. *International Journal of Molecular Sciences*, *26*(7), 3091.

Reddy, P., Hemsworth, J., Guthridge, K. M., Vinh, A., Vassiliadis, S., Ezernieks, V., Spangenberg, G. C., & Rochfort, S. J. (2020). Ergot alkaloid mycotoxins: Physiological effects, metabolism and distribution of the residual toxin in mice. *Scientific Reports*, *10*(1), 9714.

Ross, A. B., et al. (2017). Alkylresorcinols in cereal foods: Biological activity and potential health effects. *Nutrition Reviews*, *75*(8), 504–517. https://doi.org/10.1093/nutrit/nux020

Roszkowska, A., Pawlicka, M., Mroczek, A., Bałabuszek, K., & Nieradko-Iwanicka, B. (2019). Non-celiac gluten sensitivity: A review. *Medicina*, *55*(6), 222.

Sadeghi, P., Sohrabi, H., Majidi, M. R., Eftekhari, A., Zargari, F., de la Guardia, M., & Mokhtarzadeh, A. A. (2024). Mycotoxins detection in food samples through lateral flow assays (LFAs): An update for status and prospect. *TrAC Trends in Analytical Chemistry*, *176*, 117722. https://doi.org/10.1016/j.trac.2024.117722

Salvador, G. H., Pinto, Ê. K., Ortolani, P. L., Fortes-Dias, C. L., Cavalcante, W. L., Soares, A. M., Lomonte, B., Lewin, M. R., & Fontes, M. R. (2023). Structural basis of the myotoxic inhibition of the *Bothrops pirajai* PrTX-I by the synthetic varespladib. *Biochimie*, *207*, 1–10.

Scherf, K. A. (2019). Immunoreactive cereal proteins in wheat allergy, non-celiac gluten/wheat sensitivity (NCGS), and celiac disease. *Current Opinion in Food Science*, *25*, 35–41.

Schwake-Anduschus, C., Lorenz, N., Lahrssen-Wiederholt, M., Lauche, A., & Dänicke, S. (2020). German monitoring 2012–2014: Ergot of *Claviceps purpurea* and ergot alkaloids (EA) in feeding stuffs and their toxicological relevance for animal feeding. *Journal of Consumer Protection and Food Safety*, *15*, 321–329. https://doi.org/10.1007/s00003-020-01298-7

Schwingshackl, L., Knüppel, S., Michels, N., Schwedhelm, C., Hoffmann, G., Iqbal, K., De Henauw, S., Boeing, H., & Devleesschauwer, B. (2019). Intake of 12 food groups and disability-adjusted life years from coronary heart disease, stroke, type 2 diabetes, and colorectal cancer in 16 European countries. *European Journal of Epidemiology*, *34*, 765–775.

Sharma, G. M., Pereira, M., & Williams, K. M. (2015). Gluten detection in foods available in the United States – A market survey. *Food Chemistry*, *169*, 120–126.

Smits, N., Voorhuijzen-Harink, M., Cochrane, S., Robouch, P., Cubero Leon, E., Cattapan, F., Ravenhorst, M., Taverniers, I., Bortoletto, N., Taylor, C., Bruggeman, C., Tao, W., Koops, A., & Hoek-van den Hil, E. (2024). *Guidance on how to analyse unintended food allergen presence: Considerations for selection of allergen ELISA and DNA kits to analyse unintended food allergen presence* (WFSR Report 2024.007). Wageningen Food Safety Research. https://edepot.wur.nl/669866

Sørensen, C. V., Fernández, J., Adams, A. C., Wildenauer, H. H., Schoffelen, S., Ledsgaard, L., Pucca, M.B., Fiebig, M., Cerni, F.A., Tulika, T., & Laustsen, A. H. (2024). Antibody-dependent enhancement of toxicity of myotoxin II from *Bothrops asper*. *Nature Communications*, *15*(1), 173.

Tedner, S. G., Asarnoj, A., Thulin, H., Westman, M., Konradsen, J. R., & Nilsson, C. (2022). Food allergy and hypersensitivity reactions in children and adults—A review. *Journal of Internal Medicine*, *291*(3), 283–302.

Thermo Fisher Scientific. (2024). *Rye (*Secale cereale*) allergen information*. Allergen Encyclopedia. https://www.thermofisher.com/phadia/us/en/resources/allergen-encyclopedia/f5.html

Tye-Din, J. A., Galipeau, H. J., & Agardh, D. (2018). Celiac disease: A review of current concepts in pathogenesis, prevention, and novel therapies. *Frontiers in Pediatrics*, *6*, 350.

Uhlig, S., et al. (2016). Mycotoxins in cereals: Occurrence, toxicity, and exposure assessment. *Comprehensive Reviews in Food Science and Food Safety*, *15*(5), 957–977. https://doi.org/10.1111/1541-4337.12206

Vusthepalli, P. S., Vusthepalli, G. S., Manne, A. A., Nannapaneni, S., Veeravilli, S., Setti, R., & Vemuri, P. K. (2022). Comprehensive study on key pollen allergens. *Journal of Pure and Applied Microbiology*, *16*(1), 110–115.

Walkowiak, S., Taylor, D., Fu, B. X., Drul, D., Pleskach, K., & Tittlemier, S. A. (2022). Ergot in Canadian cereals – Relevance, occurrence, and current status. *Canadian Journal of Plant Pathology*, *44*(6), 793–805.

Warren, C. M., Jiang, J., & Gupta, R. S. (2020). Epidemiology and burden of food allergy. In *Current allergy and asthma reports* (Vol. 20). Springer.

WHO (World Health Organization). (2024). *Mercury and health.* https://www.who.int/news-room/fact-sheets/detail/mercury-and-health

WHO/FAO. (2020). *Code of practice on food allergen management for food business operators.* Codex Alimentarius International Food Standards; p. 20. https://iawg-fsq.org/docs/a19b101e-9b15-41a6-97e3-aa888f72455d

Wiczkowski, W., et al. ((2019). Phenolic acids in cereal-based foods. *Open Chemistry, 17*(1), 860–875. https://doi.org/10.1515/chem-2019-0103

Yu, J., & Pedroso, I. R. (2023). Mycotoxins in cereal-based products and their impacts on the health of humans, livestock animals and pets. *Toxins, 15*(8), 480.

Zhan, F., Wang, T., Iradukunda, L., & Zhan, J. (2018). A gold nanoparticle-based lateral flow biosensor for sensitive visual detection of the potato late blight pathogen, phytophthora infestans. *Analytica Chimica Acta, 1036*, 153–161.

Zhang, R., Li, Y., Gao, Y., Wang, F., Zhou, P., Zhang, Y., Li, Y., Wang, K., Rui, T., Zhu, G., & Fang, Z (2025). Study on the improvement of rye bread quality by bacterial laccase derived from streptomyces coelicolor. *Journal of Cereal Science, 123*, 104166.

10 Future Aspects of Rye Grains Quality and Role of Technologists in Its Management

Sidharth Pradeep, Johann Koshy Thampan, Dayanand Peter, Thottiam Vasudevan Ranganathan, Aanchal Sharma and Gulzar Ahmad Nayik

10.1 INTRODUCTION

Rye grain is a crop that has long been known to civilizations and its importance in a temperate climate. The nutritional aspects of rye, specifically, are high in dietary fiber, vitamins, and minerals, making it a nutritional asset. *Secale cereale* is the scientific name of rye, a kind of cereal grass grown primarily to produce grain, which is widely used to prepare bread, whiskey and animal feed. It is particularly famous for doing better compared to most of the other cereal varieties in regions experiencing poor agricultural situations because of its ability to grow in extreme, freezing climates and in soils of poor quality. Rye has long been a part of the diet of Eastern European and Northern European inhabitants, and its utility has ensured its relevance to agriculture in the future.

The overall performance and market value of *S. cereale*, or rye grain, is determined by a confluence of superior nutritional attributes and its unique agronomic resilience. Nutritionally, rye is distinguished by its robust composition, boasting the highest dietary fiber content among common grains (15.1 g/100 g), a high antioxidant capacity (60 mg/100 g) and a low glycemic index (GI) of 34, which collectively enhance digestive health, provide protection against oxidative stress and aid in blood sugar management. Agronomically, rye is a remarkably strong cereal crop known for its ability to grow successfully in extreme, freezing climates and poor-quality soils, performing better than many other cereal varieties in regions experiencing poor agricultural situations. However, a key processing dynamic that defines its performance is its naturally lower gluten content compared to wheat, which necessitates specialized methodologies—such as enzyme supplementation or fermentation—to improve dough rheology, enhance volume and achieve the desired texture in rye-based products. These collective attributes—superior nutrition, environmental

DOI: 10.1201/9781003585817-10

TABLE 10.1
Nutritional Superiority of Rye: A Comparative Analysis

Nutrient	Rye	Wheat	Rice	Corn
Fiber (g/100 g)	15.1	12.2	0.4	7.3
Protein (g/100 g)	9.0	12.6	2.7	9.4
Antioxidants (g/100 g)	60.0	45.0	5.0	20.0
Glycemic index	34.0	54.0	73.0	58.0

adaptability and unique processing needs—determine the overall quality and market relevance of rye grain. The nutritional profile of rye grains as compared to that of other cereals is given in Table 10.1

10.2 FUTURE CHALLENGES IN RYE QUALITY MANAGEMENT

Rye grain (*S. cereale*) is a cereal grass mainly cultivated for its grain, which is utilized to make bread, whiskey and animal feed. Known to civilizations for centuries, this species is especially important in temperate climates. Rye is a robust cereal crop that is tolerant of poor soils and cool environments. Due to its ability to grow in extreme and freezing climates as well as in soil which is poor quality and with low agricultural potential, rye performs comparatively better than many cereals in less-than-ideal locations. Rye has always been a significant component of food in the diets of Eastern European and Northern European denizens, and its functional uses can easily transfer as an initial use for rye grain in future agricultural progress.

Rye is a strong cereal crop that is highly resistant to poor soils and cold climates. Overall, environmental and agricultural changes present serious difficulties for its production and quality. Climate change faces a serious challenge in its growth and quality. Erratic weather conditions, such as drought, excessive rain and unanticipated frosts, disrupt the optimal germination temperature of 8–12°C, leading to lower yields, generally ranging from 2.5–4 tons per hectare. Furthermore, extreme weather stress conditions lower protein synthesis of rye and, consequently, its nutritional value. Water enhances this risk of mycotoxin contamination, especially of the ergot fungus by *Claviceps purpurea*. It grows well under a temperature range of 15–25°C with moisture levels above 85% under heavy rainfall during flowering, enhancing chances of contamination. Besides grain quality being affected, the crop becomes hazardous to eat and hence detrimental to food safety (Németh & Tömösközi, 2021).

Supply chain and storage are dangers that have been posed to the rye quality. Degeneration of quality is achieved in the presence of pests, mold and aflatoxins resulting from moisture and temperature changes and poor conditions of storage.

The first can be diminishing the value of the market, and the second one is a health risk (Rajarammanna et al., 2008).

Rye quality is an exciting prospect in the future due to technological innovations and sustainability. Climate-receptive agriculture, potentially including precision agriculture, may help to reduce some of the negative factors in crop production caused by changing climates. The identification of bioactive compounds can open the door to functional foods and nutraceuticals through nutritional enhancement. Sustainability in production and processing can be relatively friendly to the environment and efficient. Finally, the innovations, focused on consumers, will be aimed at the design of the rye-related products that would shape the new preferences. Collaboratively, technologists, farmers and researchers will play a significant role in making rye a long-term food source (Ghafoor et al., 2024).

The need for a future-focused strategy in managing rye quality is driven by significant, compounding threats: climate volatility and evolving market demands. This proactive approach is not merely beneficial but essential for the long-term viability of rye as a high-quality food resource. Key challenges include that unpredictable weather due to climate change can reduce yield, lower the grain's nutritional value and increase the risk of dangerous mycotoxin contamination, notably from the ergot fungus (*C. purpurea*). Furthermore, current issues like the low genetic diversity of rye make it susceptible to new pests, diseases and environmental stress. Post-harvest, poor storage conditions continue to pose dangers, leading to quality decline via pests, mold and aflatoxins. Mitigating these threats requires a concerted effort to develop climate-resilient rye varieties, improve processing techniques and leverage new technologies to ensure the grain remains safe, high-quality and sustainably produced for the global community.

The second major problem is that the rye lacks genetic diversity. Monoculture farming depends on a few high-yielding varieties, which become susceptible to pests, diseases and environmental changes, making it unsustainable in the long run. Loss of genetic diversity limits a crop's capacity to adapt to new challenges like climate stress and new emerging pathogens. Control of this problem requires appropriate breeding programs to produce varieties of rye with increased resistance to disease, drought and stresses under diverse environmental conditions.

Between the years 2000 and 2020, rye yields declined consistently because of unprecedented weather, periods of extended drought and climate-driven soil degradation, making it difficult to maintain conventional agricultural practices in this country's growing climate change circumstances (Figure 10.1). As farmers have resolved to address these challenges, climate-smart processes have grown in popularity among growers and researchers, as evidenced by the breed selection of drought-resistant rye varieties and commercialized precision agriculture solutions to assess the soil, irrigate the crop with coverage and apply nutrients more efficiently. These processes conserve valuable resources and improve efficiency while offering practical functions to support return in rye, which will help farmers adapt and succeed with rye production in unpredictable climates (Rajarammanna et al., 2008).

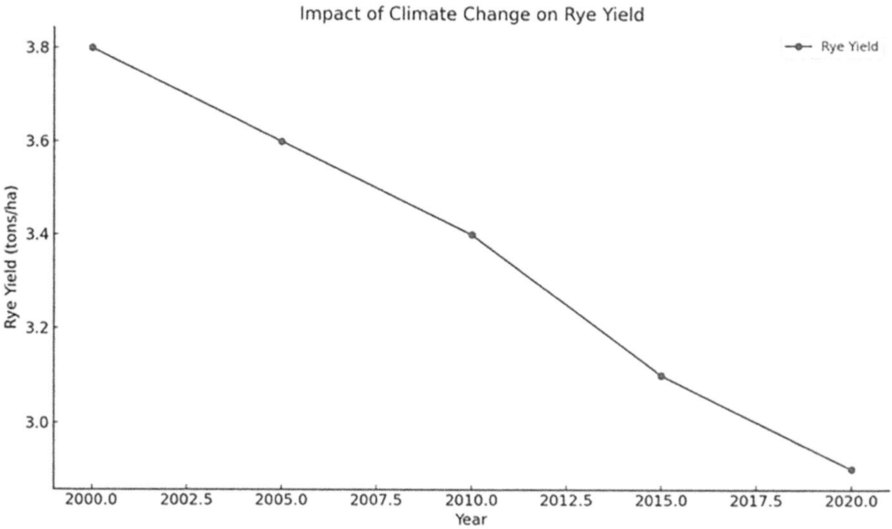

FIGURE 10.1 Impact of climate change on rye yield (2000–2020).

10.3 FUTURE CHALLENGES IN RYE QUALITY MANAGEMENT

In the coming years, the quality of rye will increasingly be challenged by climate change, changing consumer preferences and technological advances in agricultural production systems. Increasing temperatures, unpredictable rainfall patterns and soil degradation pose a threat not only to yields, but also to the nutritional and processing quality of the grain, while consumers will demand products that are of high quality and sustainably produced. Genetic and breeding factors will also impact the ability to produce adapted rye varieties that retain desirable quality traits under expected stresses. Finding a balance between environmental pressures, market expectations and scientific innovation will be important for the future quality and competitiveness of rye production.

10.4 CLIMATE CHANGE IMPACTS

Climate change will likely be an important factor affecting the future quality of rye, as changing temperatures and precipitation patterns will have direct effects on crop growth and yield stability. Higher temperatures can interfere with the physiological performance of rye plants, often resulting in reduced grain quality and increased susceptibility to insects and disease (Ghafoor et al., 2024). At the same time, changing precipitation patterns would result in longer periods of drought or increased periods of wetness, both of which negatively impact the development of the grain, its storage capability, stability and post-harvest quality (Ghafoor et al., 2024).

10.5 THREATS FROM PESTS AND DISEASES

The new pests and pathogens are also posing a significant threat to ensuring the quality and yield of the rye production in the coming years. According to recent studies, diseases like ergot may significantly impair crop productivity and even grain safety due to contamination, and high humidity and fluctuations in temperature favor the development of staphylologic pathogens. Climate change also increases risk because the geographic scope and severity of these threats can increase, and integrated pest management tactics do not necessarily rely on single-unit responses. The farmers and researchers are required to utilize multifaceted management systems to maintain the quality of the rye under such conditions, such as integrated pest management programs and crop rotations.

10.6 PROCESSING AND STORAGE CHALLENGES

Rye quality management includes the immediate issues surrounding processing and storage. Rye grains boast of unique traits that preserve the quality and yield of the grain, which require a unique milling and manipulation procedure.

Due to its extreme vulnerability to ergot fungi, in particular, an ergot fungus (*C. purpurea*), rye is among the biggest threats to both the historical and the modern.

Ergotism is a deadly disease that results from the consumption of grain contaminated with this fungus. Although the contemporary grain cleaning and milling processes have considerably reduced this risk in the food supply, it is a constant menace in the industry that must be keenly checked. Another biotic problem of rye is the recurrence of stem rust across the whole of Europe, and this constitutes a serious challenge to future farmers. Its height is somewhat high, which gives the organism a certain level of weed suppression, but agronomically speaking, the organism is susceptible to lodging, which is the bending and breaking of the stem, especially when the nitrogen fertilization is high and this makes the mechanical harvesting task more complex (Grains Research and Development Corporation, 2018). The lighter endosperm typically results in lower extraction rates from milling, and the ways we store our grain under varying conditions, temperatures and moisture can promote spoilage, germination and mycotoxins to further degrade the grain's overall integrity and usage (Németh & Tömösközi, 2021). Improved storage systems, along with direct quality assessments, can help mitigate these issues.

10.7 CHANGING CONSUMER DEMANDS

Consumer preferences in the rye market are changing rapidly due to the increase in demand for food items made with dietary fiber, health-related ingredients and alternative foods made with gluten-free products. These market shifts may require changes to the specialty, quality characteristics of rye, and thus, the growing practices required for them to comply with market standards rooted in current demand.

10.8 GENETIC IMPROVEMENT AND BREEDING CHALLENGES

The recent efforts to genetically enhance rye (*S. cereale*) have not been cumulative. Some of the objectives for the breeding include improved resistance to diseases, increased nutritional quality and resistance to biotic and abiotic stress. Achieving these outcomes will require an integration with the preservation of biodiversity and adaptation to its site-specific agronomic environment. The amassing of genomics development provides optimism to create the possibility of integrating these priorities, but it must be validated with multiple field trials before we can have certainty of system efficacy and ensure we can hastily translate genomic advantage into viable planting systems (Brzozowski et al., 2023).

10.9 EMERGING TECHNOLOGIES IN RYE QUALITY MANAGEMENT

Recent advancements in technology are revolutionizing rye management from planting through to harvest. Breeding technologies, such as CRISPR and genomic tools, will help scientists identify rye that are more disease resistant and nutritionally diverse in shorter timeframes. Furthermore, these tools are quicker and more accurate than traditional breeding, which will enable growers to produce rye that provides market avenues while facilitating the sustainability practices they need to enact. In addition, smart farming technology, such as drones and sensors, can provide real-time data on soil moisture, nutrients in soil, and the health of the crop. Thus, growers can then use water, fertilizer and pesticide only when and where they are needed, increasing profitability by eliminating waste and increasing yields and quality. Overseeing these developing agricultural practices will illustrate a sustainable, efficient, profitable growing environment for rye growers.

Post-harvest improvement technology is of substantial importance because it directly contributes to the quality maintenance of rye grain. After the grain is harvested, the drying system dries it mostly to the same moisture content and prevents moisture spoilage. Damage and foreign materials in the grain can be processed with automated machines to enhance the quality of the grain. Controlled atmosphere storage technology—which controls humidity and oxygen levels—allows for the longevity of rye to be increased by maintaining grain value while minimizing waste. Quality testing technologies such as near-infrared (NIR) spectrometer technologies allow for very rapid and non-destructive analysis of grain quality parameters, some of interest would be protein, moisture etc. To meet food safety requirements for now, real-time sensors have been developed to detect various contaminants such as mycotoxins. The culmination of these post-harvest technologies represents completeness in terms of rye grain quality control, consumer confidence and adding value to the supply chain. Having this combination of technology provides the basis for efficient, sustainable quality creation of rye.

10.10 ROLE OF FOOD TECHNOLOGISTS IN RYE QUALITY MANAGEMENT

Food technologists serve an important function in enhancing and prolonging rye grain quality to meet modern food production needs of health-concerned consumers, food safety and sustainability. The following gives a detailed description of the key spheres of involvement of food technologists with rye quality.

Enhancement of Functional Foods: Food technologists are key players in fortifying rye-based products with bioactive compounds to promote health benefits. Rye itself contains phenolic compounds, dietary fiber and antioxidants, and these same functional compounds can be concentrated through new processing technologies. These bioactive compounds are known to be beneficial for health through beneficial health properties such as antioxidant effects, lowering cholesterol and enhancing gut health. Food technologists are working on various methods, such as fermentation, germination or enzymatic treatment, to facilitate the bioavailability of these health-promoting bioactive compounds to consumer-ready rye-based products that are considered functional foods or nutritionally superior foods. In addition to the health and nutrition aspects, product development using rye-based ingredients in baked goods is critical to offer to health-conscious consumers. The low GI and high-fiber content of rye make it an attractive functional ingredient for the development of bread, crackers and other bakery products. Food technologists are also focusing on reducing or altering baking methods using rye-based ingredients to keep or enhance the health benefits of rye, while enhancing means such as texture, shelf-life and effects on taste. For example, when they added sourdough fermentation, it not only enhanced the digestibility of rye but also added value in the antioxidants of bread. The food technologies bring additional value to further the needs in the consumer market of healthier, whole-grain product substitutes.

Improvement in Processing Methodologies: Food technologists play a significant part in improving rye dough rheology and bread-making properties. Rye flour contains less gluten than wheat flour, and this may lead to a bad texture and structure in bread. Some of the processing methodologies used by technologists are enzyme supplementation (i.e., xylanases and amylases). The techniques enhance the extensivity of the dough and volume of bread. This yields better chewiness, moistness and volume of rye-based products, which is preferred by consumers. The influence of single additives (i.e., emulsifiers or hydrocolloids) is also investigated by technologists in order to enhance the bread-making properties of rye flour and to preserve the bread dough structure of rye dough. In the case of consumers who have gluten-free diets, consumers are interested in gluten-free rye formulations developed by food technologists. When compared to wheat, the level of gluten in rye is naturally lower, but there is still enough gluten in it to cause coeliac disease. Recent studies are directed at gluten-free substitutes of rye flour with the combination of it with certain other gluten-free grain products (rice, corn, sorghum) to taste and feel like rye.

Sustainability has become a focus in current food technology, with food technologists advocating for organic farming processes to produce rye. Organic farming practices promote soil health and biodiversity while reducing the environmental effects of using pesticides and fertilizers. Technologists advocate sustainable processes by creating varieties of rye that are resistant to insects and crop diseases, thereby reducing the amount of chemical applications. Supporting crop rotation with rye as a crop element of sustainable farming systems helps promote soil yield and reduces reliance and use of synthetic fertilizers. The other sustainability option is minimizing waste by using by-products such as rye bran. Rye bran is a valuable milling by-product that is high in fiber, antioxidants and phenolic compounds. Food technologists look at the usability of rye bran to develop functional food products, including dietary supplements, fiber-enriched snacks and animal feed. By using rye bran in food formulations, food technologists are reducing waste in the production process and increasing the nutritional quality of finished products. The research using rye straw for bioenergy production also offers additional sustainability outcomes by using this by-product.

Rye stands out among grains due to its superior nutritional profile. With 15.1 g of fiber per 100 g, rye surpasses wheat (12.2 g), corn (7.3 g) and rice (0.4 g), making it an excellent choice for improving digestive health and promoting satiety. Its antioxidant content, measured at 60 mg per 100 g, is significantly higher than wheat (45 mg), corn (20 mg) and rice (5 mg), providing protection against oxidative stress and supporting overall wellness. Additionally, rye's low GI of 34 makes it ideal for blood sugar management, especially compared to wheat (GI 54), corn (GI 58) and rice (GI 73). This makes rye particularly valuable for individuals managing diabetes or seeking sustained energy release.

New technologies are reshaping the management of rye quality, enabling better quality monitoring, analysis and improvement approaches. These technologies support decision-making through production and processing to ensure quality standards are achieved and maintained.

10.10.1 IoT and Smart Farming

The Internet of Things (IoT) is critical to modern rye quality management through monitoring crop and environmental conditions on site in real time. Variables like soil moisture, soil temperature and nutrient levels can be assessed through monitoring sensors. The monitoring data can be processed to improve irrigation and fertilization, which would lead to a healthier rye crop. Smart farming processes and automated choices enable producers to make decisions that will not only yield more but also lead to better-quality grain.

10.10.2 Precision Agriculture

The precision agriculture technologies utilize data analytics and geographic information systems (GIS) to drive farming practices adjusted for localized conditions in the field. Using drone imagery and satellite data can provide insights for farmers to determine where particular areas of their field may need adjustments due to nutrient deficiencies, pest damage etc. This targeted input service allows producers to utilize and apply nutrients such as fertilizers or pesticides effectively, minimizing wasted inputs and improving overall quality of grain (Hudson, 2024).

10.10.3 Advanced Breeding Techniques

New breeding methods, including CRISPR and other gene editing technologies, have excellent potential for improving rye quality. All of the advanced breeding technologies permit the rye genome to be accurately modified, thus enhancing traits including disease resistance, yield and nutritional quality. A modernized breeding system will enable the farmers to adapt to the environments and meet their needs more easily and preserve the quality of high standards.

10.10.4 Blockchain for Traceability

More commonly, blockchain technology is being used to increase traceability in the rye production supply chain. The blockchain will enable a digital chain of information about the production processes that is impossible to alter and connects all actors with each other to underline transparency and responsibility. Blockchain will also offer quality guidance and tracking of the field to fork and promote consumer confidence. Blockchain might also assist producers in complying with regulatory requirements and organizing regulatory audits.

10.10.5 AI and Machine Learning

Algorithms in machine learning and artificial intelligence (AI) have the potential to drastically change quality management in rye production by supplying predictive analytics about potential quality issues (Peter, 2024). The technologies can take into account a vast amount of input from numerous data sources and identify patterns that the human inspector would not recognize. For example, the AI will predict a possible disease in rye, even before it occurs, and take measures to prevent any decline in quality (Peter, 2024). Additionally, machine learning might provide greater efficiencies within the world of the machine for grading and sorting rye with automated measures that will always demonstrate quality.

10.10.6 Big Data Analytics

Rye quality management through big data analytics enables producers to evaluate and manage the massive amounts of data they receive from many different sources (such as weather patterns, soil health and market patterns). Analytics can inform the farmers to modify their farming practices to be able to achieve better quality results; for example, understanding climate trends could help inform the best times to plant and which crop varieties would be most likely to produce high-quality rye that results in a market advantage.

This type of technology also provides producers with efficiencies not only in the quality management of rye but as a component of the sustainability of the rye production system. These technologies allow stakeholders to respond to problems and ensure the robustness of quality and consumer requirements in a context of change in agriculture.

Advances in post-harvest technologies improve the safety, quality and value of rye grain while generating less waste. Smart storage technology (i.e., hermetic silos,

IOT capability on humidity and temperature sensors) reduces damage. Modified atmosphere storage (carbon dioxide enrichment and vacuum packaging) minimizes microorganisms and pests. Digital tags enabled by blockchain technologies improve traceability and transparency in the supply chain. Cutting-edge sorting technologies (e.g., optical sorting, magnetic separators) achieve better quality. Drying technologies (e.g., continuous flow dryers, microwave-assisted drying) reduce moisture without degrading grain quality. Fumigation technologies that employ no chemicals, such as ozonation and heat treatment, are eco-friendly pest solutions. Spectroscopy analyzers and electronic noses keep track of grain quality and sensory characteristics in real time. Advanced milling, dehulling and extrusion treatments add value during processing advantages, while preserving nutrients. Overall, these technologies support the integrity of grains and greater efficiency in terms of storage and processing operations.

10.11 FUTURE PROSPECTS AND CONCLUSION

The future outlook for rye is very optimistic, driven largely by increasing consumer recognition of health and wellness, which is contributing to its advance as a functional food. Given increased consumer education relating to the nutritional benefits of whole grains, you would expect the interest in rye-based functional foods from consumers to increase. As has been demonstrated in the literature, rye has substantial amounts of fiber, antioxidants and phenolic compounds, and has been shown to contribute to digestive health, lower risks of cardiovascular diseases and aid in blood sugar control. The changes in consumer interest have created an opportunity for food manufacturers to consider a new range of innovative rye-based functional foods for health-conscious consumers. In the future, you can anticipate that rye-based functional foods will continue to expand across multiple categories, including baked goods (breads etc.), cereals, snacks and beverages that provide healthier alternatives to traditional refined food products. Technologists will carry on with the development of processes to utilize the health benefits of rye to ensure that food products continue to meet consumers with a good quality, nutrient-rich food product that helps consumers adapt to a modern world. Another key aspect of the future of rye is to expand research on bioactive compounds. Rye contains soluble fibers along with some antioxidants such as lignans and phenolic acids that give rye additional health benefits. Current research will continue to assess bioactive compounds and their utilization in functional food ingredients and dietary supplements. Together, technologists, nutritionists and researchers could conduct further research to enhance bioactive compounds extractions, making them more bioavailable and efficacious in health interventions. This could lead to increased production of functional food groups, which could lead to a novel therapeutic method to address chronic disability, which would in turn increase demand for rye.

Rye's future is likewise dependent on collaboration among technologists, agronomists, and policymakers. These individuals can collaborate to help address some of the most pressing sustainability and quality issues. The agronomists will be focused on the development of rye varieties that will resist adverse conditions and limit the application of chemical inputs, food technologists will be key to improving processing techniques while maintaining the protein quality and safety attributes of the product, and policymakers can create the political environment that benefits sustainable cropping practices through policies and regulations that encourage and support organic rye growing and food safety policies. A collaborative effort can be made to ensure rye is a crop that has the capacity to produce for future generations while minimizing environmental impacts to enhance further food production. Rye has significant potential to help foster the above-mentioned areas (i.e., sustainability, global food security, health trends) as it is environmentally adaptable and nutrient-dense, making it a viable, sustainable and healthy alternative to the more widely produced grains. Rye has the potential to immensely improve food systems in most parts of the world where climate change poses a threat to the production of farm crops. Additionally, its health benefits, including being good for heart health and assisting in weight management, will continue to resonate with the global trend toward preventive health, where nutrition and diet will continually be a primary means of prevention against health maladies.

It is likely that technology leaders will stay at the forefront of innovation in rye production and processing in the name of improving the quality and sustainability of rye. They will help improve and develop new technologies for rye processing that will provide functional food products with a longer shelf life and improved preservation of the healthy components of rye. Technology leaders will serve an important purpose within their area of expertise by advancing food safety and mitigating food waste. They will translate this information to complement training and education for food product relevant stakeholders within the food product supply chain.

In conclusion, the rye has significant potential to fulfill the needs of an increasing world population interested in health and sustainability. By being able to embrace its whole range of nutritional and functional attributes, coupled with the support of the technologist, rye has an opportunity to support food security and wellness in the world. In addition, agronomists, food technologists and policy makers will need to work together to establish that rye maintains its place in our agriculture systems of the future for healthier diets and more sustainable food systems. One of the more innovative ideas that will make a difference may be blockchain technology applied to traceability for rye-based foods. This would allow the consumer to receive during the entire food continuum not only real-time information about the place in which the rye foods were grown but also information on the quality of the rye foods and the sustainability practices that were used on the farm to manage the rye, leading to greater transparency and trustworthiness within the food system. Innovations used on the farm, including development of smart-storage capabilities, natural, non-chemical insect control, real-time monitoring of product quality and the use of effective processing approaches, can all reduce post-harvest losses, while also preserving the nutritional and functional value of rye grains. Furthermore, traceability system implementations lead to a fully transparent supply chain, which fortifies consumer trust and meets global standards. With an increasing demand for high-quality grains, embracing such technological interventions will be critical to sustainable production, food security and long-term viability of the grain sector.

REFERENCES

Brzozowski, L. J., Szuleta, E., Phillips, T. D., Van Sanford, D. A., & Clark, A. J. (2023). Breeding cereal rye (*Secale cereale*) for quality traits. *Crop Science*, *63*(4), 1964–1987

Ghafoor, A. Z., Karim, H., Studnicki, M., Raza, A., Javed, H. H., & Asghar, M. A. (2024). Climate change and rye (*Secale cereale* L.) production: Challenges, opportunities and adaptations. *Journal of Agronomy and Crop Science*, *210*(4), e12725.

Grains Research and Development Corporation. (2018, March). *Cereal rye: GrowNotes™ (Northern Region)*. Grains Research and Development Corporation. https://grdc.com.au/_data/assets/pdf_file/0019/315316/GRDC-GrowNotes-Cereal-Rye-Northern.pdf

Hudson, A. (2024). *Embracing data and emerging technologies for quality management excellence – Dataversity*. DATAVERSITY. https://www.dataversity.net/embracing-data-and-emerging-technologies-for-quality-management-excellence/

Németh, R., & Tömösközi, S. (2021). Rye: Current state and future trends in research and applications. *Acta Alimentaria*, *50*(4), 620–640.

Peter, S. (2024). *The role of artificial intelligence (AI): Machine learning in modern quality management*. Quality Magazine. https://www.qualitymag.com/articles/98259-the-role-of-artificial-intelligence-ai-machine-learning-in-modern-quality-management

Rajarammanna, R., Jayas, D. S., & White, N. D. G. (2008, July). *Comparison of deterioration of rye under different storage regimes*. Canadian Society for Bioengineering/SCGAB Annual Conference.

Rye. (n.d.). *Wikipedia*. Retrieved September 27, 2025, from https://en.wikipedia.org/wiki/Rye

U.S. Department of Agriculture, Foreign Agricultural Service. (n.d.). *Rye*. Retrieved September 27, 2025, from https://www.fas.usda.gov/data/production/commodity/0451000

Index

Note: Locators in *italics* represent figures and **bold** indicate tables in the text.

For Product Safety Concerns and Information please contact our EU representative GPSR@taylorandfrancis.com
Taylor & Francis Verlag GmbH, Kaufingerstraße 24, 80331 München, Germany

www.ingramcontent.com/pod-product-compliance
Ingram Content Group UK Ltd.
Pitfield, Milton Keynes, MK11 3LW, UK
UKHW020159200926
475702UK00004B/533
9781032956244